TEA

Medicinal and Aromatic Plants – Industrial Profiles

Individual volumes in this series provide both industry and academia with in-depth coverage of one major medicinal or aromatic plant of industrial importance.

Edited by Dr Roland Hardman

TEA
Bioactivity and Therapeutic Potential

Edited by

Yong-su Zhen
Department of Oncology
Institute of Medicinal Biotechnology
Chinese Academy of Medical Sciences
and Peking Union Medical College
Beijing, China

Associate Editors

Zong-mao Chen
Chinese Academy of Agricultural Sciences
Zhejiang, China

Shu-jun Cheng
Chinese Academy of Medical Sciences
and Peking Union Medical College
Beijing, China

Miao-lan Chen
Chinese Academy of Medical Sciences
and Peking Union Medical College
Beijing, China

CRC Press
Taylor & Francis Group
Boca Raton London New York

CRC Press is an imprint of the
Taylor & Francis Group, an **informa** business

A TAYLOR & FRANCIS BOOK

CRC Press
Taylor & Francis Group
6000 Broken Sound Parkway NW, Suite 300
Boca Raton, FL 33487-2742

First issued in paperback 2019

© 2002 by Taylor & Francis Group, LLC
CRC Press is an imprint of Taylor & Francis Group, an Informa business

No claim to original U.S. Government works

ISBN-13: 978-0-415-27345-9 (hbk)
ISBN-13: 978-0-367-39620-6 (pbk)

Every effort has been made to ensure that the advice and information in this book is true and accurate at the time of going to press. However, neither the publisher nor the authors can accept any legal responsibility or liability for any errors or omissions that may be made. In the case of drug administration, any medical procedure or the use of technical equipment mentioned within this book, you are strongly advised to consult the manufacturer's guidelines.

British Library Cataloguing in Publication Data
A catalogue record for this book is available from the British Library

Library of Congress Cataloging in Publication Data
A catalogue record has been requested

Visit the Taylor & Francis Web site at
http://www.taylorandfrancis.com

and the CRC Press Web site at
http://www.crcpress.com

CONTENTS

PREFACE TO THE SERIES

There is increasing interest in industry, academia and the health sciences in medicinal and aromatic plants. In passing from plant production to the eventual product used by the public, many sciences are involved. This series brings together information which is currently scattered through an ever increasing number of journals. Each volume gives an in-depth look at one plant genus, about which an area specialist has assembled information ranging from the production of the plant to market trends and quality control.

Many industries are involved such as forestry, agriculture, chemical, food, flavor, beverage, pharmaceutical, cosmetic and fragrance. The plant raw materials are roots, rhizomes, bulbs, leaves, stems, barks, wood, flowers, fruits and seeds. These yield gums, resins, essential (volatile) oils, fixed oils, waxes, juices, extracts and spices for medicinal and aromatic purposes. All these commodities are traded world-wide. A dealer's market report for an item may say "Drought in the country of origin has forced up prices."

Natural products do not mean safe products and account of this has to be taken by the above industries, which are subject to regulation. For example, a number of plants which are approved for use in medicine must not be used in cosmetic products.

The assessment of safe to use starts with the harvested plant material which has to comply with an official monograph. This may require absence of, or prescribed limits of, radioactive material, heavy metals, aflatoxins, pesticide residue, as well as the required level of active principle. This analytical control is costly and tends to exclude small batches of plant material. Large scale contracted mechanized cultivation with designated seed or plantlets is now preferable.

Today, plant selection is not only for the yield of active principle, but for the plant's ability to overcome disease, climatic stress and the hazards caused by mankind. Such methods as *in vitro* fertilization, meristem cultures and somatic embryogenesis are used. The transfer of sections of DNA is giving rise to controversy in the case of some end-uses of the plant material.

Some suppliers of plant raw material are now able to certify that they are supplying organically-farmed medicinal plants, herbs and spices. The European Union directive (CVO/EU No 2092/91) details the specifications for the **obligatory** quality controls to be carried out at all stages of production and processing of organic products.

Fascinating plant folklore and ethnopharmacology leads to medicinal potential. Examples are the muscle relaxants based on the arrow poison, curare, from species of *Chondrodendron*, and the antimalarials derived from species of *Cinchona* and *Artemisia*. The methods of detection of pharmacological activity have become increasingly reliable and specific, frequently involving enzymes in bioassays and avoiding the use of laboratory animals. By using bioassay linked fractionation of crude plant juices or extracts, compounds can be specifically targeted which, for example, inhibit blood platelet aggregation, or have antitumour, or antiviral, or any other required activity. With the assistance of robotic devices, all the members of a genus may be readily screened. However, the plant material must be **fully** authenticated by a specialist.

The medicinal traditions of ancient civilizations such as those of China and India have a large armamentarium of plants in their pharmacopoeias which are used throughout South East Asia. A similar situation exists in Africa and South America. Thus, a very high percentage of the world's population relies on medicinal and aromatic plants for their medicine. Western medicine is also responding. Already in Germany all medical practitioners have to pass an examination in phytotherapy before being allowed to practice. It is noticeable that throughout Europe and the USA, medical, pharmacy and health related schools are increasingly offering training in phytotherapy.

Multinational pharmaceutical companies have become less enamoured of the single compound magic bullet cure. The high costs of such ventures and the endless competition from "me too" compounds from rival companies often discourage the attempt. Independent phytomedicine companies have been very strong in Germany. However, by the end of 1995, eleven (almost all) had been acquired by the multi-national pharmaceutical firms, acknowledging the lay public's growing demand for phytomedicines in the Western World.

The business of dietary supplements in the Western World has expanded from the health store to the pharmacy. Alternative medicine includes plant based products. Appropriate measures to ensure the quality, safety and efficacy of these either already exist or are being answered by greater legislative control by such bodies as the Food and Drug Administration of the USA and the recently created European Agency for the Evaluation of Medicinal Products, based in London.

In the USA, the Dietary Supplement and Health Education Act of 1994 recognized the class of phytotherapeutic agents derived from medicinal and aromatic plants. Furthermore, under public pressure, the US Congress set up an Office of Alternative Medicine and this office in 1994 assisted the filing of several Investigational New Drug (IND) applications, required for clinical trials of some Chinese herbal preparations. The significance of these applications was that each Chinese preparation involved several plants and yet was handled as a **single** IND. A demonstration of the contribution to efficacy, of **each** ingredient of **each** plant, was not required. This was a major step forward towards more sensible regulations in regard to phytomedicines.

My thanks are due to the staff of Harwood Academic Publishers who have made this series possible and especially to the volume editors and their chapter contributors for the authoritative information.

Roland Hardman

PREFACE

Tea as a beverage is of great importance to humans. Tea consumption has a long history of over 2,000 years. Originated in China, drinking tea as a habit of daily life has spread all over the world. Currently, tea is one of the most popular beverages globally. Because tea is widely consumed by hundreds of millions of people in a perpetual manner, the possible effects of tea on human health is of particular importance in the field of medical, agricultural, and food research. The general view of tea has experienced a series of changes. Originally in ancient China, tea was taken as a medicine to detoxify or to cure diseases. Later on, tea was recognized as a tonic, which is beneficial to human health. In the course of development, tea is widely accepted as a beverage. Despite those changes, tea remains a kind of medicine, at least in part, in traditional Chinese medicine, in which tea is used alone or in most cases used in combination with other herbs to treat a variety of disorders. Modern medical research has found that tea and tea products display a wide spectrum of bioactivity and show therapeutic effectiveness in a number of experimental disease models. The subject of bioactivity and therapeutic potential of tea and tea products has drawn much more attention.

As a plant product, tea has a highly complicated composition. Fresh tea leaves and the processed tea consist of a great number of substances which can be roughly divided into two categories, the non-volatile compounds and the volatile aroma compounds. The non-volatiles that constitute the major part of the tea solids include polyphenols, flavonols, amino acids, carotenoid and other pigments, carbohydrates, organic acids, caffeine and purine derivatives, enzymes, vitamins, and many others. The physiological effect of caffeine is well known and documented. The biological activity and pharmacological effect of tea polyphenols have been under intensive investigation in recent decades. Tea polyphenols are noted for a variety of biological activities, such as antioxidative activity, antimicrobial activity, antipromotion activity in carcinogenesis, and antitumor activity. The volatile aroma compounds are hundreds in number but relatively low in quantity. The bioactivity of most aroma compounds remains to be further elucidated. The composition of tea may be different in the tea leaves from various cultivars and geographic areas. Furthermore, the composition of tea undergoes a series of changes in manufacture processing, leading to a great diversity of the composition among green tea, black tea or semi-fermented tea. Taking all the above-mentioned into account, tea is a highly complicated object to be evaluated for its bioactivity and the potentially therapeutic applications. However, due to its high complexity, tea is a rich resource of bioactive compounds from which new drugs may be discovered. In addition to the evaluation of the bioactivity of tea as a whole, it is of importance to search for new active separated principles from tea products.

The goal of this book *Tea: Bioactivity and Therapeutic Potential* is to cover all relevant aspects including botanical identification, processing and major tea categories, composition and the chemistry of constituents, biological activities, physiological and pharmacological effects, and experimental therapeutic effects. The therapeutic applications of tea based on traditional Chinese medicine are also included. The contributors of this book are renowned experts from botanical, agricultural, chemical, biochemical, pharmacological, and medical circles. In the following chapters we have integrated substances with their activity in order to cover the major advances comprehensively. A detailed discussion on the chemistry of tea may provide a broad basis for the elucidation of related bioactivity. An extended discussion on those major effects of tea may provide some clues leading to determination of its therapeutic uses.

Yong-su Zhen

CONTRIBUTORS

Miao-lan Chen
Chinese Academy of Medical Sciences
and Peking Union Medical College
9 Dong Dan San Tiao
Beijing 100730
China

Zong-mao Chen
Tea Research Institute
Chinese Academy of Agricultural
Sciences
1 Yunqi Road
Hangzhou
Zhejiang 310008
China

Shu-jun Cheng
Cancer Institute
Chinese Academy of Medical Sciences
and Peking Union Medical College
Panjiayuan
Beijing 100021
China

Shen-de Li
Cancer Institute
Chinese Academy of Medical Sciences
and Peking Union Medical College
Panjiayuan
Beijing 100021
China

Zhen-yu Li
Laboratory of Systematic and
Evolutionary Botany
Institute of Botany
Chinese Academy of Sciences
20 Nanxinchun
Xiangshan
Beijing 100093
China

Wei-bo Lu
Institute of Basic Theory
China Academy of Traditional Chinese
Medicine
18 Beixincang, Dongzhimennei
Beijing 100700
China

Pei-zhen Tao
Institute of Medicinal Biotechnology
Chinese Academy of Medical Sciences
and Peking Union Medical College
1 Tiantan Xili
Beijing 100050
China

Hua-fu Wang
Tea Research Institute
Chinese Academy of Agricultural
Sciences
1 Yunqi Road
Hangzhou
Zhejiang 310008
China

Pei-gen Xiao
Institute of Medicinal Plant Development
Chinese Academy of Medical Sciences
and Peking Union Medical College
Xibeiwang
Haidian
Beijing 100094
China

Ning Xu
Tea Research Institute
Chinese Academy of Agricultural
Sciences
1 Yunqi Road
Hangzhou
Zhejiang 310008
China

Xiao-qing You
Tea Research Institute
Chinese Academy of Agricultural
Sciences
1 Yunqi Road
Hangzhou
Zhejiang 310008
China

De-chang Zhang
Institute of Basic Medical Sciences
Chinese Academy of Medical Sciences
and Peking Union Medical College
5 Dong Dan San Tiao
Beijing 100005
China

Yong-su Zhen
Institute of Medicinal Biotechnology
Chinese Academy of Medical Sciences
and Peking Union Medical College
1 Tiantan Xili
Beijing 100050
China

1.   TEA AND HEALTH – AN OVERVIEW

MIAO-LAN CHEN

Chinese Academy of Medical Sciences and Peking Union Medical College, 9 Dong Dan San Tiao, Beijing 100730, China

Tea is one of the most popular beverages in the world. Drinking a cup of tea for pleasure or in times of stress is a part of daily life for millions of people all over the world. It is estimated that one-half of the population in the world consumes tea. The production and trade of tea has become an important business for centuries. In 1990, the tea growing area in the world reached 2.45 million hectares and total output reached 2.51 million tons. World tea consumption has increased steadily. China, for example, has experienced an increase in tea consumption of 6.0% annually for the period 1961–1984 and is expected to maintain the momentum at a rate of 10% to the year 2000. In the United Kingdom, annual tea consumption is expected to increase 1.4% annually until the year 2000. In terms of annual tea consumption per capita, Ireland has the highest value at 3.07 kg (triennial average over the period 1986–1988), followed by Iraq (2.95 kg), Qatar (2.91 kg), UK (2.84 kg) and Turkey (2.73 kg) for the same period (Robinson and Owuor, 1992; Chen and Yu, 1994). The daily consumption of tea is approximately 3 billion cups all over the world. About 80% of consumers prefer black tea and the rest consume green tea and semi-fermented oolong tea. Green tea is preferred in China, Japan, and Middle East countries, the oolong tea is mainly consumed in the eastern part of China and in Japan. In Great Britain, tea is drunk by more than 80% of the population. The average intake for those tea drinkers considering tea as a healthy drink was estimated to be 0.8 litre per day (Marks, 1992). Tea is effective for quenching thirst. However, the ability of tea to quench thirst is not the main reason for its popularity as a beverage. This relies much more on its sensory properties, customs, prices, availability and apparent health benefits. Because of the high popularity of tea, the relationship between tea and health has come as one of the most attractive topics in biomedical sciences.

1.   HISTORY OF TEA CONSUMPTION

The discovery and use of tea has a long history. It may originate during the "Shen-Nong" era of ancient China, around 5000 to 6000 years ago (Chen, 1994). The probable center of origin of tea is in southwest China. The tea plant, *Camellia sinensis*, has been known to and cultivated by Chinese people for a very long time. The word "cha" meaning tea in Chinese is the first word that was coined for tea in the world (Yao, 1992). Originally, tea was used as medicine for various illnesses in China. The first literary mentions of tea all agree that its taste was remarkably bitter. At that time it was also called "bitter tea". Bitterness of tea provides an important clue permitting the assumption that early drinkers decocted freshly picked tea leaves. The development from tea as "medicine" to tea as a "drink" began in late Zhou Dynasty (1124 B.C.–222 B.C.) and tea drinking gradually rose in popularity in Qin Dynasty (221–206 B.C.) (Yao and Chen, 1995). Since then, tea has been recognized both as a beverage and, in some occasions, as a medicine. In Qin Dynasty, great changes took place in the way tea was perceived. Medicinal tea had become tonic tea. The difference between medicines and tonics is that the former cured disorders whereas the latter kept one fit. Chinese medicine had always stressed prevention; therefore, doctors recommended tea to healthy persons to keep them that way. Certainly, the market of tea for healthy people wanting to maintain good health was far greater than that solely for curing illnesses. This change in attitude toward tea resulted in a great rise in popularity (Evans, 1992a).

Tea drinking continued to spread widely and rapidly during the Han Dynasty (206 B.C–220 A.D.) and the effects of tea was documented. As listed in the *Materia Medica (Ben Cao Jing)*, tea as a medicine acts as "an antidote to herbal poisons, as a cure for swelling and abscesses in the head and as a sleep inhibitor." The famous Han surgeon Hua Tuo summed up the medical viewpoint of that period in his dissertation, "To drink bitter tea constantly makes one think better." The word "constantly" by Hua Tuo is significant because it shows that people by the end of the dynasty were already drinking tea all day long, evidently as a tonic of longevity. There were records stating that "tea sobers one after drinking alcohol." In the era of Three Kingdoms (220–280 A.D.), people utilized tea to offset the effects of inebriation caused by the exaggerating consumption of alcohol. Tea had become a popular remedy for drunkenness and its ill effects. There were also records indicating that "drinking tea induces sleeplessness" and "tea keeps one awake." People of that era knew that drinking too much tea and too frequently may result in chronic sleeplessness (Evans, 1992b).

In Tang Dynasty (618–906 A.D.), tea production including cultivation, harvesting and processing had developed rapidly and the popularity of tea drinking reached a new level. Green tea was first invented as steamed green tea cake instead of directly cooking the tea leaves for drinking brew. Although tea was still considered a healthy drink and often qualified as "invigorating", it had now become a prestigious social drink, the so-called national drink of China. There were meticulous and luxurious tea banquets, tea parties and tea competitions held in the Imperial Palace, in the circle of high-ranking officials and social elite. Tea was also widely consumed by the average person. There existed a variety of tea specialities. Even different classes of Tang society, rich and poor, urban and peasant, as well as ethnic minorities drank tea differently. The book *The Classic of Tea* written by Lu Yu of the Tang Dynasty is the first monograph on tea in the world. In the book Lu Yu presented a comprehensive description and discussion on tea including production, preparation, quality assertion etc. Lu Yu stated that "tea growing wild is superior, garden tea takes second place." In general, Tang tea cultivators planted hill-tea on shady land along mountain slopes subject to cloud cover, exposed to fogs or frequent mists. The time for harvesting is important. Tea leaves were picked "in March-April prior to the spring rains" when new shoots had appeared and the leaves were young and tender. Tea leaves picked thereafter are less desirable, because tea leaves grow faster after rains and are consequently larger. Having good quality water is important in making a tea drink. A good cup of tea depends upon the quality of the water as much as the quality of tea. Lu Yu listed selected tea-water sources and chose mountain spring water over all others. People in Tang Dynasty paid very much attention to the preparation of tea drinks. As Lu Yu warned, "Tea improperly prepared can cause sickness." He mentioned: "The first cup of tea should have a haunting flavor, strange and lasting. When you drink tea, sip only, otherwise you will dissipate the flavor. Moderation is the very essence of tea." A Tang poet described how pleasant and joyful people felt on tea physically and mentally. In the so-called "The seven cups of tea", he wrote that tea could moisten the lip and throat, break loneliness, and penetrate one's barren entrails; moreover, could call one up to the realm of immortals and the feeling to ride a sweet breeze and waft away (Evans, 1992c).

Tea continued to gain popularity in China after Tang Dynasty. Teahouses first appeared in Song Dynasty (960–1279 A.D.) and quickly spread throughout the country. Teahouses were known as places where one could relax and have a good time. Black tea which the Chinese called "red tea" was manufactured and consumed in Ming dynasty (1368–1644 A.D.). Most of the manufactured black tea was exported and the majority of Chinese remained consuming green tea. Drinking of tea was considered beneficial to health. In the book *Tea Manual (Cha Pu)* written in Ming Dynasty, the author concluded that "Drinking genuine tea helps quench the thirst, aids digestion, checks phlegm, wards off drowsiness, dispels boredom and dissolves greasy foods."

In Japan the first tea was brought from China in the early 9th century. China started supplying Russia with small quantities of tea toward the end of the 17th century, and the trade was first carried overland by caravans. The first tea to reach Europe went by way of the Dutch who brought the first consignments to Holland in the early part of the 17th century. The early supplies of tea entering England were brought over from Holland. In London the first tea was served to the public in 1657. By the mid 1750s tea houses and tea gardens were appearing in and around London. Tea was soon to become the national drink in the British Isles (Weatherstone, 1992). An author in the late 18th century described the difference in the way of tea drinking between Chinese and the Europeans. He mentioned that Chinese drank tea without sugar; however, almost everyone in Europe added sugar to tea. Since then, great changes have taken place, and the difference, at least in some regions, seems to be less prominent in the present time.

In China, from Tang Dynasty to Qing Dynasty (618–1911 A.D.), there had existed a great number of books contributed to tea. Those included mainly 3 categories, namely, books on herbal medicines, tea manuals and general historic publications. As summarized by Lin (1992), tea was reported to exhibit 24 kinds of physiological and therapeutic effects, such as causing less sleep, calming down, clearing sight, relieving headache, dispelling thirst, dissipating fever, detoxification, helping digestion, reducing obesity, diuresis, as a pectoral for chest diseases, invigorating, strengthening teeth, and more. In addition to the probable applications as medicine, tea used as a daily beverage has made great contributions to human health in at least two major aspects (Zhu, 1992). Firstly, tea drinking changes the habit of how people consume water. In ancient times, when people felt thirsty they would simply drink natural, unprocessed water that might contain pathogenic microbes. Since the adoption of tea drinking, people had used boiling water to make tea infusion. In fact this practice helped people avoid a variety of infectious diseases. Secondly, tea appears to be a good substitute for alcoholic beverages. Those people who very much enjoyed tea drinking might avoid alcohol over-consumption that causes severe damage to the human body.

2. COMPOSITION OF TEA AND THE ACTIVE CONSTITUENTS

For a better understanding of the bioactivity of tea and its physiological and pharmacological effects, it is essential to scrutinize the chemical composition of tea and its bioactive constituents. There exist volatile and non-volatile compounds. Generally speaking, the tea aroma is mainly dependent on the volatile compounds it contains, while the color and the taste of tea are mainly dependent on the non-volatile compounds.

2.1. Chemical Composition of Tea Flush

Tea flush is generally a reference to young shoots of tea that consist of the terminal bud and two adjacent leaves. In fresh tea flush there exist a wide variety of non-volatile compounds: polyphenols, flavonols and flavonol glycosides, flavones, phenolic acids and depsides, amino acids, chlorophyll and other pigments, carbohydrates, organic acids, caffeine and other alkaloids, minerals, vitamins, and enzymes. The chemical composition of the tea leaves depends upon leaf age, the clone being examined, soil and climate conditions, and agronomic practices.

The total polyphenols in tea flush ranges from 20% to 35%. Tea polyphenols include mainly six groups of compounds. Among them, the flavonols (mainly the catechins) are the most important group and occupy 60–80% of the total amount of polyphenols. The catechins have been widely, and intensively investigated for their bioactivity and utilization. Four major catechins, namely (–)-epigallocatechin-3-gallate (EGCG), (–)-epigallocatechin (EGC), (–)-epicatechin-3-gallate (ECG), and (–)-epicatechin (EC), constitute around 90% of the total catechin fraction; and (+)-catechin (C) and (+)-gallocatechin (GC) are present about 6% of the fraction. There are some minor catechins that constitute less than 2% of the total catechins. The catechins that are water-soluble, colorless compound contribute to astringency and bitterness in green tea.

There are three major flavonol aglycones in the fresh leaf, kaempferol, quercetin and myricetin. These substances occur both as free flavonols and as flavonol glycosides. The glycosidic group may be glucose, rhamnose, galactose, arabinose, or rutinose.

These compounds are considered to contribute to bitterness and astringency in green tea (McDowell & Taylor, 1993).

Amino acids constitute around 4% in tea flush. The most abundant amino acid is theanine (5-N-ethylglutamine) which is apparently unique to tea and found at a level of 2% dry weight (50% of free amino acid fraction). The precursors in the biosynthesis of theanine in tea plant have been identified as glutamic acid and ethylamine. The site of theanine biosynthesis is the root and from there it is transferred to younger leaves; thus the roots have the highest concentration in the tea plant.

Free sugars constitute 3–5% of the dry weight of tea flush. It consists of glucose, fructose, sucrose, raffinose and stachyose. The monosaccharides and disaccharides contribute to the sweet taste of tea infusion. The polysaccharides present in tea flush can be separated into hemicellulose, cellulose and other extractable polysaccharide fraction. Some investigations have demonstrated that polysaccharides extracted from manufactured tea showed decreasing effect on blood-glucose level.

Caffeine is the major purine alkaloid present in tea. The content of caffeine in tea flush is approximately 2–5% (dry weight basis). Theobromine and theophylline are found in very small quantities. Traces of other alkaloids, e.g. xanthine, hypoxanthine and tetramethyluric acid, have also been reported.

Many volatile compounds, collectively known as the aroma complex, have been detected in tea. The aroma in tea can be broadly classified into primary or secondary products. The primary products are biosynthesized by the tea plant and are present in the fresh green leaf, whilst the secondary products are produced during tea manufacture (Sanderson & Graham, 1973). Some of the aroma compounds, which have been identified in fresh tea leaves, are mostly alcohols including Z-2-penten-1-ol, n-hexanol, Z-3-hexen-1-ol, E-2-hexen-1-ol, linalool plus its oxides, nerol, geraniol, benzylalcohol, 2-phenylethanol, and nerolidol (Saijo & Takeo, 1973). The aroma complex of tea varies with the country of origin. Slight changes in climate factors can result in noticeable changes in the composition of the aroma complex. Notably, teas grown at higher altitudes tend to have higher concentrations of aroma compounds and superior flavor, as measured by the flavor index (Owuor *et al.* 1990). Growing tea in a shaded environment may change the aroma composition and improves the flavour index. The aroma complex also varies with season and these variations appear to be larger under temperate or sub-tropical climates (Gianturco *et al.* 1974).

2.2. Chemical Composition of Made Tea

A series of changes occur in the process of manufacturing. There are three basic types of tea manufacture, resulting in the production of green, semi-fermented, and black teas. They differ mainly in the degree of fermentation. Green tea undergoes little or no fermentation and black tea is produced as a result of a full fermentation. Semi-fermented tea (oolong tea) is a product of partial fermentation. The major steps for the manufacturing of green tea include spreading-out, fixing, rolling and drying. For black tea manufacture the major steps include withering, rolling, fermentation and drying.

The total content of polyphenols decreases to some degree during green tea processing. In comparison with fresh tea leaves, polyphenol contents of manufactured green tea generally decrease by around 15%. Polyphenols undergo marked changes during black tea processing. As a result of enzymatic oxidation of the catechins by polyphenol oxidase, two groups of polyphenol compounds, theaflavins and thearubigins, are formed which are thought to be unique to black tea. The enzyme oxidizes the catechins to their respective o-quinones, which rapidly react with each other and other compounds to form theaflavins and thearubigins. Theaflavins account for between 0.3 and 1.8% of the dry weight of black tea and between 1 and 6% of the solids in tea liquor. Various theaflavins and their respective precursors are as follows: theaflavin (EC + ECG), theaflavin-3-gallate (EC + EGCG), theaflavin-3'-gallate (ECG + EGC), theaflavin-3,3'-digallate (ECG + EGCG) and isotheaflavin (EC + GC). Determined by Coxon *et al.* (1970a,b), the approximately relative proportions of the theaflavins in black tea were, theaflavin (18%), theaflavin-3-gallate (18%), theaflavin-3'-gallate (20%), theaflavin-3,3'-digallate (40%), and isoflavin together with the theaflavic acids approximately 4%. The exact levels of different theaflavins vary with the fermentation conditions applied (Robertson, 1983). Theaflavins are bright red pigments giving the tea liquor the characteristics described by tasters as "brightness" and "briskness". The contribution made by these compounds to tea quality differs with individual theaflavins. It is believed that the digallate contributes most while theaflavin itself contributes least.

The thearubigins constitute between 10 and 20% of the dry weight of black tea and represent approximately 30–60% of the solids in tea liquor (cited from Robertson, 1992). Unlike the theaflavins, the thearubigins have still not been characterized. They are diverse in their chemistry and possibly their molecular size which may range in molecular weight from 700 to 40,000 Da. The level of free amino acids may increase during tea processing. Because of an increase in the activity of proteolytic enzymes, proteins are hydrolyzed with a concurrent increase of free amino acids. Accordingly, there is a change in the qualitative composition of free amino acids in the manufactured tea as compared with the fresh leaf. Twenty-five amino acids are reportedly found in made tea; among those theanine is the highest in quantity. The contents of theanine in made tea averaged 1.37% in the estimation of 100 g tea samples. Theanine has two enantiomers: L- and D-theanine. The average level of D-theanine was around 1.85% of the total theanine. The relative amounts of D-theanine display inverse correlation to tea quality. The ratio of D-theanine to L-theanine increases under high temperature of storage. Therefore, the ratio of theanine enantiomers might be used as an indicator for long-term storage or as a tool in the grading of tea. Theanine is considered to be important in the taste of green tea. Recently, theanine has become a substance of interest for its bioactivity. As reported (Yokogoshi, 1995), theanine showed a lowering effect on blood pressure in hypertensive rats. Theanine may also act as biochemical modulator to enhance the antitumor effect of doxorubicin (Sadzuka *et al.* 1996).

The level of caffeine may increase during withering, an early step of tea processing, due to the losses of other components. However, caffeine content may decrease during the firing process of tea manufacturing. Caffeine accounts for some 5–10% of the solid material extracted from tea when mixed with boiling water. Theobromine accounts for about a further 0.3% of the total extracted solids, but contrary to common belief, there is no theophylline in tea as drunk ordinarily (Scott *et al.* 1989). The pharmacological properties of caffeine have been recognized for many years. Caffeine has marked stimulatory effects upon the central nervous system and has been employed therapeutically for this purpose (Aranda *et al.* 1977). Caffeine relaxes smooth muscle of the bronchi, making it of value for the treatment of asthma and the bronchospasm of chronic bronchitis. In humans, caffeine can increase the capacity of muscular work. This may, at least in part, be due to its stimulatory effect upon the nervous system. There has been some concern about the untoward effects of caffeine. A number of investigations on caffeine intake have been reported. According to data cited by Scott *et al.* (1989), the estimated average daily caffeine intakes per capita were 50 mg worldwide, 186–325 mg in USA, and 359 in UK. The no-effect dosage of caffeine was recommended as 40 mg/kg/day (Elias, 1986). As reported, the average caffeine content of tea (mg/cup) was 55 mg/cup for bagged tea or leaf tea (Scott *et al.* 1989). Notably, the caffeine content goes up with the infusion time. The contents of caffeine per cup were 48 mg (2 min) and 80 mg (5 min) for bagged tea as well as 38 mg (2 min) and 60 mg (5 min) for leaf tea (Starvic *et al.* 1988). Apparently, the quantity of caffeine ingested by average or moderate tea consumers, taking 3–6 cups a day, is below the above-mentioned dosage level unless significant additional amounts of caffeine come from other sources.

The contents of tea aroma in manufactured tea differ from those of the fresh tea leaves. During the process of tea manufacturing, the amounts of various aroma compounds are changing differently; some increase, while some others decrease. While some of the aroma compounds of black tea are present in the fresh leaf, most of them are formed during tea manufacture via enzymatic, redox or pyrolytic reactions. So far, over six hundred aroma compounds have been found. Those include hydrocarbons, alcohols, aldehydes, ketones, acids, esters, lactones, phenols, nitrogenous compounds, sulfur compounds, and miscellaneous oxygen compounds (Robinson & Owuor, 1992). The majority of aroma compounds are probably derived from carotenes, amino acids, lipids and terpene glycosides. The compounds produced from carotenes have a major effect on the aroma of tea. Flavor teas are normally produced from green leaf with high carotene contents. It is of interest to investigate the bioactivity of the aroma compounds. Some of them have been reported to display a variety of biological effects. Geraniol, for example, is an inhibitor of mevalonate biosynthesis, causing reduction of cholesterol. Geraniol inhibits the proliferation of cultured tumor cells and exerts inhibitory effect on the growth of transplanted hepatoma in rats and melanoma in mice (Yu *et al.* 1995). Geraniol also shows antifungal activity (Mahmoud, 1994). Since tea aroma composed of a complicated group of compounds, the bioactivity of the majority of tea aroma compounds remains to be further explored.

As well known, there are a great variety of teas that differ in the origin places and countries, the plant cultivars, and the manufacturing practices. Different kinds of tea may vary in their compositions, thus the biological effects may vary to some degree. As estimated by Gill (1992), there were over 250 different specialty tea products included in the grocer price list. At the present time in China more than 200 varieties exist. These include 138 green teas, 10 black teas, 13 semi-fermented teas, and many miscellaneous teas (Cheng, 1992).

3.   POTENTIAL MEDICAL APPLICATIONS OF TEA

It has been known for a long time that tea may exert a number of physiological effects on humans. Caffeine, one of the noticeable components in tea, has been under a wide range of investigations and thought to be responsible for many of the tea effects, beneficial or undesirable. In addition to caffeine, other components of tea, especially the polyphenols, have been proved by modern biomedical research to possess a variety of physiological and pharmacological effects. The understanding on the biological effects of tea and its components and the related mechanism of action has been rapidly expanded, particularly over the last two decades.

3.1.   Antimicrobial Activity of Tea

Medical books written as early as in the Song Dynasty (960–1279 A.D.) in China mention that green tea in combination with ginger can effectively cure dysenteric disorders, including those so-called red and white in appearance (Lin, 1992). Modern medical research has demonstrated that tea and tea products are active against a wide range of microorganisms, implying that tea may be potentially useful for treatment of some infectious illnesses.

A number of reports indicated that green tea and black tea can inhibit the growth of a wide spectrum of pathogenic bacteria including *Staphylococcus aureus, Shigella dysenteriae, Salmonella typhosa, Pseudomonas aeruginosa, Vibrio cholerae, and others*. Both tea powder and tea infusion are active. In a comparison of the activity of green tea and black tea against various bacteria known to cause diarrheal diseases, the Gram-positive bacteria were more sensitive than Gram-negative bacteria. In the case of *Staphylococcus aureus*, black tea showed stronger bactericidal activity than green tea and coffee (Toda, 1989). Tea polyphenols are major components responsible for the antibacterial activity of various tea products. The active tea polyphenols include EC, ECG, EGC, EGCG, and theaflavins and their MIC (minimum inhibition concentration) values were estimated in the range of 100–800 ppm (Hara *et al.* 1989). In addition, tea aroma compounds, such as linalool, geraniol, nerolidol, *cis*-jasmone and caryophyllene, also display antibacterial activities. It is of importance that tea can inhibit methicillin-resistant *Staphylococcus aureus* (MRSA) which poses severe problems in clinical chemotherapy. As reported, aqueous extracts of different types of tea were bactericidal to staphylococci at well below "cup of tea" concentration and the active compounds were determined to be EGC, EGCG, and ECG in green tea as well as theaflavin and its gallates in black tea (Yam *et al.* 1998). In addition to the direct antibacterial activity, tea extracts can reverse the methicillin resistance in methicillin-resistant *Staphylococcus aureus* and, to some extent, the penicillin resistance in beta-lactamase-producing ones. There exists a synergy between beta-lactam antibiotics and tea extracts (Yam *et al.*, 1998).

The anticaries activity of tea has drawn much attention. Tea products have shown inhibitory activity against mutans streptococci and glucosyltransferases. Mutan streptococci that are known to synthesize glucans have been implicated as primary causative agent of caries (Hamada *et al.* 1984). Two types of glucans, water-soluble and water-insoluble are synthesized by two different groups of glucosyltransferases in these bacteria. The water-insoluble glucan is highly adhesive to tooth surface resulting in the formation of dental plaque. The bacteria grow in dental plaque, metabolize various sugars there and produce organic acids, especially lactic acid, which retains in the plaque, eventually to decalcify the tooth enamel and develop dental caries (Koga *et al.* 1986). An earlier investigation conducted at primary schools over a year has found that the incidence of dental caries among children who took a cup of tea immediately after lunch was found to be significantly lower than that among children who did not (Onisi *et al.* 1981). Various tea extracts have shown bactericidal activity against mutan streptococci. Moreover, several catechins, the components from green tea, are active against cariogenic bacteria. As reported, GC and EGC completely inhibited the growth of three strains of cariogenic bacteria at 250 and 500 μg/ml respectively. The MIC values of EGCG ranged between 500 and 1000 μg/ml. The inhibitory activity of GC and EGC was stronger than that of C and EC; and EGCG was more active than ECG. In addition, tea polyphenols inhibit the water-insoluble glucan synthesis catalyzed by glucosyltransferase from *Streptococcus mutans*. Tea aroma compounds also display inhibitory effects against *Streptococcus mutans*. Nerolidol, an aroma compound, is active against this microorganism with an MIC of 25 μg/ml. Preliminary clinical trials on the preventive effect of tea on tooth caries have provided positive results (Elvin-Lewis *et al.* 1986; Ooshima *et al.* 1994). Evidently, tea and tea products may be useful for caries control.

It is of interest to assess the antiviral activity of tea. As early as 1949, Green reported the inhibitory effect of black tea extracts against influenza virus A in embryonated eggs. Tea polyphenols, especially EGCG, have been found to be active against a series of viruses under experimental conditions. The antiviral activity of tea polyphenols seems to be attributed to interference with virus adsorption. Notably, reverse transcriptase of human immunodeficiency virus type 1 (HIV-1 RT) is highly sensitive to the inhibitory effects of tea polyphenols such as EGCG and ECG. As determined, EGCG was a very strong HIV-1 RT inhibitor with an IC_{50} of 6.6 nM (Tao *et al.* 1992). However, the inhibitory effects of tea polyphenols on the enzyme may be counteracted by bovine serum albumin or other agents (Moore *et al.* 1992). Although active against the enzyme, EGCG and ECG were reported to show no inhibitory effects against HIV-1 in cell cultures. The potential usefulness of tea and products for antiviral therapy needs further investigation.

3.2. Effects of Tea on Cardiovascular Disorders

The physiological effects of tea and tea products on the cardiovascular system and their potential uses for the prevention and treatment of cardiovascular disorders have drawn a great deal of interest. In traditional Chinese medicine, tea, especially green tea, can be used as a major component in a composed prescription for treatment of hypertension or coronary heart diseases. Epidemiological and experimental investigations on the effects of tea on cardiovascular disorders have been concentrated mainly on the effects on blood pressure, on blood lipids, and on atherosclerosis.

3.2.1. Effects on blood pressure

Epidemiological investigations have shown that tea consumption may exert a lowering effect on blood pressure. A survey of adults in China showed that the average rate of hypertension in the group who drank tea as a habit was lower than that in the group who did not. A clinical investigation on patients with hypertension revealed that a 10 g daily intake of green tea for half a year resulted in reduction of the blood pressure by 20–30% (Chen, 1994). Oral administration of tea polyphenols also yielded a decrease of blood pressure in patients. A number of studies also revealed the blood pressure decreasing effects of tea in experimental animal models. The reported active substances include tea extracts, polyphenols, and tea tannin. In addition, theanine, the amino acid component of tea, was found to be effective in decreasing the blood pressure of spontaneous hypertensive rats. Caffeine may exert a variable effect on blood pressure. Intravenous administration of caffeine may cause an initial fall in blood pressure and then a secondary rise. So, caffeine should be taken into account if whole tea preparation instead of tea component is used in studying the effect on blood pressure.

3.2.2. Effects on blood lipids

Lipids exist ubiquitously in the living body and they can be classified into three major categories, namely, simple lipids (mainly triglyceride), compound lipids (such as phospholipids and glycolipids), and derived lipids (cholesterol and fatty acids). Since lipids account for much of the energy expenditure, the transport of lipids through blood circulation is of particular importance in the organism. Many classes of lipids are transported in the blood as lipoproteins and the blood levels of lipids may serve as an indicator for metabolic disturbances and abnormal status. Apart from free fatty acid, some major lipoproteins in the blood have been identified that are important physiologically and in clinical diagnosis. Those include chylomicrons, very low density lipoproteins (VLDL), low density lipoprotein (LDL), and high density lipoproteins (HDL).

Several surveys on populations reveal that there is an inverse relationship between tea drinking and the blood level of cholesterol. An increase in green tea consumption may substantially decrease serum total cholesterol and triglyceride concentration; in addition, it may associate with an increased HDL level. However, there are discrepancies as some surveys found no statistically significant relationship between tea drinking and serum lipid levels. During the past decade, a series of investigations have been conducted in experimental animals including mice, rats and rabbits. The animals fed with high fat/high cholesterol diet had reduced plasma cholesterol level when given tea instead of water to drink or given a diet supplemented with tea (Matsuda et al. 1986; Matsumoto et al. 1995). The hypocholesterolemia activity of tea catechins can be attributed to the inhibition of intestinal cholesterol absorption as well as the enhancement of cholesterol excretion through feces.

3.2.3. Effects on atherogenesis

Atherosclerosis is one of the most frequently occurring cardiovascular diseases. The changes of the arterial vessel include a deposit of lipid at the intima (inner layer of the blood vessel) accompanied with thrombosis and the proliferation of smooth muscle cells and the connective tissue. These lesions usually involve large and medium-sized arteries, namely the aorta and the major branches to vital organs such as the heart, the brain, and the kidneys, leading to severe consequences. As well known, hyperlipidemia is one of the risk factors for atherogenesis. As known, the plasma level of LDL-cholesterol is positively related to atherogenesis whereas the HDL-cholesterol level is negatively related to the disorder. The oxidation of LDL cholesterol may play an essential role in the pathogenesis. Epidemiological studies have shown that tea consumption may occasionally have a beneficial effect. Tea consuming may result in a lower risk of atherosclerosis, coronary heart disease, and stroke (Stensvold *et al.* 1992; Hensrud & Heimburger, 1994). Experimental research also provides evidence that tea and tea products are active in suppression of atherogenesis by lowering lipidemia and inhibiting platelet aggregation that leads to the formation of thrombus.

3.3. Effects of Tea on Cancer

Cancer is a serious problem of health that causes global concern. According to the WHO 1998 report, in the year of 1997 there were 57.45 million cancer patients including 9.24 million newly detected cancers and 6.23 cancer deaths. The possible effectiveness of tea and tea products on cancer prevention and treatment attracts great attention in the medical circle.

3.3.1. Tea and cancer prevention

The topic on the relationship between tea consumption and cancer incidence has been under investigation for decades. It could be evaluated in two aspects, harmful or beneficial. As to the possible harmful effect, there is inadequate evidence for the carcinogenicity of tea in humans and in experimental animals. According to the IARC Working Group, the overall evaluation is that tea is not classifiable as to carcinogenicity to humans (WHO International Agency for Research on Cancer, 1991). However, on the other hand, a number of epidemiological studies have shown that there exists an inverse correlation between tea consumption and the incidence of certain kinds of cancer in humans. Intensive research has demonstrated the inhibitory effect of tea infusion and its components, especially polyphenols, on chemical carcinogenesis of various cancers in experimental animals (Cheng *et al.* 1991; Yang & Wang, 1993). Tea extracts and tea polyphenols have shown a variety of biological effects that may be related to carcinogenesis, the development of cancer. Those effects include the anti-mutagenicity, the inhibitory effect on promotion (Cheng *et al.* 1989), the blocking effect on the process of cell transformation, the scavenging effect on free radicals as well as suppressing the precancerous changes during carcinogenesis such as edema and hyperplasia induced by TPA (Tong *et al.* 1992). Both epidemiological and experimental investigations have provided sound basis for the potential utilization of tea and tea products in cancer prevention.

3.3.2.   *Potential use of tea in cancer therapy*

In addition to cancer prevention, the possible use of tea in cancer treatment is also a topic of interest. Tea extracts, containing mainly polyphenols, display cytotoxicity to cancer cells. Tea extracts or tea polyphenols are active in suppressing the proliferation of cultured cancer cells. The inhibition is dose-dependent. In a comparison of the component activity, EGCG, GC and EGC are more potent than EG, ECG, catechin and caffeine (Valcic *et al.* 1996). Tea polyphenols can reduce the surviving ability of cancer cells and exert a variety of biological effects on cancer cells such as inducing programmed cell death (Hibasami *et al.* 1996), blocking cell cycle progression (Yan *et al.* 1990), and inhibiting the activity of telomerase (Naasani *et al.* 1998). A number of investigations have shown that tea extracts or tea components are of therapeutic efficacy against cancers in experimental animals. They are active against both benign tumors and malignant tumors. Tea extracts or tea polyphenols can reduce the size of papillomas, benign tumors in nature, in skin carcinogenesis as well as inhibit the growth of transplanted tumors, which are malignant in nature (Wang *et al.* 1992; Oguni *et al.* 1988). Notably, tea polyphenols may play a role as biochemical modulators to enhance the antitumor activity of cytarabine and methotrexate (Zhen *et al.* 1991). Tea polyphenols exerted modulating effects to render doxorubicin-resistant cancer cells sensitive to doxorubicin (Stammler & Volm, 1997). Moreover, tea catechins may reduce the toxic effect of cisplatin. Theanine, an amino acid component of tea, may reduce the toxicity of doxorubicin with no decrease of its antitumor activity (Sugiyama *et al.*, 1997).

4.   FUTURE TRENDS

Tea is one of the most popular beverages in the world. Tea consumption has been experienced over a long period of time by a huge population. Therefore, tea consumption is highly relevant to health. The effect of tea on human health has drawn great attention for centuries, particularly in recent decades, and will continue to remain an attractive issue in biomedical research.

Tea consumption has a long history of more than two thousand years. Originating in China, then spreading to Japan and Europe and other areas, tea has been consumed by thousands of millions of people and the effects of tea on humans has been documented in a huge amount of literature and records. As it is well known, tea is safe and may be beneficial to humans' health; and no substantial untoward effects occur, if the intake of tea remains at the average level.

Tea may play an active role in the prevention of certain kinds of disorders, especially chronic diseases in humans. Epidemiological and experimental investigations have provided evidence that tea may exert a blocking effect on the development of various disorders, such as microbial infections, cancers, cardiovascular diseases and others. The utilization of tea and tea products for prevention is promising. What needs to be further investigated includes the determination of the active constituent, elucidation of the basic mechanism of action, and the evaluation of clinical effectiveness.

The potential of tea used as therapeutic agent is of particular interest. In traditional Chinese medicine, tea, mostly in combination with other herbal medicines, has been applied to the treatment of various diseases. Modern medical research has provided a wide range of evidence that tea may be effective in therapy. For example, tea and its polyphenolic components display cytotoxicity to cancer cells and show therapeutic efficacy against tumor growth in experimental animals. Another example is that due to its antimicrobial activity tea seems to be useful for treating certain kinds of infections. Moreover, tea may be used as biochemical modulator to enhance the therapeutic effectiveness of other drugs. For the purpose of therapeutic application, it is essential to identify and isolate the active constituent, to evaluate the therapeutic efficacy with relevant models, and to detect possible toxic effects before entering clinical trials.

REFERENCES

Aranda, J.V., Gorman, W., Bergsteinsson, H., Gunn, T. (1977) Efficacy of caffeine in treatment of apnea in the low-birth-weight infant. *J. Pediatrics*, **90**, 467–472.

Chen, Z.M. (1994) The physiologically modulating function of tea to humans. *Tea Abstacts*, 8 (1), 1–8; 8(2), 1–8.

Chen, Z.M. and Yu, Y.M (1994) Tea. In C.J. Arntzen and E.M. Ritter, (eds), *Encyclopedia of Agricultural Science*, Academic Press, San Diego, vol. 4, pp. 281–288.

Cheng, Q.K, (1992) Contemporary famous tea. In Z.M. Chen, (ed), *China Tea Book*, Shanghai Culture Publishers, Shanghai, pp. 128–258.

Cheng, S.J., Ho, C.T., Huang, M.T., Wang, Z.Y., Liu, S.L., Gan, Y.N., Li, S.Q. (1989) Inhibitory effect of green tea extracts on promotion and related action of TPA. *Acta Acad. Med. Sin.*, **11**, 259–264.

Cheng, S.J., Ding, L., Zhen, Y.S., Lin, P.Z., Zhu, Y.J., Chen, Y.Q., Hu, X.Z. (1991) Progress in studies on the antimutagenicity and anticarcinogenicity of green tea epicatechins. *Chin. Med. Sci. J.*, **6**, 233–238.

Coxon, D.T., Holmes, A., Ollis, W.D. (1970a) Isotheaflavin. A new black tea pigment. *Tetrahedron Lett.*, pp. 5241–5246.

Coxon, D.T., Holmes, A., Ollis, W.D. (1970b) Theaflavic and epitheaflavic acids. *Tetrahedron Lett.*, pp. 5247–5250.

Elias, P.S. (1986) Current biological problems with coffee and caffeine. *Cafe Cacao The*, **30**, 121–138.

Elvin-Lewis, M., Steelman, R. (1986) The anticariogenic effects of tea drinking among Dallas children. *J. Dent. Res.*, **65**, 198.

Evans, J.C. (1992a) *Tea in China, the History of China's National Drink*, Greenwood Press, New York, pp. 19–20.

Evans, J.C. (1992b) Ibid, pp. 32–34.

Evans, J.C. (1992c) Ibid, pp. 47–48.

Gianturco, M.A., Biggers, R.E., Riddling, B.H. (1974) Seasonal variations in the composition of the volatile constituents of black tea. A numerical approach of the correlation between composition and quality of tea aroma. *J. Agric. Food Chem.*, **22**, 758–764.

Gill, M. (1992) Speciality and herbal teas. In K.C. Willson and M.N. Clifford (eds), *Tea: Cultivation to Consumption*, Chapman and Hall, London, pp. 517–534.

Green, R.H. (1949) Inhibition of multiplication of influenza virus by extracts of tea. *Proc. Soc. Exp. Biol. Med.*, **71**, 84–85.

Hamada, S., Koga, T., Ooshima, T. (1984) Virulence factors of *Streptococcus mutans* and caries prevention. *J. Dent. Res.*, **63**, 407–411.

Hara, Y., Ishigami, T. (1989) Antibacterial activities of tea polyphenols against foodborne pathogenic bacteria. *J. Jpn. Soc. Food Sci. Technol.*, **36**, 996–999.

Hensrud, D.D. and Heimburger, D.C. (1994) Antioxidant status of fatty acids and cardiovascular disease. *Nutrition*, **10**, 170–175.

Hibasami, H., Achiwa, Y., Fujikawa, T., Komiya, T. (1996) Induction of programmed cell death (apoptosis) in human lymphoid leukemia cells by catechin compounds. *Anticancer Res.*, **16**, 1943–1946.

Koga, T., Okahashi, N, Asakava, H., Hamada, S. (1986) Adherence of *Streptococcus mutans* to tooth surfaces. In S. Hamada *et al.* (eds), *Molecular Microbiology and Immunobiology of Streptococcus Mutans*, Elsevier Science Publishers, Amsterdam, pp. 111–120.

Lin, Q.L. (1992) Pharmacological properties of tea. In Z.M. Chen, (ed), *China Tea Book*, Shanghai Culture Publishers, Shanghai, pp. 91–101.

Mahmoud, A.L. (1994) Antifungal action and antiaflatoxigenic properties of some essential oil constituents. *Lett. Appl. Microbiol.*, **19**, 110–113.

Marks, V. (1992) Physiological and clinical effects of tea. In K.C. Willson and M.N. Clifford (eds), *Tea: Cultivation to Consumption*, Chapman and Hall, London, pp. 707–739.

Matsuda, H., Chisaka, T., Kubomura, Y., Yamahara, J., Sawada, T., Fujimura, H., Kimura, H. (1986) Effects of crude drugs on the experimental hypercholesterolemia. I. Tea and its active principles. *J. Ethnopharmac.*, **17**, 213–224.

Matsumoto, N. and Hara, Y. (1995) Blood-cholesterol depressing activity of tea catechin. *Food Chem.*, **13**, 81–84.

McDowell, I. and Taylor, S. (1993) Tea: types, production and trade. In R. Macrae, R.K. Robinson and M.J. Sadler, (eds), *Encyclopaedia of Food Science, Food Technology and Nutrition*, Academic Press, London, Vol. 7, pp. 4521–4533.

McDowell, I. and Taylor, S. (1993) Tea: Chemistry. In R. Macrae, R.K. Robinson and M.J. Sadler, (eds), *Encyclopaedia of Food Science, Food Technology and Nutrition*, Academic Press, London, Vol. 7, pp. 4527–4533.

Moore, P.S., Pizza, C. (1992) Observations on the inhibition of HIV-1 reverse transcriptase by catechins. *Biochem. J.*, **288**, 717–719.

Naasani, I., Seimiya, H., Tsuruo, T. (1998) Telomerase inhibition, telomere shortening, and senescence of cancer cells by tea catechins. *Biochem. Biophys. Res. Commun.*, **249**, 391–396.

Oguni, S., Nasu, K., Yamamoto, S., Nomura, Y. (1988) On the antitumor activity of green tea leaf. *Agric. Biol. Chem.*, **52**, 1879–1880.

Onisi, M., Shimura, N., Nakamura, C., Sato, M. (1981) A field test on the caries preventive effect of tea drinking. *J. Dent. Hlth.*, **31**, 13.

Ooshima, T., Minami, T., Aono, W., Tamura, Y., Hamada, S. (1994) Reduction of dental plaque deposition in humans by Oolong tea extract. *Caries Res.*, **28**, 146–149.

Owuor, P.O., Obaga, S.O., Othieno, C.O. (1990) Effects of altitude on the chemical composition of back tea. *J. Sci. Food Agric.*, **50**, 9–17.

Robertson, A. (1983) Effects of physical and chemical conditions on the *in vitro* oxidation of tea leaf catechins. *Phytochem.*, **22**, 897–903.

Robertson, A. (1992) The chemistry and biochemistry of black tea production – the non-volatiles. In K.C. Willson and M.N. Clifford (eds), *Tea: Cultivation to Consumption*, Chapman and Hall, London, pp. 555–601.

Robinson, J.M. and Owuor, P.O. (1992) Tea aroma. In K.C. Willson and M.N. Clifford (eds), *Tea: Cultivation to Consumption*, Chapman and Hall, London, pp. 603–647.

Sadzuka, T., Sugiyama, T., Miyagishima, A., Mozawa, Y., Hirota, S. (1996) The effects of theanine, as a novel biochemical modulator, on the antitumor activity of adriamycin. *Cancer Lett.* **105**, 203–209.

Saijo, R., and Takeo, T. (1973) Volatile and non-volatile forms of aroma compounds in tea leaves and their changes due to injury. *Agric. Biol. Chem.*, 37, 1367–1373.

Sanderson, G.W. and Graham, H.N. (1973) On the formation of black tea aroma. *J. Agric. Food Chem.*, **21**, 576–585.

Scott, N.R., Chakraborty, J., Marks, V. (1989) Caffeine consumption in the United Kingdom: a retrospective survey. *Food Sci. Nutr.*, **42F**, 183–191.

Stammler, G. and Volm, M. (1997) Green tea catechins (EGCG and EGC) have modulating effects on the activity of doxorubicin in drug-resistant cell lines. *Anticancer Drugs*, 8, 265–268.

Starvic, B., Klassen, R., Watkinson, B. *et al.* (1988) Variabilities in caffeine consumption from coffee and tea: possible significance for epidemiological studies. *Food Chem. Toxicol.*, **26**, 111–118.

Stensvold, I., Tverdal, A., Solvoll, K., Foss, O.P. (1992) Tea consumption: relationship to cholesterol, blood pressure and coronary and total mortality. *Prev. Med.*, **21**, 546–553.

Sugiyama, T., Sadzuka, Y., Hirota, S. (1997) Effects of theanine, a tea leaf component, on antitumor activity and inhibition of tumor metastasis of adriamycin, *Proc. Am. Cancer Res.*, **38**, 611.

Tao, P.Z., Zhang, T., Zhou, P., Wang, S.Q., Chen, S.J., Jiang, J.Y., Guo, J.T., Chen, H.S. (1992) The inhibitory effects of catechin derivatives on the activities of human immunodeficiency virus reverse transcriptase and DNA polymerases. *Acta Acad. Med. Sinica*, **14**, 334–338.

Toda, M., Okubo, S., Hiyoshi, R., Shimamura, T. (1989a) The bactericidal activity of tea and coffee. *Lett. Appl. Microbiol.*, **8**, 123–125.

Tong, T., Cheng, S.J., Li, S.Q., Bai, J.F., Hara, Y. (1992) Inhibitory effect of (–)-epigallocatechin gallate on TPA-induced activities. *Carcinogenesis, Mutagenesis, Teratogenesis*, **4**, 1–4.

Valcic, S., Timmermann, B.N., Alberts, D.S., Waechter, G.A., Krutsch, M., Wymer, J., Guillen, J.M. (1996)Inhibitory effect of six green tea catechins and caffeine on the growth of four selected human tumor cell lines. *Anticancer Drugs*, 7, 461–468.

Wang, Z.Y., Huang, M.T., Ho, C.T., Chang, R., Ma, W., Ferraro, T., Reuhl, K.R., Yang, C.S., Conney, A.H. (1992) Inhibitory effect of green tea on the growth of established skin papillomas in mice. *Cancer Res.*, **52**, 6657–6665.

Weatherstone, J. (1992) Historical introduction. In K.C. Willson and M.N. Clifford (eds), *Tea: Cultivation to Consumption*, Chapman and Hall, London, pp. 1–23.

WHO International Agency for Research on Cancer. (1991) Coffee, tea, mate, methylxanthines and methylglyoxal. *IARC Monographs on the Evaluation of Carcinogenic Risks to Humans*, **51**, 207–271.

Yam, T.S., Shah, S., Hamilton-Miller, J.M. (1997) Microbiological activity of whole and fractionated crude extracts of tea (*Camellia sinensis*), and of tea compounds. *FEMS Microbiol. Lett.*, **152**, 169–174.

Yam, T.S., Hamilton-Miller, J.M., Shah, S. (1998) The effect of a component of tea (*Camellia sinensis*) on methicillin resistance, PBP2' synthesis, and beta-lactamase production in *Staphylococcus aureus*. *J. Antimicrob. Chemother.*, **42**, 211–216.

Yan, Y.S., Zhou, Y.Z., You, L.Q., Tian, Z.Q. (1990) Effect of Chinese green tea extracts on the human gastric carcinoma cell *in vitro*. *Chin. J. Prevent. Med.*, **24**, 80–82.

Yang, C.S. and Wang, Z.Y. (1993) Tea and cancer. *J. U.S. Natl. Cancer Inst.*, **85**, 1038–1049.

Yao, G.K. (1992) Tea history. In Z.M. Chen, (ed), *China Tea Book*, Shanghai Culture Publishers, Shanghai, pp. 1–9.

Yao, G.K. and Chen, P.F (1995) *Tea Drinking and Health*, Shanghai Culture Publishers, Shanghai, pp. 6–7.

Yokogoshi, H. (1995) Reduction effect of theanine on blood pressure and brain 5-hydroxyindoles in spontaneous hypertensive rats. *Biosc. Biotech. Biochem. J.*, **59**, 615–618.

Yu, S.G., Hildebrandt, L.A., Elson, C.E. (1995) Geraniol, an inhibitor of mevalonate biosynthesis, suppresses the growth of hepatomas and melanomas transplanted to rats and mice. *J. Nutr.*, **125**, 2763–2767.

Zhen, Y.S., Cao, S.S., Xue, Y.C., Wu, S.T. (1991) Green tea extract inhibits nucleoside transport and potentiates the antitumor effect of antimetabolites. *Chin. Med. Sci. J.*, **6**, 1–5.

Zhu, Z.Z. (1992) Tea affairs in ancient time. In Z.M. Chen, (ed), *China Tea Book*, Shanghai Culture Publishers, Shanghai, pp. 11–15.

2. BOTANICAL CLASSIFICATION OF TEA PLANTS

PEI-GEN XIAO AND ZHEN-YU LI*

Institute of Medicinal Plant Development, Chinese Academy of Medical Sciences and Peking Union Medical College, Research Center on Utilization and Conservation of Chinese Materia Medica, Beijing 100094, China

1. INTRODUCTION

Linneus (1753) in his "Species Plantarum" (1st ed.) nominated *Camellia* cultivated in Japan and the tea plants cultivated in China and Japan as *Camellia japonica* and *Thea sinensis*, respectively. The nomenclature of *Camellia* was for the memory of G.J. Kamel (1661–1706), or Camellus, a Moravian Jesuit traveller in Asia, while Thea from the transliteration of Fijian's dialect of tea. In 18th century, *Camellia* and *Thea* were widely accepted as two separated genera, but latterly were united for the first time by Sweet (1818), who selected the name *Camellia* L. for the combined genera. Until the fifth decade of the twentieth century, the concept of the genus *Camellia sensu lato* was more popularized, and *Thea sinensis* L. was recombined by O. Kuntze (1887) as *Camellia sinensis* (L.) O. Kuntze.

The genus *Camellia* with some 100 species was found mainly in eastern and southeastern Asia. Ming and Zhang (1996) classified *Camellia* into Subgen. *Thea* (L.) H.Y. Chang and Subgen. *Camellia*, and all the tea plants have been concentrated within Subgen. *Thea* Sect. *Thea* (L.) Dyer.

As early as three thousand years ago, Chinese people started to use tea, initially as remedy for detoxification. Starting from West Han period (206 BC to 23 AD) tea has become a daily drink. Today tea is accepted as one of the three major beverages worldwide, and is widely cultivated in the warm regions.

Apart from *Camellia sinensis* (L.) O. Kuntze, local people in the remote mountainous districts of southwestern China collect several kinds of spontaneous tea plants to use as tea. Those include *C. taliensis* W.W. Smith Melchior, *C. grandibracteata* H.T. Chang et F.L. Yu, *C. crassicolumna* H.T. Chang, *C. remotiserrata* H.T. Chang, H.S. Wang et P.S. Wang, and *C. gymnogyna* H.T. Chang. The above-mentioned tea plants have different characteristics in morphology, phytochemistry as well as in their usage from those of the genuine tea; therefore, it is necessary to make a distinct classification of tea plants for the purpose of promoting further investigation, conservation and utilization of this important economic plant.

* Laboratory of Systematic and Evolutionary Botany, Institute of Botany, Chinese Academy of Sciences, Beijing 100093, China

2. THE TRUE TEA PLANT – *CAMELLIA SINENSIS*

At present, the widely cultivated tea plant worldwide belongs solely to one species, that is *Camellia sinensis* (L.) O. Kuntze, although it has been further divided into many small species, subspecies or varieties by different authors.

2.1. Taxonomic History

Linneus named the tea plant as *Thea sinensis* L. in his "Species Plantarum" (1st ed. 1753) based on several prior references, including an illustration from E. Campfire's "Amoenitatum Exoticarum" (1712). In the second edition (1762), however, Linneus nominated *Thea bohea* L. for *Thea sinensis* L., in the meantime, another species *Thea viridis* L. was added. The name "bohea" is a tea variety "Bai-hao" in Fijian's dialect and "viridis" means green tea, the former with 6 and the latter with 9 petals.

Many authors have considered tea plants using a small species concept. They are Loureiro (1790), Salisbury (1796), Sweet (1818), Hayne (1821), Rafinesque (1838), Griffith (1838, 1854), Masters (1844), Makino (1905, 1918), Chang (1981, 1984, 1987, 1990), Tan *et al.* (1983), and Zhang *et al.* (1990). Various tea plants including cultivated varieties and several spontaneous ones should be mostly at the species level.

On the contrary, Dyer (1874), Seemann (1869), Kuntze (1887), Brandis (1911), Rehder and Wilson (1916), Cohen-Stuat (1916), and Kingdon-Ward (1950) claimed that tea plants should be considered on the basis of a broad polymorphic species and not necessarily divided into infraspecific taxa.

There exists a third viewpoint in which tea plant is treated as one species and further classified into many infraspecific taxa. Aiton (1789), Ventenat (1799), Sims (1807), De Candolle (1824), Loddiges (1832), Koch (1853), Choisy (1855), Miquel (1867), Pierre (1887), Watt (1889, 1908), Pitard (1910), Kitamura (1950) and Sealy (1958) all followed the same principle, but their concept and scope of true tea plant varied. To date, the synonyms under *Camellia sinensis* (L.) O. Kuntze have reached more than sixty.

The wild population of *Camellia sinensis* shows abundant biodiversity. The variation of chromosome structure reveals heterozygosity and polymorphism, the variability of morphology expresses mainly by the differentiation on the form, size, texture, color and vesture of leaves, sepals and petals. The styles vary from free to united (Ming, 1992).

Many scholars like Hadfield (1974), Satyanarayana and Sharma (1986), Banerjee (1987, 1992), Mohanan and Sharma (1981) and Ming (1992) carried out systematic investigations on the morphological variation of true tea plants. Since the true tea plant has had at least 2700 years of cultivation history in China, there are a large number of cultivated races, and some new races appear through individual mutation or hybridization. In remote mountainous districts of southern Yunnan province, China, several primitive races of true tea plant were found, and their characteristics are very closely similar to the wild ones. The karyotypes of chromosomes between wild and cultivated teas revealed no obvious difference (Chen *et al.* 1983).

There is no reproductive interruption between big leaf form (Assam types) and small leaf form (China types) and hence they can be hybridized (Carpenter, 1950; Wight and Barua, 1957, Wight 1959).

Among the true tea plants there exist many chemically differentiated races. In the buds of *Camellia ptilophylla* H.T. Chang, there is no caffeine but abundant theobromine (Chang *et al.* 1988). Furthermore, in the buds of *C. sinensis* var. *pubilimba,* there is no caffeine but abundant flavonoid–camellianin B. There may be even some difference between their contents of purine alkaloids and some other constituents; however, no significant morphological differentiation occurs. Thus, in our point of view, *Camellia sinensis*, the unique polymorphic species, can be regarded as the genuine and true tea plant.

2.2. Taxonomic Characters

Habits   Evergreen trees or small trees with single bole in nature, or usually shrub-like in red sandstone field in Mid-China (Rehder and Wilson, 1916). The plants in cultivation develop into very short shrubs possessing densely arranged branchlets, with more buds and smaller leaves, due to artificial branchlet-cut and bud-pick. The age of cultivated tea plants is usually between 40 to 50 years. When older, the plant will be cut down. Hence tea plants in cultivation are usually shrub-like.

Branches   Erect to spreading; buds silky-pubescent.

Internodal length   1.5–7 cm.

Leaf pose   Erect, horizontal to decurved.

Lamina   Elliptic, oblong, obovate-oblong to oblanceolate, (2.5-) 5–20 (–30) × (1.5-) 3–6 (–9) cm, thin-coriaceous to coriaceous or papery, densely pubescent or villous beneath, or glabrous, light-green to dark-green, occasionally pigmented with anthocyanin, matted to glossy above; apex acute, acuminate to cuspitate, acumen obtuse rounded; lateral nerves 7–14 (–16) pairs.

Petioles   (2-) 4–8 (–10) mm long, pubescent, villous or glabrous.

Sclereids   Branched or unbranched (Barua & Dutta, 1959; Barua & Wight, 1958).

Pedicels   Spreading at anthesis, decurved in fruit, 4–9 mm long; bracteoles 2 (–4), deciduous.

Flowers   Terminal or axillary, solitary or 2–3 (–5), (1.8-) 2.5–4 cm in diameter, fragrant.

Sepals   5–8, late-ovate to suborbicular, 2–5 (–6) mm long, glabrous, ciliate or inside or outside silvery pubescent or villous, persistent.

Petals   5–11, obovate, obovate-oblong or suborbicular, concave, 8–18 mm long, slightly united at the base, glabrous or finely pubescent outside (Brandis, 1911), white, occasionally pink-tinged.

Stamens   Numerous (100–300) in many rows, 7–15 mm long, glabrous; filaments white, outer ones united at the base, adherent to the base of the petals, inner ones free; anthers dorsifixed, yellow.

Ovary   Globose to void, (2-) 3-loculed, sparsely pilose or silvery villous; each locule with 2–5 ovules.

Styles   (2-) 3, 8–12 mm long, filiform, united at the base, or connate to near apex, arms ascending to spreading horizontally.

Capsules   Compressed globose, (2-) 3-cornered, (2) –3-loculed, woody, 1.8–3 cm in diam., with 1 or 2 seeds in each locule, loculicidally dehiscent, valves 1–2 mm thick.

Seeds   Usually 1 (–2) in each cell, subglobose or semiglobose, smooth, 1.2–1.8 cm in diam., pale brown.

Pollen grains   Subspheroidal-oblate, 3-colporate, (40.6–53.1) 47.7 × 53.5 (46.9–62.5) μm, exine sculpture rugulate with beaded muri, exine thickness 2–2.5 μm. (Wei et al. 1992).

Chromosome   Most of tea plants are diploid (2n = 30; x = 15) (Morinaga et al. 1929; Kato & Simura, 1971; Bezbaruah, 1971; Kondo, 1979; Li & Liang, 1990; Liang et al. 1994), triploid (2n = 45) was only found in a cultivated type, *Camellia sinensis* f. *macrophylla* Kitam. (Karasawa, 1935; Kitamura, 1950). Karyotype formulae 2n = 30 = 20m + 8sm + 2 st, 2n = 30 = 22 + 8 sm, 2n = 30 = 18 m + 12 sm, 2n = 30 = (10–24) m + (6–20) sm + (2–4) st (Chen et al. 1983; Liang et al. 1994).

2.3.   Origin and Dispersal

Various hypotheses concerning the native area of tea plants have been proposed. These could be summarized as follows: (1) China and Japan (Linneus, 1753, 1762); (2) China (De Candolle, 1824; Zhang, 1988; Ming, 1992; Liang et al. 1994); (3) Upper Assam (Seemann, 1869); (4) Assam to South China (Cohen-Stuat, 1916; Rehder and Wilson, 1916; Sealy, 1958); and (5) Central Asia (Kingdon-Ward, 1950). The third hypothesis of the Upper Assam origin was negated by Henry (1897) and Kingdon-Ward (1950). and the fifth one was not widely accepted because of the lack of evidence. In our opinion, the original area of tea plants should be determined by the following four principles: (1) discovery of wild old plants, (2) records from ancient books, (3) discovery of archaeology and ethnobotany studies; and (4) acknowledgment of the center of variation and diversity of Sect. Thea.

2.3.1.   Origin of tea plants

In 1826 Scott sent a specimen with a couple of leaves, collected from Munipur in India, to Calcutta. Wallich named it as *Camellia scottiana* Wall. (nom. nud.) and included in his Catalogue no. 3668., which was later treated as a synonym of *C. theifera* Griff. (Dyer, 1873, 1874). In his paper entitled "Discovery of the genuine Tea plants in Upper Assam", Wallich (1835) regard this species as "Assam Tea". After careful investigation of the tea plants of Upper Assam in the spring of 1836, Griffith (1838) wrote in his "Report on the tea plant of Upper Assam"; he said "The largest plants exist in the Kufoo locality, one being observed to measure 43 feet in length, with a diameter near the base of six inches. Occasionally the plant reaches to a height of 47 or 50 feet". He named this species *Camellia theifera* Griff. with an exquisite illustration in the report. Other investigations, however, insisted that these plants must have been introduced to Upper Assam long ago (Kingdon-Ward, 1950; Bor, 1953; Chang, 1981). Seemann (1869) speculated that tea plants in China had been introduced by Buddhist priests from India.

Fortunately, Henry collected a specimen of wild tea from Hainan, China, which was determined to be *Thea bohea* L. (= *Camellia sinensis*) by Hance in 1885 and thought to be "really wild". Wild tea plants were also found in southern Yunnan, eastern Sichuan, and middle Taiwan, of China (Henry, 1897; Rehder et Wilson, 1916; Masamune et Suzuki, 1936). However, less attention has been paid to them.

Since 1980 wide and thorough investigations have been carried out in southern China on resources of tea plants. As a result, many old tea plants were found in Yunnan, Sichuan, Hunan, Guangxi, Guangdong, Hainan, and Fujian provinces. Only in Yunnan province does the number of trees with a diameter of more than 1 meter exceed 20.

In Yunnan Menghai Mount-Daihei, 1900 m above sea level, six aged tea plants have been found in the forest, with the biggest one up to 30 m high and 1.3 m in diameter. In Jinping Fenshuilaoling natural preserved area, 2000–2500 m above sea level, wild tea plants are found growing in the primitive forest with their trunks measuring as much as 2 m in diameter (Hsueh & Jing, 1986). In Yunnan Mt-Gaoligong, more than one hundred big tea trees have been discovered. The trunk diameters range from 80 to 138 cm and the highest one as tall as 20.7 m. The spontaneous living ancient trees of Assam tea are found growing mostly in primitive forest, 1800–2500 m above sea level, and the trees are characterized by a long growing period, extended crown, scattered branches, long internodes, and larger leaves with acuminate apex. They taste bitter. Their texture is thin and fragile, their flowers and fruits are larger, and the cuttings are difficult to generate roots from; while most of the cultivated tea plants in eastern and central China are clones, their cuttings are easier to grow roots from.

According to the "*Hua-Yang-Guo-Zhi*", in the garden of the Sichuan imperial kinsman tea plant was already cultivated, and in the articles of tribute to the Emperor Zhouwuwang there was a kind of "fragrant tea". This demonstrated that tea cultivation in Sichuan has had at least 2700 years of history (Zhuang, 1988).

In Tang dynasty Lu Yu in his "*Tea Classic*" (7th century) recorded: "Tea, which is a fine southern tree, one foot, two feet to dozens of feet high, in Bashan and Shanchuan the tree trunk could reach two men's arms around". In 1910, Wilson collected wild tea in red sandstone ravine of Sichuan Pachou (Ba county), the plant was up to 5 m high (Rehder & Wilson, 1916). Moreover, in Central Chongqing county Mt-Jinxia, an area of higher latitude, wild tea plant could reach over 10 m high (Chen, 1990). More recently, in Tongzi county of northern Guizhou province, some specific live ancient tea plants were also discovered. The wild tea growing in the forest of the boundary between southern Yunnan, Laos and northern Vietnam belongs to Assam tea. The Assam tea growing in central Hainan Bawangling forest could reach up to 15 m high and 40 cm in diameter (Chang, 1980).

In southern Yunnan province of China, the minority nationalities have had their traditional tea civilization. The Hani minor nationality living in Menghai Mt-Nannuo has a specific "Nuobo (meaning tea) culture", respecting tea as totemism. According to the folklore of Jinuo minor nationality, the history for using tea could be traced back to the period of matriarchal society (Wang, 1992).

Assam tea was initially nominated by Wallich in 1835. According to Griffith's description, the plant feature would be judged as no more than 500 years old even up to now, while the same race "Pu-er tea" had been cultivated in Yunnan for more than 1700 years. In Yunnan, the spontaneous Assam tea plant is found in the region south of the latitude of 25 degrees north; but in the northwestern and northern parts of Yunnan which are at the same latitude as Upper Assam, only China tea grows (Ming, 1992). Southeastern Tibet neighboring to Assam grows only cultivated tea; notably, no wild tea can be found there (Chang, 1986). Upper Assam is located along the route "from Sichuan to India" – a folk communication route which started from the 4th century BC, known as the passageway of "Chengdu – Dianchi – Dali – Baoshan – Tengchong – Upper Burma – Assam". Therefore, it is likely that Assam tea was disseminated by the local people in the region south of the latitude of 25 degrees north; starting there, the spread went through the communication route to Upper Assam.

All species under Sect. *Thea* are distributed over a wide area from southwestern to southern parts of China, among which two species extended further south to the northern part of Indochina Peninsula. In Yun-Kui (Yunnan-Guizhou) plateau where all these species and their subspecies of Sect. *Thea* are densely gathered, each species in this area has obviously replaced distribution and infraspecific differentiation, forming this section's differentiated center and diversity center. According to Ming (1992) the Sect. *Thea* may be evolved from Sect. *Archecamellia* and originated from subtropical mountains of Yunnan, Guizhou and Guangxi. The population differentiation of *Camellia sinensis* in Yun-Kui plateau has some regular features. In general, two regions can be roughly divided along the latitude of 25 degrees north. In the south tea differentiated as Assam types, while in the north as China types. In between, there existed various intermediate types. The variations of leaves, sepals, petals and vesiture are manifold and the style fusion varies from united merely at the base to a much greater part. In addition, the color of petals varied from white to pink tinged. For the above reasons, we share the same viewpoint of Zhuang (1988) and Liang *et al.* (1994) that the Yun-Kui plateau seems to be the center of variation diversity of the tea, where the wild tea and all of the allied plants exist.

2.3.2. *Dispersal*

The seeds of tea are big and glossy, and are unlikely to be carried by animals for a long distance. Experiments showed that exposure of tea seeds to sunshine caused dehydration and loss of germination (Liu *et al.* 1989). Therefore, most of the tea plants in Sect. *Thea* have only a narrow distributed area, particularly those species not yet introduced and utilized. The reason for the widespread distribution of *Camellia sinensis* ought to be attributed to the success of human introduction and cultivation.

According to ancient literature, the cultivation of tea in eastern Sichuan of China has nearly three thousand years of history. In Han dynasty the function of tea was gradually turned from remedy to popular beverage, and its cultivation area was expanded enormously, reaching Zhejiang Mt-Tianmu in east China. Surprisingly, a box of tea sample was even found in the grave of a noble-man buried in West Han Wen Emperor period (179–165 BC) in Jiangling Hubei. In Tang dynasty, the production area of tea was expanded further, and the fashion of tea drinking even evolved as a branch of Chinese culture. When the merchandise tea-brick reached Tibet, the nomads there enjoyed tea heartily and it became indispensable in their daily diet. At present, the Tibetan dialect for tea has still remained the same dialect of "Jia" which was introduced from Chang-an as early as 641 AD (Wang, 1992).

Japan is one of the earliest countries to which tea was introduced. In 805 AD, a Japanese monk first introduced Chinese tea to the near river district. In 828 AD, tea was introduced to Korea. In the beginning of 17th century, the Dutch seagoing vessels started to transport tea products from China and introduced tea plant to Sri Lanka. In 1727, tea plants were introduced to Indonesia from China and Japan. In 1768, J. Ellis introduced the tea plant to Kew Garden (Aiton, 1811). According to the record of J. Loureiro (1790), a part of cultivated tea races in Vietnam came from Guangzhou, China. In 1780, the British East-India Company initially imported tea seedlings from Guangzhou to India. In 1796, tea plants were introduced from Fujian to Taiwan. The Japanese later introduced Assam tea to Taiwan in 1895. To date, tea plants have been widely cultivated in many tropical and subtropical regions worldwide, in which the latitudes range from 40°N to 33°S.

3. CLASSIFICATION OF SECT. THEA

Many classification systems within the genus *Camellia* or *Thea* have been proposed. Here are the major items:

1). With sections in Camellia L.
    Dyer (1874), 2 sections   Sealy (1958), 12 sections
    Cohen-Stuart (1916), 5 sections   Chang (1981, 1982), 4 subgenera, 20 sections
    Melchior (1925), 5 sections   Ming and Zhang (1996), 2 subgenera, 14 sections
2). With sections in Thea L.
    Pierre (1887), 6 sections
3). With more genera of narrow sense besides Camellia and Thea of Linneus (1753)
    Rafinesque (1830, 1838)   Blume (1825)
    Nees (1833–34)   Nakai (1940)
    Hallier (1921)   Hu (1956, 1965)

However, we prefer the systems of Ming and Zhang (1996):

Subgen. 1. Thea (L.) H.T. Chang   Subgen. 2. Camellia
    Sect. 1. Piquetia (Pierre) Sealy   Sect. 9. Heterogenea Sealy
    Sect. 2. Archecamellia Sealy   Sect. 10. Stereocarpus (Pierre) Sealy
    Sect. 3. Cylindrica Ming   Sect. 11. Tuberculata H.T. Chang
    Sect. 4. Corallina Sealy   Sect. 12. Camellia
    Sect. 5. Longissima H.T. Chang   Sect. 13. Paracamellia Sealy
    Sect. 6. Thea (L.) Dyer   Sect. 14. Calpandrica (Bl.) Pierre
    Sect. 7. Theopsis Cohen Stuart
    Sect. 8. Eriandria Cohen Stuart

Sect. Thea (L.) Dyer includes tea plants and their allied species.

3.1. Sect. Thea (L.) Dyer

Dyer (1874) divided the genus *Camellia* into two sections, Sect. Camellia and Sect. Thea (L.) Dyer. This was followed by Sealy (1958), Chang (1981) and Ming (1992), despite of the differences that some species are included.

Camellia Sect. Thea (L.) Dyer
Type species: *Camellia sinensis* (L.) O. Kuntze
Evergreen trees or shrubs. flowers 1–3 (–5), terminal or axillary, white, occasionally pink-tinged; pedicles erect, spreading to decurved; bracteoles 2 (–4), small, deciduous, rarely persistent at anthesis; sepals 5 (–8), small, persistent; stamens numerous in many rows, glabrous, outer filaments shortly united at the base; ovary (2-) 3-5-loculed; styles (2-) 3–5, united at the base, or connate to near apex. Capsules compressed globose or globose, woody, with a persistent columella axis. Seeds glabrous.

Seven species, naturally distributed in northern Indochina Peninsula, Yunnan-Guizhou Plateau and southern China, ranging from Upper Burma eastwards to western Guangdong, China, and from southern Sichuan, China, southwards to northern Thailand (Ken, 1972; Nagamasu, 1987). Yunnan-Guizhou Plateau is the center of variation and diversity of Sect. Thea.

3.2. Subdivision of Sect. Thea

Chang (1981, 1984) divided Sect. Thea (L.) Dyer into four series, that is, Ser. Quinquelocularis H.T. Chang, Ser. Pentastylae H.T. Chang, Ser. Gymnogynae H.T. Chang, and Ser. Sinensis. The major characters are listed as follows:

> Ovary 5-loculed, styles 5-lobed or 5-parted
> > Ovary glabrous – Ser. Quinquelocularis
> > Ovary hairy – Ser. Pentastylae
> Ovary 3-loculed, styles 3-lobed
> > Ovary glabrous – Ser. Gymnogynae
> > Ovary hairy – Ser. Sinensis

In fact, 4-loculed ovaries can be found in the 5-loculed ovary groups (*C. crassicolumna* and *C. taliensis*), similiar to 2-loculed ones in 3-loculed ovary group (*C. sinensis*) (Pierre, 1887; Ken, 1972). In *C. sinensis* the hairs on the ovary show much variation, from glabrous via hirsute to pubescent, or from base via middle to top. Furthermore, there is no obvious relation to any other characters. So, in our opinion, the hairs on the ovary should not be used as a criterion in dividing lower ranks in the section and Ming's (1992) system, abandoning the rank of series, is more acceptable.

In Sect. Thea, the number of sepals or petals and the existence of hairs or not on the ovary are all unstable characters and should be abandoned. For example, the petals of *Camellia sinensis* are usually glabrous, but Brandis (1911) and our own observation of a piece of specimen from Kunming, Yunnan (B.Y. Qiu 55254, PE), has found hairs on the back of petals. However, the hairs on leaf buds, instead of drop or not of hairs on mature leaves, in our opinion, are very stable characters and could play an important role in the subdivision of Sect. Thea.

Key to species and subspecies:
1. Ovary (4-) 5-loculed, styles (4-) 5, united at the base, or connate to near apex.
   2. Leaf buds hairy.
      3. Capsules globose or ovoid, valves (4-) 5–8 mm thick.
         4. Ovary hairy......................1a. *C. crassicolumna* subsp. *crassicolumna*
         4. Ovary glabrous.................. 1b. *C. crassicolumna* subsp. *kwangsiensis*
      3. Capsules compressed globose, valves ca. 1–2.5 mm thick
         5. Lower surface of mature leaves pubescent; pedicles 6–7 mm long, pubescent; bracteoles persistent at anthesis.........2. *C. grandibracteata*
         5. Mature leaves glabrous; pedicels 10–15mm long, glabrous; bracteoles early deciduous...4. *C. remotiserrata*
   2. Leaf buds glabrous; capsules compressed globose, valves 1–2.5 mm thick.
      6. Ovary hairy...3a. *C. taliensis* subsp. *taliensis*
      6. Ovary glabrous...............................3b. *C. taliensis* subsp. *tachangensis*
1. Ovary (2-) 3-loculed, styles (2-) 3, united at the base, or connate to near apex.
   7. Leaf buds hairy; capsules compressed globose.
      8. Flowers 5–7 cm in diam.; sepals 6–8 mm long; ovary glabrous, stylar arms recurved; valves 4–7 mm thick......................................4. *C. gymnogyna*
      8. Flowers (1.8-) 2.5–4 cm in diam.; sepals 2–5 (–6) mm long; ovary hairy or glabrous, stylar arms ascendent or spreading horizontally; valves 1–2 mm thick.
         9. Ovary hairy.. 5a. *C. sinensis* subsp. *sinensis*
         9. Ovary glabrous............................5b. *C. sinensis* subsp. dehungensis
   7. Leaf buds glabrous; ovary glabrous; capsules globose, valves 1.5 mm thick.....
      .... 6. *C. costata*

3.2.1. *Camellia crassicolumna* H.T. Chang

Type: Yunnan, Xichou, C.P. Tsien 644 (holotype, PE).
   la. subsp. *crassicolumna* (Figure 1: 4–6.)
   DISTRIBUTION: Southeastern Yunnan, China. In evergreen broad-leaf forests at altitudes from (1300-) 1600 to 2300 m.
   2n = 30 (*C. makuannica, C. purpurea*) (Gu *et al.* 1988).
   1b. subsp. *kwangsiensis* (H.T. Chang) Z.Y. Li et Hsiao, comb. et stat. nov.
Type: Guangsi, Tianlin, Y.K. Li 560 (holotype, SCBI; isotype, PE)
   DISTRIBUTION: Southeastern Yunnan and northwestern Guangxi, China. In evergreen broad-leaf forests at altitudes from 1500 to 1900 m.

3.2.2. *Camellia grandibracteata* H.T. Chang et F.L. Yu

Type: Yunnan, Yunxian, Y.J. Tan A 10001 (holotype, SYS; isotype, ZJTI).
   DISTRIBUTION: Western Yunnan of China. In evergreen forests at altitudes from 1750 to 1805 m.
   2n = 30 (Li, 1989).

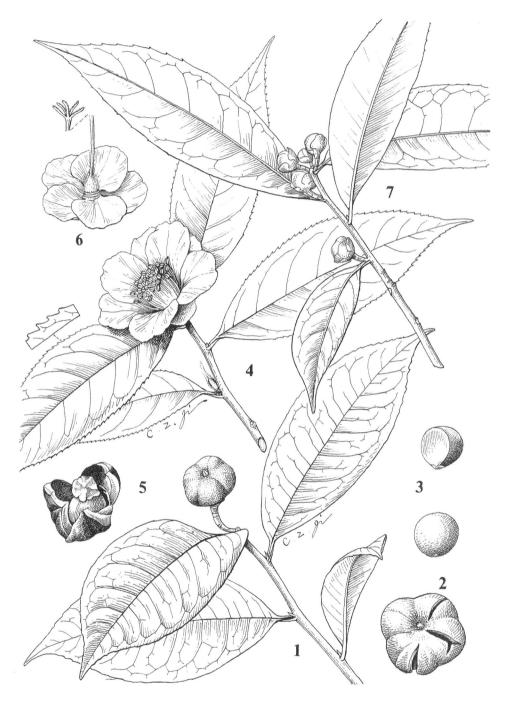

Figure 1. *Camellia taliensis* subsp. *taliensis*: 1. fruiting branch. *C. taliensis* subsp. *tachangensis* 2. fruit; 3. seeds. *C. crassicolumna* subsp. *crassicolumna* 4. flowering branch; 5. fruit; 6. ovary and sepals. *C. costata* 7. flowering branch

3.2.3. *Camellia taliensis* (W.W. Smith) Melchior

Types: described from Tali, based on three specimens; lectotypes, Yunnan, Tali, G. Forrest 13477 (selected by Ming in Acta Bot. Yunn. 14 (2): 119. 1992) (hololectotype, E; isolectotype, K).

3a. subsp. *taliensis* (Figure 1:1)

DISTRIBUTION: Southeastern Yunnan, China; Upper Irrawaddy River Basin (Shan to S. Myitkyina), Burma; and Chiang Mai, Thailand. In evergreen forests, scattered on slopes and ridges or by streams at altitudes from (800-) 1300 to 2400 (–2700) m.

2n = 30 (*C. taliensis, C. irrawadiensis*)

3b. subsp. *tachangensis* (F.C. Zhang) Z.Y. Li et Hsiao, comb. et stat. nov. (Figure 1: 2–3).

Type: Yunnan, Shizon, F.C. Zhang 005 (holotype, YAU; isotype, KUN).

DISTRIBUTION: Northeastern Yunnan, western Guangxi, and southwestern Guizhou, China. In evergreen broad-leaf forests at altitudes from 1500–2350 m.

2n = 30 (*C. quinquelocularis, C. tetracocca*) (Gu *et al.* 1988; Liang *et al.* 1994)

3.2.4. *Camellia remotiserrata* H.T. Chang, F.L. Yu et P.S. Wang

Type: Yunnan, Weixing, F.L. Yu et P.S. Wang A35005 (holotype, SYS; isotype, ZJTI)

DISTRIBUTION: Northeastern Yunnan, northern Guizhou, and southern Sichuan, China. In evergreen broad-leaf forests and exposed situations at altitudes from 920 to 1350 m.

2n = 30 (*C. nanchanica* and *C. gymnogynoides*) (Li, 1985; Liang *et al.* 1994)

3.2.5. *Camellia gymogyna* H.T. Chang *(Figure 2: 5–6.)*

Type: Guangxi, Lingyun, C.C. Chang 11123 (holotype, SCBI).

DISTRIBUTION: Southeastern Yunnan, Guangxi, and southeastern Guizhou, China. In evergreen broad-leaf forests at altitudes from 1000 to 1600 m.

2n = 30 (Liang *et al.* 1994)

3.2.6. *Camellia sinensis* (L.) O. Kuntze

Typification: Linneus (1753) gave several prior references, including an illustration from Kaempfer (1712); lectotype not designated.

6a. subsp. *sinensis* (Figure 2: 1–4.)

DISTRIBUTION: Yunnan, Guizhou, Guangxi, and southern Sichuan, China; nouthern Vietnam; northern Laos; and northeastern Burma. In evergreen broad-leaf forests at altitudes from (800-) 1000 to 2500 m. Now widely cultivated and naturalized in tropical or subtropical regions from near sea level to an altitude of 2000 m.

2n = 30 (Morinaga *et al.* 1929; Kato & Simura, 1971; Bezbaruah, 1971; Kondo, 1979; Li & Liang, 1990; Liang *et al.* 1994); 2n = 45 (*C. sinensis* f. *macrophylla*) (Karasawa, 1935; Kitamura, 1950).

6b. subsp. *dehungensis* (H.T. Chang et B.H. Chen) Z.Y. Li et Hsiao, comb. et stat. nov.

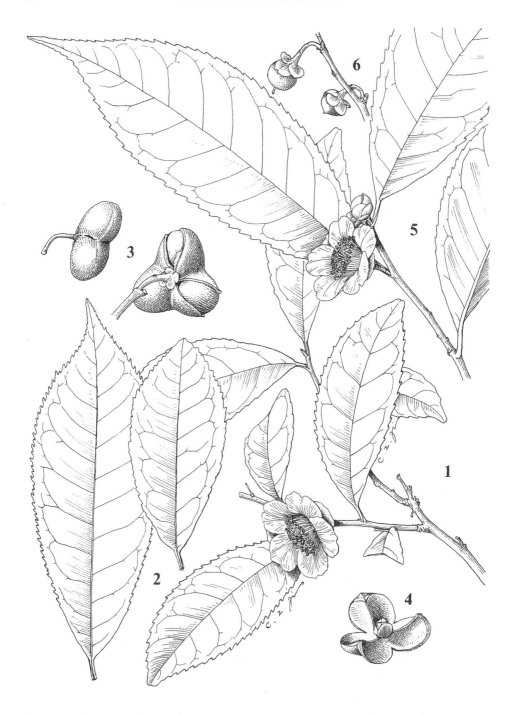

Figure 2. *Camellis sinensis* subsp. *sinensis*: 1. flowering branch; 2. leaves; 3. fruits; 4. capsule valves and persistent columella. **C. gymnogyna**: 5. flowering branch; 6. fruits

Type: Yunnan, Luxi, B.H. Chen *et al.* A04003 (SYS, ZJTI).

DISTRIBUTION: Southern Yunnan and western Guangxi, China. In evergreen broad-leaf forests or brushes in mountain slopes from low altitudes to an elevation of 2000 m.

3.2.7. *Camellia costata* Hu et S.Y. Liang ex H.T. Chang *(Figure 1: 7)*

Type: Guangxi, Zhaoping, S.Y. Liang 6505169 (holotype, SYS; isotype, PE).

DISTRIBUTION: Northern Guangxi, northwestern Guangdong, and southeastern Guizhou, China. In evergreen broad-leaf forests at altitudes from 700 to 1100 m.

2n = 30 (*C. yungkiangensis*) (Liang *et al.* 1994).

3.3. Economic Uses

The genus *Camellia*, Sect. *Thea* is very important economic ally. This is because it consists of the true tea plant and its closely related taxa. Along with cultivated *Camellia sinensis*, the wild plants and related species in southwestern mountainous districts of China have been used as tea drink for a long time. True tea itself as a major beverage and health drink is historically regarded to possess many beneficial effects, such as stimulant, diuretic, digestive, "dispelling summer heat and dampness", anti-aging, anti-dysenteric, anti-diarrhea and slimming etc. In addition, *C. taliensis* in its type locally called "Gan-tong tea". Both its subsp. *taliensis* and subsp. *tachangensis* grow glabrous buds and contain low levels of caffeine. The latter subspecies has been introduced and cultivated in Yunnan and features anti-freezing properties, even in −8°C it could survive for 10 days, and its yield is high and good quality. *C. gymnogyna* suffers fewer diseases and insect pests, with higher yields and a specific flavor. *C. grandibracteata*, *C. remotiserrata*, *C. crassicolumna* and *C. crassicolumna* subsp. *kwangsiensis* have all been used as tea drinks locally. The above-mentioned tea taxa apparently possess their characteristic germplasm, which can be utilized for improving the properties of cultivated tea such as adversity-resistance, yields and races.

The chromosome numbers in Sect. *Thea* are usually 2n = 30, the hybrid after species intercross could be reproductive (Wight and Barua, 1957).

The kernels of *C. sinensis* contain 28.4–35.6% oil in which 16.3–28% is palmic acid, while in *C. taliensis* contains 26.2% oil with palmic acid up to 31% (Chia and Zhou, 1987).

Tea is an evergreen woody plant with fragrant and beautiful flowers and planted as a good ornamental tree. In horticulture there is an ornamental race – *C. sinensis* cv. "*variegata*" as well.

4. DISCUSSION

Camellia sinensis has been associated with people's daily life for a long historical period. Its cultivation together with breeding and free intercrossing have resulted in a number of species. Furthermore, species of wild population added more varieties with geographical, ecological and chemical features. There is, in general, no reproductive interruption between cultivated and wild teas. Thus, we agree that the true tea plant could be treated as one species in a broad sense, i.e. *Camellia sinensis* (L.) O. Kuntze.

After 1980, the survey of wild tea resources has resulted in gratifying achievements in China, such as the discovery of a wild tea in southwestern Yun-Kui high plateau with a life span of more than 1700 years old and the discovery of several wild species closely related to true tea. All of these, we believe, would lead us to a better understanding about the origin and differentiation of the Sect. *Thea* and *Camellia sinensis*.

Taking advantage of modern technologies related to molecular biology, chemo-taxonomy, numerical taxonomy, experimental biology and their integration, further investigations may clear up inter-specific and infra-specific taxonomic affinities of tea plants. We hold that it is essential to intensify the search for tea germplasm in different geographical areas. This will be beneficial not only for the identification and preservation of genotypes in respect to the further utilization of tea breeding programs, but also for developing a new concept of tea classification on the basis of their biogeography, convergence and divergence in the course of their evolution.

REFERENCES

Aiton, W. (1789) Hortus Kewensis, Ed. 1, George Neal, London, 2: 230.

Aiton, W. (1811) Hortus Kewensis, Ed. 2, George Neal, London, 3: 303.

Banerjee, B. (1987) Can leaf aspect affect herbivory? A case study with tea. *Ecology*, **68**, 839–43.

Banerjee, B. (1992) Botanical classification of tea, in *Tea, Cultivation to Consumption* (eds. K.C. Wilson and M.N. Clifford), Chapman & Hall, London, pp. 25–49.

Barua, D.N. and Dutta, A.C. (1959) Leaf sclereids in taxonomy of *Thea Camellias* II. Camellia sinensis L., *Phytomorphology*, 9, 372–382.

Barua, D.N. and Wight, W. (1958) Leaf sclereids in taxonomy of *Thea Camellias* I. Wilson's and related Camellias. *Phytomorphology*, 8, 257–264.

Bezbaruah, H.P. (1971) Cytological investigations in the family Theaceae I. Chromosome chambers in some *Camellia* species and allied genera. *Caryologia*, 24, 421–426.

Blume, C.L. (1825) Bijdragen tot de Flora van Nederlandsch Indie, Lands Drukkerij, Batavia.

Bor, N.L. (1953) Manual of Indian Forest Botany, Oxford University Press.

Brandis, D. (1911) Indian Trees, Constable and Company LTD, London, p. 64.

Carpenter, P.H. (1950) The wealth of India (article on *Camellia sinensis*), CSIRI, Delhi.

Chang, H.T. (1981) Thea – a section of beveragial Tea-trees of the Genus *Camellia*, *Acta Sci. Nat. Univ. Sunyats.* (1), 87–99.

Chang, H.T. (1981) A Taxonomy of the Genus *Camellia*, Sunyatsen University Press, Guangzhou, pp. 1–179.

Chang, H.T. (1984) A revision on the tea resource plants, *Acta Nat. Univ. Sunyats.* (1), 1–12.

Chang, H.T. (1986). Theaceae, in Flora Xizangica (ed. C.Y. Wu), Science Press, Beijing, 3: 259.

Chang, H.T. (1990) New species of Chinese Theaceae. *Acta Sci. Univ. Sunyats.*, 29, 85–97.

Chang, H.T. and Bartholomew, B. (1984) Camellias, Timber Press, Oregon, pp. 137–153.

Chang, H.T. and Ye, C.X. (1988) A discovery of new tea resource-Cocoa Tea, *Acta Sci. Nat. Univ. Sunyats.* (3), 131–133.

Chen, H. (ed.) (1990) *Handbook of Natural Resources in China*, Science Press, Beijing, pp. 308–311.

Chen, R.Y. (1938) Cytology studies on *Camellia* 1. Karyotype analysis of cultivated Tea, *C. sinensis* and Bada wild Tea, in *Rep. and Abstr. Pres. 50th Anniv. Bot. Soc. China*, p. 531.

Chia, L.C. and Chou, J. (1987) *Oil Plants in China*. Science Press, Beijing, pp. 382–84.

Choisy, J.D. (1855) Mémoire sur les Familles des Ternstroemiacées et Camelliacées, Jules-Guillaume, Genève, p. 67.

Cohen-Stuat, C.P. (1916) Voorbereidende onderzockingen tea dienst van de Selektie der Thee Plant, v. h. Proefstation voor Thee, XL.

De Candolle, A.P. (1824) Prodromus Systematis Naturalis Regni Vegetabilis, Argentorati et Londini, 1: 530.

Dyer, W.T.T. (1873) On the determination of three imperfectly known species of Indian Ternstroemiaceae, *Journ. Linn. Soc. Bot.*, **13**, 328–331.

Dyer, W.T.T. (1874) Ternstroemiaceae, in *The Flora of British India* (ed. J.D. Hooker), Reeve, London &t Ashford, **1**, 292.

Griffith, W. (1838) Report on the Tea plant of Upper Assam. *Trans. Agr. Hort. Soc. Calcutta*, **5**, 104.

Griffith, W. (1854) Notulae ad Plantas Asiaticas, Calcutta, **4**, 558.

Gu, Z.J., Xia, L.F, Xie, L.S., Kondo, K. (1988) Report on the chromosome numbers of some species of *Camellia* in China, *Acta Bot. Yunn.* **10**, 291–296.

Hadfield, W. (1974) Shade in north-east India tea plantations. II Foliar illumination and canopy characteristics, *J. Appl. Ecol.*, **22**, 179–199.

Hallier, H. (1921) Beiträge zur Kenntnis der Linaceae (DC. 1819), *Beih. Bot. Centrlbl.* 39(2): 162.

Hance, H.f. (1885) Spicilegia florae sinensis: diagnoses of new, and habitats of rare or hitherto unrecorded, Chinese plants, *J. Botany*, **23**: 321.

Hayne, F.G. (1821) Getreue Darstellum und Beschreibung der in der Arzneykunde Gebräuchlichen Gewächse, 7: t. 27.

Henry, a. (1897). Botanical exploration in Yunnan, *Kew Bull. Misc. Inform.* 1897: 100.

Hoffmannsegg, J.C. (1824). Verzeichnis der Pflanzenkulturen. *Gräflich Hoffmannseggischen Gärten zu Dresden und Rammenau,* Dresden, p. 117.

Hsueh, J.J. and Jiang, H.Q. (eds.) (1986). *Yunnan Forests*, Yunnan Science and Technology Press, Kunming, and China Forestry Publishing House, Beijing, p. 478.

Hsu, B.R. (1993) Age-old tea plants in Gaoligongshan, Yunnan, *Plants*, 1993 (6): 11.

Hu, H.H. (1956) *Sinopyrenanria* and *Yunnanea*, two new genera of Theaceae from Yunnan, *Acta Phytotax. Sin.*, **5**, 279–283.

Hu, H.H. (1965) Glyptocarpa, a new genus of Theaceae, *Acta Phytotax. Sin.*, **10**, 25.

Karasawa, K. (1935) On the somatic chromosome number of triploid Thea, *Jpn. J.. Genet.*, **2**, 320.

Kato, M. and Simura, T. (1971) Cytogenetical studies on *Camellia* species I. *Jpn. J.. Breed.*, **21**, 265–268.

Ken, H. (1972) Theaceae. In T. Smitinand and K. Larsen, (eds.), *Flora of Thailand,* ASRCT Press, Bangkok, **2**, 148.

Kingdon-Ward, F. (1950) Does Wild tea exist? *Nature*, **165**, 297–299.

Kitamura, S. (1950) On Tea and Camellias. *Acta Phytotax. Geobot.* Kyoto, **14**(2), 56–63.

Koch, K.H.E. (1853) Hortus Dendrologicus, F. Schneider et Comp., Berlin, p. 69.

Kondo, K. (1979) Cytological studies in cultivated species of *Camellia* V., *Jpn. J. Breed.* 29, 205–210.

Li, G.T. and Liang, T. (1990) Karyotype studies on six taxa of *Camellia* in China, *Guihaia* 10 (3): 189–197.

Liang, G.T., Zhou, C.Q., Lin, M.J., Chen, J.Y. and Liu, J.S. (1994) Karyotype variation and evolution for Sect. Thea in Guizhou, *Acta Phytotax. Sin.* **32**, 308–315.

Linnaeus, C. (1753) Species Plantarum Ed. 1, Laurentii Salvii, Stockholm, p. 515.

Linnaeus, C. (1762) Species Plantarum Ed. 2, Laurentii Salvii, Stockholm, p. 735.

Liu, X.Y., Huang, Y.X. Wang, H.Y. and Liang, X.Y. (1989) Illustrations of Seeds and fruits, Guangxi Science and Technology Press, Nanning, p. 77.

Loddiges, C. (1832) The Botanical Cabinet, John and Arthur Arch, London, 19: t. 1828.

Loureiro, J. (1790) Flora Cochin-Chinensis, Typis et Expensis Academicis, Lisbon, pp. 338–9.

Makino, T. (1905) Observations on the Flora of Japan, *Bot. Mag. Tokyo*, 19: 135.

Makino, T. (1918) A contribution to the knowledge of the flora of Japan, *Journ. Jap. Bot.* 1: 39.

Masters, J.W. (1844) The Assam tea plant compared with tea plant in China, *J. Agri. Hortic. Soc. India,* **3**, 63.

Melchior, H. (1925) Theaceae, In H.G.A. Engler & K.A.E. Prantl, (eds), *Die Natrlichen Pflanzenfamilien,* 2nd ed., Leipzig, Berlin, 21, 109–154.

Ming, T.L. (1992) A revision of Camellia Sect. Thea, *Acta Botan. Yunnan.*, **14**, 115–132.

Ming, T.L. (1997) Theaceae, In Z.Y. Wu, (ed), *Flora Yunnanica*, Sciences Press, Beijing, 8, 263–382.

Ming, T.L. and Zhang, W.J. (1996) The evolution and distribution of genus *Camellia*, *Acta Botan. Yunnan.* **18**, 1–13.

Miquel, F.A.W (1867) Prolusio florae Iaponicae, *Ann. Mus. Lugd. Bat.* **3**, 17.

Mohanan, M. and Sharma, V.s. (1981) Morphology and systematics of some tea (*Camellia* spp.) cultivars, *Proceedings of the Fourth Symposium on Plantation Crops*, **4**, 391–400.

Morinaga, T., Fukushima, E., Kano, T., Maruyama, Y. and Yamasaki, Y. (1929) Chromosome numbers of cultivated plants II, *Bot. Mag. Tokyo*, 43: 591.

Nagamasu, H. (1987) Note on a *Camellia* species, Sect. Thea, of Doi Chang, North Thailand, *Acta Phytotax. Geobot.*, **38**, 210.

Nakai, T. (1940) A new classification of the Sino-Japanese genera and species which belong to the tribe Camellieae (II), *J. Jap. Botan.* **16**(11), 27–35.

Nees, von E. (1833–34) Sasanque, in P.F. von Siebold, Nippon, 4: 13.

Pierre, J.B L. (1887) Flore Forestière de la Cochinchine, Octave Doin, Paris, 2: pl. 113–4.

Pitard, C.J. (1910) Ternstroemiacees, In M.H. Lecomte, (ed), *Flore Generale L'Indo-Chine*, Masson et Cie, Éditeurs, Paris, **1**, 340–344.

Rafinesque, C.S. (1830) Medical Flora, Atkinson *et al.*exander, Philadelphia, 2: 267.

Rafinesque, C.S. (1837) Flora Telluriana, The Author, Philadelphia, 1: 17.

Rafinesque, C.S. (1838) Sylva Telluriana, The Author, Philadelphia, pp. 139–40.

Rehder, A. and Wilson, S.H. (1916) *Thea sinensis* Linnaeus, In C.S. Sargent, (ed.), *Plantae Wilsonianae* Cambridge, The University Press, **2**, 391–392.

Salisbury, R.A. (1796) Prodromus Stirpium in Horto ad Chapel Allerton Vigentium, London, p. 370.

Satyanarayana, N. and Sharma, V.S. (1986) Tea (Camellia L. spp.) germplasm in south India. In H.C. Srivastava, B. Vatsya and K.K.G. Menon, (eds.), *Plantation Crops: Opportunities and Constraints,* Oxford IBH Pubishing Co., New Delhi, pp. 173–9.

Sealy, J.P. (1958) A Revision of the Genus *Camellia*, Royal Horticultural Society, London, pp. 111–31.

Seemann, B. (1869) Synopsis of the genera *Camellia* and *Thea* in Trans. Linn. Soc. London, 22: 349–50.

Sims, J. (1807) *Thea chinensis* var. *bohea*, Curtis's Bot. Mag. 25: sub t. 998.

Steudel, E.G. (1841) Nomenclator Botanicus Ed. 2, Typ. Et Sumtibus J.G. Cottae, Stuttgart, p. 678.

Sweet, R. (1818) Hortus Suburbanus Londinensis, J. Ridgway, London, p. 157.

Tan, Y.J., Chen, B.H. and Yu, F.L. (1983) New species and new varieties of tea trees in Yunnan, China, *Tea Sci. Res. J.* (China), 1–20.

Ventenat, E.P. (1799) Tableau du Règne Végétal, Selon la Méthode de Jussieu, Drisonnier, Paris, 3: 158.

Wallich, N (1835) Discovery of the genuine tea plant in Upper Assam, *Jour. Asiat. Soc. Bengal*, **4**, 42–49.

Wang, L. (1992) Tea Culture in China, China Books, Beijing, pp. 1–349.

Watt, G. (1889) Dictionary of the Economic Products of India, Calcutta, p. 79.

Watt, G. (1908) Tea and the tea plant, *J. Roy. Hort. Soc.* (London), **32**, 76–80.

Wei, Z.X., Zavada, M.S., Ming, T.L. (1992) Pollen morphology of *Camellia* (Theaceae) and its taxonomic significance, *Acta Botan. Yunnan.* **14**, 275–282.

Wight, W. (1959) Nomenclature and classification of the tea plant, *Nature,* **183**, 1726–1728.

Wight, W. and Barua, D.N. (1957) What is tea? *Nature,* **179**, 506–507.

Zhang, F.C., Ding, W.R., Huang, Y. (1987) Opinions on study and use of Yunnan tea resources, in J.Q. Xiao *et al.* (eds.), *Selected Papers of Rational Exploitation and Utilization of Yunnan's Biological Resources.* Yunnan Science and Technology Press, Kunming, pp. 143–151.

Zhang, F.C., Ding, W.R., Huang, Y. (1990) Three new species of the genus Camellia from Yunnan. *Acta Botan. Yunn.* **12**, 31–34.

Zhuang, W.F. (1988) An Introduction to the Tea History of China, Science Press, Beijing, pp. 24–47.

3. GREEN TEA, BLACK TEA AND SEMI-FERMENTED TEA

NING XU AND ZONG-MAO CHEN

*Tea Research Institute, Chinese Academy of Agricultural Sciences,
1 Yunqi Road, Hangzhou, Zhejiang 310008, China*

1. INTRODUCTION

Tea is one of the three major non-alcoholic beverages in the world. Tea plant has been cultivated for several thousand years in China. The tea plant *Camellia sinensis* or *Camellia assamica* is believed to originate in the mountainous region of southwestern China as many species and endogenously wild tea trees have been discovered in primitive forests in Yunnan province (Yu and Lin, 1987). The words *te, qia* or *cha* denote tea in various Chinese dialects, and in one or other of these forms has been transposed into other languages. At the very beginning, leaves from wild tea plants in forests were picked and steeped in boiling water and the brew was drunk. During the second and third century AD, tea leaves were pressed into cakes with rice gruel and then dried. The dried tea cake was then ground for brewing. From the 10th to 14th centuries, the Chinese adopted the pan-firing method to process green tea. At the end of the 14th century, dark green tea came into being. This was followed by the development of processing for production of Oolong tea, white tea, black tea, and scented tea. Classification of made tea is based on the degree of fermentation and oxidization of polyphenols in fresh tea leaves. Up to now, there are six types of teas: green, black, Oolong, dark green, white, and yellow tea. Green, yellow and dark green tea are non-fermented tea as polyphenols in fresh leaves are less oxidized, green tea belongs to absolutely non-fermented tea; however, polyphenols in yellow and dark green tea are non-enzymatically oxidized during the processing period. White tea, Oolong tea and black tea are fermented teas, while green tea is non-fermentation and black tea the most fermentation, thus, Oolong tea is also called semi-fermented tea. Each kind of tea has its characteristic flavor and appearance. Besides the above six teas, flower scented tea, compressed tea, instant tea and herbal teas are classified as reprocessed teas. In this chapter, special emphasis will be placed on the classification, manufacture, processing biochemistry and factors affecting the quality of these teas.

2. GREEN TEA

2.1. Classification of Green Tea

Green tea, a non-fermented tea, has been known in China from very early times. Even today, green tea is produced and consumed in the greatest quantities in

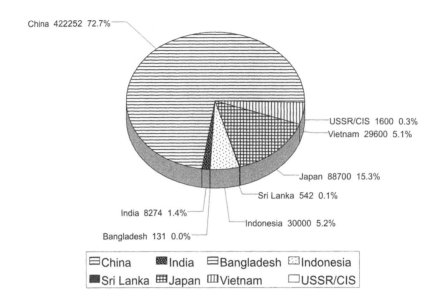

China 422252 72.7%

USSR/CIS 1600 0.3%
Vietnam 29600 5.1%

Japan 88700 15.3%

Sri Lanka 542 0.1%

India 8274 1.4%

Indonesia 30000 5.2%

Bangladesh 131 0.0%

⊟China   ▓India   ⊟Bangladesh   ⊡Indonesia
▓Sri Lanka   ⊞Japan   ⊞Vietnam   ☐USSR/CIS

Figure 3.1   Green Tea Production in the World (1996) (in metric tons) (ITC, 1997)

China. In 1996, 72.7% of the world total green tea production is produced in China. Besides China mainland, it is also produced in Japan, Indonesia, China Taiwan, and other Asian countries (ITC, 1997). Nowadays, people have paid special attention to its healthy effect on the human body.

China is the birthplace of green tea. In Tang Dynasty (618–907 AD), green tea was first invented as steamed green tea cake instead of directly cooking tea leaves for drinking brew. In Song Dynasty (960–1127 AD), pan-fixed green tea was developed. Nowadays, there are various green teas classified according to their manner of processing and appearance (Table 1).

Japan is the second largest producer of green tea in the world. Steamed green teas are the main tea product. The types of green tea produced in Japan are listed in Table 2.

2.2. Processing of Green Tea

Generally, there are no limitations on leaf shape processing green tea. Many cultivars are suitable for green tea processing. However, the quality of green tea is determined by the degree of tenderness of the fresh green leaf. Made green tea shape can be needle, twisted, flat, round, compressed shape or even as ground powder. Various steps are required for the manufacturing of green tea: fresh green leaf→spreading-out (or

Table 1 Classification of Green Tea.

Fixing manner	Drying manner	Shape	Representative
Pan-fixed	Pan-dried	Twisted shape	Xiumei, Gongxi
		Round shape	Gunpowder
		Flat shape	Longjing, Qiqiang
		Needle shape	Yuhuacha
	Basket dried	Twisted shape	Maofeng tea, Raw materials for processing scented tea
		Flat shape	Taipinghoukui
		Flaky shape	Luanguanpian
	Roasted	Twisted shape	Mengdingganlu
		Needle shape	Anhuasongzhen
		Flat shape	Jingtingluxue
	Sun dried	Compressed	Puer Fangcha
		Bowl shape	Tuo cha
		Twisted shape	Raw materials for processing dark green tea
Steam fixed		Needle shape	Gyokura
		Twisted shape	Sencha
		Ground tea	Matcha

not)→fixing (panning or steaming)→ rolling→drying (in pan, basket, machine or by sunlight).

First, green leaves are plucked from small or medium leaf type cultivars. The plucking standards are different according to green tea types. Tender and uniform tea leaves are

Table 2 Major Types of Japanese Green Tea.

Tea type	Description
Gyokura (pearl dew)	The finest of Japanese green tea. The tea plant is cultivated under shade (60–90% darkness) for about 2 weeks before plucking the flush
Sencha	The most popular Japanese green tea, mostly produced from the clone Yabukida
Kabusecha	Made from plants kept under 50–80% darkness for 1–2 weeks.
Tencha	The tea flush is the same as that for Gyokuro, but the processing differs slightly from that used in the making of Gyokuro or Sencha because the rolling process is omitted.
Matcha	Made from Tencha by grinding in a stone tea-rolling mortar. This tea is well known because it is used in tea ceremonies.
Bancha	A low-grade Shencha, made from mature and coarse leaves.
Hojicha	Made by roasting Bancha at around 180°C
Kamairicha	A pan-fired tea, made from young leaves as Sencha. The first processing step is parching, instead of steaming, as in Sencha.
Tamaryokucha (Guricha)	The tea leaves are steamed first; then manufactured into comma-shaped tea, like Kamairicha, by a rolling dryer

commonly required for processing top quality green tea such as Longjing, Biluochun, Gyokura etc. Other green teas such as Xiumei, and gunpowder require mature tea shoots with 1 bud and two or three leaves. The plucked shoots generally need to be spread out in bamboo trays or directly on the ground for 1–3 hrs to emit some grass-like odor and to lose some water in order to improve green tea quality. Fixing of green leaf is the next key step in processing green tea. The purpose of fixing green leaf is to halt the activities of enzymes in tea shoots in order to prevent "fermentation" and to keep the green color of the leaves. In China, pan-firing is the most commonly used fixing method although a small amout of green tea is fixed by steaming method. However, most green teas (steamed green tea) in Japan are fixed by steaming method. The pan-fixing temperature is usually higher than 180°C and steam-fixing temperature is usually 100°C. Fast and complete fixing of green leaves is very important. If the temperature is too low, the leaves turn red; but if the temperature is too high, the leaves scorch. The water content of fixed leaves is usually controlled at 60% in China. Long fixing times and overdrying are unfavorable for the following rolling step. Some top quality green teas in China are made by hand. The fixing temperature is between 100°–200°C. Machine fixing temperatures are about 220°–300°C. After fixing, the next step in processing green tea is rolling. During rolling, leaf cells are broken and leaf juices are liberated and the rolled leaves show a twisted shape. Successful rolling depends on the degree of pressure applied, rolling time, rolling method and leaf temperature. In China tender leaves are generally rolled under light pressure for a shorter time, but mature leaves are rolled under heavier pressure, for longer and are even rolled several times. Roll-breaking is usually followed by the rolling process in order to loosen compressed leaves for better drying. Drying is usually repeated several times. For hand-made tea such as Longjing tea, long and repeated drying by hand is needed to remove the moisture, create the shape and produce its special flavor. For common green tea, the drying step is usually conducted two times. Drying can be done in a pan, basket, machine or by sunlight. Pan-drying tea will produce a tight shape in appearance and fragrant tea aroma; however, sunlight drying creates loose shaped green tea and poor taste quality.

Japan is the largest producer of steamed green tea. Gyokura (pearl dew) and Sencha are two representatives of Japanese steamed green tea. Steamed green tea is also produced in China, India and China Taiwan. For processing steamed green tea, the fixation of green leaves is done by steaming the leaf for a few minutes in perforated drums supplied with a steam blast. Most of steamed green teas are needle-shaped, with dark green dry leaf, bright or deep green liquor, green infused leaf and typical aroma. Steamed green tea contains high amounts of vitamin C while black and Oolong teas contain less vitamin C due to the oxidation during the fermentation process.

2.3. Biochemistry of Green Tea Processing

Fast and high temperature fixing of the leaf enzyme activity is the key treatment for processing green tea. Once the enzyme activities in fresh leaves are deactivated, the fixed leaves are subjected to the following rolling and drying steps. The proper fixed leaves always keep green during the whole processing. Various biochemical changes occur during green tea processing.

2.3.1. Enzyme activity

Living tea shoots contain many enzymes responsible for various biochemical metabolic pathways. Once tea leaves are picked, those enzymes still have high activity. Polyphenol oxidase (PPO), catalase (CAT), peroxidase (PO), and ascorbic acid oxidase (AAO) etc are the main enzymes in tea leaves. During fixing process, enzymes in tea leaves are deactivated by high temperature. The higher the fixing temperature, the more deactivity degree of enzymes become (Cheng, 1982).

It indicates that different enzymes in tea leaves have different responses to temperature. In the range from 15° to 25°C, the activities of CAT and PO increase along with the rise of the temperature. When the temperature is higher than 35°C, the activities of CAT and PO decrease. In addition, the activity of PPO increases lineally in the range of 15°–55°C. Deactivation of PPO occurs when temperature rises up to 65°C. Therefore, it can be concluded that the fixing process of green tea finishes only when PPO activity is deactivated completely. Many investigations show that PPO activity in tea shoots varies with leaf position, plucking standard, seasons, cultivars etc (Obanda, 1992). For different plucking standards, the fixing temperature and duration are varied. For example, the tender shoots should be fixed at high temperature for longer, as the PPO activity is higher than that in mature shoots.

2.3.2. Chlorophyll

Generally speaking, chemical changes during the fixing process are actually induced by thermo-physiochemical action. Chlorophyll is the main coloring pigment in tea infusion of green tea. During manufacturing, chlorophyll contents decrease as processing completes. If relative chlorophyll content in newly plucked shoots is 100, that in the fixed shoots is 87, rolled shoots 74 and dried shoots 52. During green tea processing, chlorophyll lost Mg due to high temperature and pH changes. Furthermore, hydrolysis of chlorophyll often occurs and chlorophyll breaks down into folic acid, phytol and Mg-free chlorophyll (Xiao, 1963).

2.3.3. Tea polyphenol

The main chemical changes of tea polyphenols (TP) during green tea processing are oxidation, hydrolysis, polymerization and transformation. Consequently, the total tea polyphenol contents decrease to some extent in comparison with fresh leaves. Generally, TP content in made tea after processing will decrease by around 15%. This change plays an active role in forming the characteristic green taste and flavor (Zhou, 1976).

The autooxidative products of TP promoted by heat are yellow substances that contribute to green tea infusion color; furthermore, the substance may conjugate with protein in tea leaves to form a water-insoluble substance which contributes to the bright color of infused leaves.

TLC investigation showed that gallated catechins would be changed to non-gallated catechins through hydrolysis under humid and heating conditions during green tea processing. This process causes remarkable changes of tea catechin composition in made tea (Cheng, 1982).

2.3.4. Protein and amino acid

During green tea processing, some proteins in tea leaves hydrolyzed into free amino acids by high temperature and moist environment. Therefore, amino acid content in made tea is higher than that in fresh leaves (Table 3). The increase of amino acids will improve the freshness of tea infusion, as amino acids are the substances that create the freshness of tea infusion. As tea polyphenols are usually decreased during green tea processing, the ratio of tea polyphenols to amino acids changes after tea processing. The coordinated ratio of tea polyphenols and tea amino acids will produce a fresh and grassy taste. On the other hand, some of the amino acids will change into volatile substances. For example, isoleucine oxidizes into isopentaldehyde and phenylalanine into phenylaldehyde. These two volatiles are beneficial to green tea quality (Zhou, 1976).

Table 3   Changes of Amino Acids During Green Tea Processing.

Item	Fresh leaves	Fixed leaves	Made tea
Amino acids	1.786	1.810	1.848
Theanine	0.646	0.760	1.060

2.3.5. Carbohydrates

Soluble carbohydrates were increased in made tea leaves in comparison with fresh shoots. Under high temperature and very moist conditions, starch became hydrolyzed and produced more soluble sugars. It was reported that soluble sugars increased significantly during and after green tea processing (Lin, 1962).

The increase of soluble sugars in made tea varied with green tea type, fixing temperature and drying method. In general, increases in soluble sugars appeared in baked green teas.

2.3.6. Aroma

Very complex biochemical changes of aroma occurred during green tea processing. Up to now, more than 600 kinds of tea aroma compounds have been identified. Only small amounts of aroma compounds come from fresh leaves, the majority come from other substances during green tea processing. At the early stages of green tea processing, lower boiling point (bp) volatile compounds were released by the high temperature while those higher bp volatile compounds remained in the processed leaves. The typical changes of volatile compounds are that lower bp volatile chemicals such as grass odor components are evaporated and higher bp volatile chemicals increase. Examples of lower bp volatile chemicals in fresh leaves are Z-3-hexen-1-ol, E-2-hexenal, acetaldehyde, formic acid, acetic acid, etc. Z-3-hexen-1-ol, with strong grass odor and bp 156°C accounts for 60% of the total volatile compounds in fresh leaves. Most of Z-3-hexen-1-ol was evaporated after fixation and drying process and a small amount changed into E-3-hexen-1-ol, the chemical with a fresh odor.

2.4. Factors Affecting Green Tea Quality

Many factors affect the green tea quality such as processing technology, tea plant cultivars, made tea shapes, fertilization, seasons and plucking standards, etc. Among the above factors, processing factors have the greatest influence on green tea quality. Leaf spreading for a short period before fixing is a very important step for processing of high quality green tea. Spreading out of green leaves in bamboo trays or on the ground can promote the hydrolysis of non-water soluble carbohydrates and pectins, formation and accumulation of non-gallated catechins, release of grass-like odor and loss of some moisture in fresh leaves for better fixation. The spreading height, turning numbers, and duration for a particular type of green tea differ according to green leaf and weather conditions. Generally speaking, 70% of moisture content after spreading is suitable. Deactivating the PPO is the main purpose of fixing. In China, a commonly accepted principle is "Tender leaves need heavier fixing while mature leaves need lighter fixing; the fixing temperature should be higher at the very beginning and then lower; and combination of promoting the moisture removal and inhibiting moisture removal". As tender leaves usually have higher enzyme activities, only higher temperature and long fixing time can thoroughly deactivate the enzymes. Heavier fixing of tender leaves will promote the hydrolysis of proteins. For example, the amino acid contents were proved to be increased with the longer fixing time (Cheng, 1982). However, over-fixing would scorch the leaf and results in a smoky taste and higher ratio of broken leaf. Lower temperature or shorter fixing time often produces red leaves due to the oxidation of polyphenols, which will lower the quality of green tea. Otherwise, relatively lower temperature is needed for fixing mature leaves as they have lower water content and enzyme activities in comparison with tender leaves. Rolling is also important to green tea quality. The degree of pressure and rolling time are the key technical parameters. Longer rolling and heavier pressure imposed on fixed leaves will produce yellowish leaves and more broken leaves. Furthermore, hydrolysis of chlorophyll and autooxidation of polyphenols cause poor tasting green tea as more juices are squeezed out.

Tea plant cultivars have deep influence on green tea quality. Early-sprouting and high ratio of amino acid to polyphenols cultivars such as cv, Longjing 43, cv, Fuding Dabaicha (China cultivar) and cv. Yabukida (Japan cultivar), etc. usually produce high quality green tea compared with other cultivars under the same cultivation and processing conditions. Proper shading of the plant can also produce high quality green tea such as Gyokura. The tea plant is cultivated under the shade for about 2 weeks (60–90% darkness) before plucking the flush. Higher rates of nitrogen application has been shown to produce high quality green tea as amino acid levels in the leaves increases in comparison with none or lower rates of nitrogen application. Fine plucking standard is undoubtedly favorable to green tea quality.

3. BLACK TEA

3.1. Classification of Black Tea

Black tea is a fully fermented tea. In 1996, worldwide black tea production accounted for 72% of the total tea production. Black tea was developed in the mid 17th century in Chongan county, Fujian province of China. The first black tea was so called Xiao Zhong black tea, the withering of which was promoted indoors by pine tree smoking. In 1850 AD, Congou black tea manufacture was created in Fujian province on the basis of Xiao Zhong black tea manufacturing method. Later, black tea was disseminated into tea producing areas in other parts of China such as Keemun, Anhui province. Nowadays, Black tea is mainly divided into Xiao Zhong black tea, Congou, meaning the laborious or assiduous sort, more time and labor being expended on this than on other varieties, Broken black tea (orthodox, CTC and LTP) and brick black tea (Table 4).

3.2. Processing of Black Tea

3.2.1. *Xiao Zhong black tea*

Xiao Zhong black tea or Souchong tea, meaning small type tea, was the first black tea to be invented in China. Xiao Zhong black tea can be divided into two types: Zheng Shang Xiao Zhong, and Jia Xiao Zhong black tea (False Xiao Zhong black tea). The former is only produced in Tongmu village in Chongan County, Fujian province, China; the latter is produced in Wuyi mountain area, Fujian province. Smoky Xiao Zhong black tea, which is made from sifted broken black tea or lower grade black tea, is also categorized as Xiao Zhong black tea. Xiao Zhong black tea has unique aroma and taste of pine tree smoke, because the leaves are fumigated by the pine tree smoke. Investigation showed that the aroma of Zheng Shang Xiao Zhong consisted of high levels of phenols, furans, nitrogen compounds, cyclopentenolones and terpenoids derived from pine needles through the smoking process (Kawakami *et al.* 1995). Xiao Shong black teas are made in the Congou manner but from coarser leaf. The processing steps are as follows:

Shoots with one bud and 3–4 leaves are spread 9–10 cm in depth on a bamboo mat and withered indoors with pine tree smoke. The room temperature is kept at 30°C by release of pine tree smoke to fumigate the leaves, which are turned at intervals of 10–20 min. The withered leaves are then rolled for 60–90 min. Fermentation takes place in cloth bags or baskets for 5–6 hrs after roll breaking of the rolled leaves. The fermented leaves are panfired at 200°C for 2–3 min and then another rolling step is conducted for 5–6 min. Drying is done by the pine tree smoke for 8–12 hrs. The unique character of Xiao Zhong Black tea is mainly formed during

Table 4 Black Tea Classification and Processing

Classification	Representative	Processing	Notes
Xiao Zhong Black tea	Zhengshan Xiao Zhong	Fresh green leaf→indoor withering or sunlight withering →1st rolling →fermentation→panfiring →2nd rolling→fumigated by pine tree smoke→drying	Only produced in Tongmu village in Chongan County, Fujian province, China
	Jia Xiao Zhong (False Xiao Zhong black tea)		
	Smoked Xiao Zhong	Broken Congou tea (as raw material)→fumigated by pine tree smoke	Quality is poor
Congou Black tea	Keemun black tea	Fresh green leaf→withering→ rolling-fermentation→drying	Produced in Qimen (Keemun) County, Anhui province, China
	Yunnan Congou		Produced in Yunan province, China
	Fanning and dust		
Broken black tea	Leaf tea Borken tea Fanning tea Dust tea	Fresh green leaf→withering→maceration →fermentation-drying	Orthodox tea, CTC teas are included in this group.
Brick black tea	Mizhuan tea	Broken black tea or low grade black tea→sifting→ blending→weighting→ modeling Product size: 23.7 × 18.7 × 2.0 cm^3 Weight: 1.125 kg	Consumed by China minority people or export

the drying step as tea aroma and pine tree smoke become completely mixed. The moisture content is maintained under 8% when the processing is finished.

3.2.2. Congou black tea

Keemun black tea is one of the most famous black teas in China. The black tea has a flowery aroma. The processing step is as follows:

Fresh leaves are spread 15–20 cm deep and withered indoors at the temperature of 20°–25°C. The leaves are turned every 2 hrs and this process lasts for 12 hrs. The average water content of withered leaves is about 60%. The leaves are then rolled either by hand or by rolling machine. Fermentation is carried out at 20°–30°C at the relative humidity over 90% for about 2–5 hrs according to seasons. After fermentation, leaves are subjected to drying at 110°–120°C for 10–15 min. After spreading of the first dried leaves, another drying is carried out at 70°–90°C for 45–60 min. The moisture content of the made tea is 4–6%. The refining of raw Keemun black tea includes sifting, polishing and blending, almost the same as that of green tea.

3.2.3. Broken black tea

As other types of black tea account for only a small proportion of the total black tea production of world, on most occasions, black tea refers to broken black tea. The processing of broken black tea is similar to Congou black tea. It needs the following five stages: withering, maceration (leaf disruption), fermentation, drying, and sorting. The withering stage has two purposes. It allows certain chemical changes and leaf moisture removal in order to achieve high quality and to be easily processed in the following stages. The leaf withering is often conducted in a withering trough that allows the pressed air to pass through the leaf bed spread in the trough. Disruption of the leaves is usually performed by machines. The orthodox roller, CTC, LTP, the rotorvane and the Legg cutting machine are usually adopted in black tea production. Orthodox roller is the traditional machine for macerating tea. It consists of a hopper containing leaf under pressure, which moves over a ridged table that is moving over a different path. It is essentially a batch machine although continuous variants have been tried. Tea produced by these machines is known as orthodox tea. CTC stands for crush, tear and curl. This machine was devised in India and consists of two toothed rollers that move at different speeds in opposite directions. The leaf moves continuously between the rollers. Tea processed using this machine is known as CTC tea. It produces smaller pieces than the orthodox roller. CTC tea accounts for around 60% of the total broken black tea production in the world. LTP stands for Lawrie Tea Processor. In this machine the tea leaves are crushed and cut by a series of knives and beaters attached to a rotating shaft. Rotorvane, devised in India, has a shaft carrying a series of angled blades which rotates in a cylindrical housing. The blades are near to the exit which is partly closed. Together these generate a pressure in the machine. The rotorvane is commonly used in series with either orthodox rollers or CTC machines. The Legg Cutter cuts tea leaves by a series of knives oscillating vertically. Following the leaf disruption stage, the fermentation stage allows macerated leaves to contact the air for a period between 40 min to nearly 3 hrs depending on the air and leaf conditions. Macerated leaves are usually placed in a fermentation machine (fermenter) or directly spread into a thin layer on the floor for fermentation. The fermentation time, temperature, aeration and other parameters are dependent on processing. Drying is usually done by a drying machine such as the multi-band drier or fluid bed drier. The dried tea consists of a mixture of different particle sizes including a proportion of stalk and fiber. Sorting is the last stage for broken black tea processing. The sorting processing equipment consists of stalk and fiber removal apparatus for separating the dried tea in uniform particle sized portions. Vibrating sorting machines and electrostatic fibre-removing rollers are often used in sorting the dried black tea.

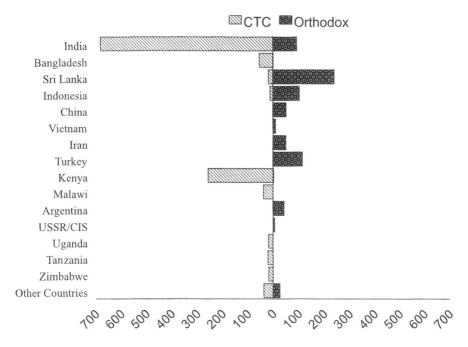

Figure 3.2 Production of black tea according to manufacture (CTC and Orthodox) (× 1000 metric tons) (ITC, 1997)

3.3. Classification of Broken Black Tea

Most of the black teas produced in the world are actually broken black tea, which are macerated either by orthodox roller, CTC machine, LTP or by other manufacturing means. Internationally, the broken black tea after sorting is classified into brokens, fannings, leaf tea and dusts according to the shape and interior quality. Pekoe, Orange. Pekoe Flowery, and various combinations of these words are used. Pekoe is often used to describe the quality and grades of different broken black teas. Those names partly indicate the fineness of the leaf constituting the class. The term Pekoe is derived from the Chines *Pek-ho* or *Bai-hao*, denoting white hair, and refers to the fine hair seen on the buds and younger leaves. When fermented tea juice is smeared on these hairs they appear yellow, orange or golden in color. Flowery Pekoe is made from the most tender buds. This class of tea consists mainly of silvery flower or tip. The four types of broken black tea can be further classified into different grades indicated by different combinations of Flowery, Broken, Orange, Pekoe and Dust. (Table 5). Each grade of black tea has particular particle sizes due to the different sizes of mesh in use. The descending order of particle size is as follows: leaf tea > brokens > fannings > dusts (Lu, 1979).

Table 5   The Grades of Black Tea.

Type	Grade	Abbreviations
Brokens	Flowery Broken Orange Pekoe	F.B.O.P.
	Broken Orange Pekoe	B.O.P.
	Broken Pekoe	B.P.
Fannings	Flowery Broken Orange Pekoe Fanning	F.B.O.P.F.
	Broken Orange Pekoe Fanning	B.O.P.F.
	Pekoe Fanning	P.F.
	Orange Fanning	O.F.
	Fanning	F.
Leaf Tea	Flowery Orange Pekoe	F.O.P.
	Orange Pekoe	O.P.
	Pekoe	P.
Dusts	Dust	D.

3.4.   Biochemistry of Black Tea Processing

The general steps of processing black tea are as follows: withering of leaves under certain plucking standards, rolling, fermenting, drying and refining. Each step has a major influence on the made tea quality and many chemical or physical changes occur.

3.4.1.   Withering

The withering of tea leaves is the first step in processing black tea. The loss of water in fresh leaves makes it easier for the subsequent rolling and fermenting. Many physical and biochemical changes take place during the withering process (Table 6) (Owuor, 1986, 1988, 1990, 1995, 1996). Generally, two aspects of withering are known to take place in this process; one is physical and the other is chemical. The physical change associated with withering is a loss of moisture from the shoot which leads to changes in cell membrane permeability. Thus, withering can be divided into chemical wither and physical wither (Owuor, and Orchard, 1989).

3.4.2.   Rolling

The rolling step in processing black tea causes the disruption of cell walls and leaves as well as the leakage of leaf juice. Once the cell walls are broken, many chemical changes take place due to the breakdown of the cell membrane and the activation of enzymes such as PPO and hydrolases. Often the enzyme activities are enhanced as suitable conditions like temperature, oxygen supply and moisture exist during the rolling process.

Table 6 Changes During the Withering Stages of Black Tea Processing.

Change type	Content	Nature
Content-decreasing type	Moisture removal	Physical
	Polysaccharides	Chemical
	Cell membrane permeability	Both physical and chemical
	Grass-like odor	Both physical and chemical
	Polyphenols	Chemical
	Lipids	Chemical
	Glycosides	Chemical
	Chlorophyll	Chemical
Content-increasing type	Free amino acids	Chemical
	Soluble protein	Chemical
	Caffeine	Chemical
	Simple carbohydrates	Chemical
	Enzyme activity	Chemical
	Flowery flavor	Chemical
	Carotenoids	Chemical
Other changes	Organic acids	Chemical
	Enzyme activities	Chemical
	Transformation of other chemicals	Chemical

3.4.3. Fermentation

The most complicated chemical changes take place during fermentation. PPO is the key enzyme responsible for the formation of theaflavins and thearubigins from catechins. The content of theaflavins and thearubigins is usually adopted as the monitoring parameter in evaluation black tea quality. Theaflavins consist of four major components, which are the oxidative products of paired catechins by PPO. The formation of theaflavins and thearubigins needs a good supply of oxygen (Owuor, 1992, 1994, and 1996).

3.4.5. Drying

Drying can cause the cessation of enzyme activity and reduces the water content of fermented leaves. It is very important to dry the fermented leaves for the correct amount of time in order to form a good quality tea. Late drying or early drying will certainly cause the deterioration of black tea quality. Furthermore, many quality-oriented flavors are formed during drying. Oxidation of leaf components continues through the drying stage while some chemical components such as amino acids and simple carbohydrates increase.

3.5. Factors Affecting Black Tea Quality

Environmental factors that affect the chemical composition of tea leaves are altitude and climate and cultural practices including fertilizer application, suitable shading, plucking standards, disease and pruning; and cultivars also affect the chemical composition of tea leaves.

An investigation on the effect of phosphorus (as single superphosphate) on black tea quality was carried out (Gogoi, *et al.* 1993). Results showed that P applications up to 50 kg P_2O_5/ha increased the major catechins and caffeine contents of tea shoots, and the thearubigin and theaflavin contents and brightness as well as total color of made teas. No additional improvements in quality were observed at higher P application rates. Different nitrogen fertilizers (ammonium sulfate) can improve theaflavin and chlorophyll levels in the processed black tea. Total chlorophyll content increased with increasing N levels up to 200 kg N/ha. Theaflavin content decreased with increasing N levels with the largest drop at 200 kg N/ha. The total color of the infusion followed an almost identical pattern to theaflavin content. Tea quality deteriorated with increasing N and was described as "grassy" for the highest levels (Lelyveld, *et al.* 1990).

An Indonesian investigation indicated that the best aroma quality appeared eg in August and September both for orthodox and CTC teas (Musalam, 1991).

Black tea type has a major influence on quality. The aroma concentrate of orthodox and CTC teas is essentially different and the flavor of CTC tea is generally inferior to that of orthodox tea. The liberation of monoterpene alcohol is favored by anaerobic conditions. The total amount of volatiles as well as their components like Z-3-hexanal, linalool and its oxides and methyl salicylate extracted from CTC teas are lower than those from orthodox teas. The less fragrant nature of CTC teas may be due to the lower amounts of essential volatile compounds, especially linalool and its oxides together with methyl salicylate (Takeo, 1983c). Generally, CTC tea has higher theaflavin value than orthodox teas. This is associated with higher activity of the oxide reductase on the catechin substrate. In turn, it inhibits the action of the hydrolytic enzyme that was reported to be responsible for producing linalool and its oxides and methyl salicylate in disrupted tea leaf tissues under anaerobic conditions (Takeo, 1981a, b, 1983a, b, c). The CTC processing resulted in higher lignin, total lipid and soluble solids contents, total color and brightness, and lower cellulose and hemicellulose contents compared with orthodox processing. An increased number of cuts during CTC processing produced higher percentages of fannings and dust. The bulk density, theaflavin, thearubigin and soluble solids contents increased as the size of the grades decreased (Mahanta *et al.* 1990).

The withering process is important in the formation of black tea aroma. The ratios of E-2-hexenal, gernaiol, benzyl alcohol and 2-phenyl ethanol are higher in non-withered tea, while that of linalool and its oxides, and methyl salicylate, are higher in withered tea. From the aroma pattern it is considered that the higher ratio of linalool and its oxides in the total aroma of withered teas, especially hard withered teas, may be one of the reasons why such teas are more fragrant than non-withered tea. It is well established that linalool and its oxides along with methyl salicylate play an important role in the flowery flavour of black tea while E-2-hexenal constitutes the grassy flavor found in non-withered teas (Takeo, 1983c).

Fermentation also influences black tea quality. The volatile flavor components are lower in non-fermented tea than semi-fermented tea. Linalool oxides are found in the volatile oil extracted from fermented leaves but not identified in the homogenates of fresh leaves. The comparison of volatile contents of semi-fermented and black tea from the same tea cultivar showed that black tea had a higher level of E-2-hexenal, Z-3-hexenal, E-2-hexenyl formate, monoterpene alcohols and methyl salicylates, while semi-fermented tea had a high content of Z-furanoid-β-ionone, uroledal, jasmine lactone and methyljasmonate. Generally, after rolling during orthodox manufacture, there is a large amount of volatile flavor components, but this falls as fermentation progresses. When the polyphenolic oxidation reaction slows down after optimum fermentation, the reaction of the residual flavor substrates may produce small increases in the amount of volatile flavor compounds (VFC) in the over-fermented tea. Rapid oxidation of polyphenols hampered VFC formation in tea leaves (Takeo, 1981a). Lowering of fermentation temperature may result in greater production of VFC polyphenolic oxidation (Hazarika et al. 1984). Thus, duration and temperature of fermentation should be optimized so that a perfect flavored tea is processed.

Pruning is an important field practice in tea production. It also influences black tea quality. The leaf from branches left after pruning produced better quality black teas than tipping leaf from the same bushes. The low quality of tipping leaf black teas was exacerbated by the application of high rates of nitrogenous fertilizers (Owuor, 1990, 1994). Total catechin content, polyphenol oxidase catechol oxidase activity, total theaflavins, brightness, total color and flavor index decreased with coarser plucking. Thearubigin content did not show a clear trend with change in plucking standard (Obanda and Owuor, 1994).

Black tea quality is also affected by shading. An analysis showed that the chlorophyll, carotenoid and anthocyanin in tea shoots grown in the shade of trees were significantly higher than those from unshaded plots. The lower accumulation of catechins and/or higher pigment contents in shaded plants resulted in tea that was less astringent and with better color and appearance. All the pigment contents, except chlorophyll, were higher in pruned tea than that in non-pruned tea, thus enhancing the quality of made tea (Mahanta and Baruah, 1992).

Blister blight disease has also influenced infusion quality in orthodox black tea. A quantitative assessment of infusion characteristics in tea made from shoots with increasingly severe infection by *Exobasidium vexans* fungus showed substantial, progressive decline in theaflavins, thearubigins, caffeine and high polymerized substances, as well as brightness, briskness and total liquor color. Similarly, total phenols and catechins, and PPO and PO activities showed a marked decline in the infected shoots (Arvind et al. 1993).

4. SEMI-FERMENTED TEA

4.1. Introduction

Semi-fermented Tea (Oolong Tea) is a unique tea with special flavor and quality. Oolong tea, which originated in Fujian province, China, derives its name from the Chinese word *Wulong*, literally meaning black dragon and symbolizing authority and nobility. The creation time of Oolong tea can be traced back to 1855. By combining green tea processing procedures with black tea processing procedures, tea farmers in Anxi county, Fujian province, invented a new tea type, Oolong tea. After more than one hundred years development, many famous Oolong teas are recognized such as Wuyi Rock tea, Phoenix Narcissus tea, Iron Buddha tea, Red Robe tea, Rougui tea, Golden Key tea, White Crown tea, etc. Oolong teas are mainly produced in Fujian and Guangdong provinces and in Taiwan. Oolong teas are widely consumed in Mainland China, Taiwan, Japan, UK, USA and Southeast Asian countries.

4.2. Classification of Semi-fermented Tea

Traditionally, the classification of Oolong teas was done according to the production locations (such as Southern Fujian Oolong, Northern Fujian Oolong and Taiwan Oolong) and special tea cultivars (such Tieguanying, Dahongpao, Fushou etc). The cultivar name of tea plant cultivated for Oolong tea making is also assigned to the processed Oolong tea. Semi-fermented teas are named by their half or nearly half fermentation degree in comparison with green tea (non-fermented tea) and black tea (fully fermented tea).

4.3. Processing of Oolong Tea

The raw material for processing Oolong tea is peculiarly plucked. The processing steps consist of plucking of green leaves→withering→rotating→fixing→rolling→ drying→refining. Details of the procedure are described as follows:

4.3.1. *Plucking of green leaves*

Usually, a fully matured shoot (a dormant banjhi bud with 2–3 leaves) is plucked for Oolong tea manufacture. On most occasions, tea cultivars are specially bred for processing Oolong tea. The name of the tea cultivar is usually the same as the related Oolong tea; for example, Tieguanyin and Fenghuang Shuixian are both referring to tea cultivar's name and oolong tea's name which were processed from the leaves of the two cultivars.

4.3.2. *Withering*

Unlike black tea, a special name called *Zuoqing*, literally meaning green-making, is given to the withering stage of Oolong tea processing. The stage is mostly done in the open air by sunlight. The successful withering depends, to a great extent, on the weather conditions. The sunny and windy weather is favorable to making Oolong. The plucked leaves are spread on bamboo mats (0.5 kg/m^2) and exposed to the sunlight for 30–60 min. During the exposure period to sunlight, leaves are turned 2–3 times. When the leaves become soft and the total moisture loss reaches 10–20%, the leaves are moved indoors for the next step. The degree of moisture loss differs according to Oolong tea types. The withering degree is as follows: Pouchong < southern Fujian Oolong tea < northern Fujian Oolong tea < red Oolong tea (Pekoe Oolong tea). After withering, leaves emit a special aroma and have the correct moisture content.

4.3.3. Rotating

Rotating is a special operation in the processing of Oolong tea. Rotating causes friction between leaves and disrupts the leaf cells. Most famous or high-grade Oolong teas are rotated by hand using a bamboo tray. A special machine designed for rotating withering leaves is now widely used for common Oolong tea. Rotating is done indoors at a temperature of 20–25°C and at a relative humidity of 75–85%. The rotating of leaves causes damage to leaf edges and fermentation takes place. The leaf edges turn red at first, gradually spreading to the inner part of leaves. The finished rotated leaves have a mosaic color picture, the fermented leaf edge becomes red and unfermented leaf around the leaf vein remains green. The rotating step in processing Oolong usually lasts for 6–8 hrs and occurs 5–6 times.

4.3.4. Fixing

The rotated leaves are immediately subjected to high temperature fixing in order to stop the fermentation at the leaf edge and to deactivate the enzyme activity in the green part. The fixing involves pan heating for 3–7 min at 180°–220°C. The exact temperature and time depend on the various trees of Oolong tea.

4.3.5. Rolling

Rolling is done at once before the temperature drops. Often, rolling is carried out 2–3 times with a suitable pressure. The detailed rolling procedure depends on the type of Oolong tea. Northern Oolong tea is rolled two times whereas southern Oolong tea three times. Most Oolong teas are rolled using a rolling mill. The cell breakage degree is lighter than green tea or black tea, about 30%. This is the reason why Oolong tea can be brewed repeatedly.

4.3.6. Drying

Drying is usually done in two stages. In the first stage, leaves are spread thinly on the bamboo basket or the drying machine and dried quickly at high temperatures. For the second drying, the temperature is lowered. Drying time and temperature varies according to different types of Oolong tea. For northern Fujian Oolong tea, slow drying at lower temperature is preferred to maximize the aroma of the Oolong tea.

Different types of Oolong tea have different requirements for processing. Table 7 lists some outlined steps of some famous Oolong teas.

4.4. Biochemistry of Oolong Tea Processing

As Oolong tea is a semi-fermented tea and has complicated processing steps, many changes take place during the procedure. Table 8 lists some major changes during each processing step. Unlike black tea and green tea, investigations on the biochemical changes during Oolong tea processing are far fewer. As Oolong tea is a peculiar tea, most studies on Oolong tea biochemistry have been focused on flavor

Table 7 Processing Steps of Some Types of Oolong Tea

Oolong tea	Processing step
Wuyi Rock tea (Rougui, Golden key, Tieguanyin, Qianlixiang)	Fresh leaves→spreading for 30–50 min→Rotating 5–7 times for 6–8 hrs→Fixing by pan or machine→Rolling→First Drying at 100°–140°C for 12–15 min→Spreading and removing stalks and broken leaves→Full Drying at 80°–86°C→Sifting, fanning and separation of leaves and stalks→Drying→Blending and packing
Anxi Tieguanyin (Iron Buddha)	Fresh leaves→Spreading→Sunlight withering→Rotating→Fixing for 7–10 min at 260°C→Alteration of rolling and drying (including first drying→First bag rolling→Second drying→second bag rolling→Full drying)→Reprocessing (including sifting→fanning→separation by hand or by machine→drying→spreading for cooling→blending and packing)
Fenghuang Shuixian (Phoenix Narcissus)	Fresh leaves→Sunlight withering→Indoor withering→Rotating→Fixing→Rolling→First drying at 50°–60°C for 10 min→Second drying at 50°–60°C for 2–3 hrs→Reprocessing (including sifting→fanning→separation of stalks and broken leaf→drying→blending and packing)
Taiwan Oolong tea	Fresh leaves→Sunlight withering for 30–60 min→Rotating 5 times at the interval of 20–60 min→Fixing at 60°–70°C→Spreading for 20 min after fixing→Rolling for 8–15 min→First drying at 105°C–115°C for 8–15 min→Second drying at 70°–90°C for 1–2 hrs→Reprocessing (including separation→fanning→sifting→blending→packing)

chemistry. During rotating treatment of Pouchong tea, aroma compounds develop: esters of Z-2-hexen-1-ol, linalool oxides, benzyl alcohol, 2-phenylethanol, nerolidol, α-farnesen, Z-jasmone, methyl jasmonate, several lactones, benzyl cyanides and indole. Recently, several investigations showed that the glycosidically bound volatiles are the precursors of the alcoholic aroma in Oolong tea (Guo W. *et al.* 1996; Moon *et al.* 1996; Ogawa *et al.* 1997).

The color substance in Oolong tea is more complicated than in black tea. The polyphenol dimers such as theaflavins and thearubgins were also identified in Oolong tea infusion; however, the forming mechanism has not been clear.

4.5. Factors Affecting Oolong Tea Quality

The quality of tea is affected by 4 major factors: cultivars, environment, cultural practices and processing techniques. Quality is more important than yield for semi-fermented tea. Soil and climate are major factors in tea production, the best quality coming from plants grown at higher altitudes where temperatures are cooler. Cultural aspects such as nutrition, weed control, irrigation, pest and disease control and harvesting have more effect on yield while leaf age and harvesting season have a marked effect on quality. Tea processing involves a series of complicated operations (withering, shaking, panning, rolling and drying) each of which can affect the final quality (Chiu, 1990)

Table 8 Major Changes During Each Processing Step

Processing step	Chemical or physical change
Spreading	Moisture removal
	Emission of grassy odor
Withering	Moisture removal
	Oxidation of catechins
	Hydrolysis
	Emission of grassy odor and transformation of flowery aroma
Rotating	Moisture removal
	Disruption of cell walls
	Fermentation
	Hydrolysis
	Oxidation of catechins
	Transformation of flowery aroma
	Breakdown of chlorophyll
	Initial shaping
Fixing	Moisture removal
	Cessation of fermentation
	Promotion of catechin oxidation
	Transformation and release of high bp aroma
	Accelerating the emission of grassy odor
	Hydrolysis
Rolling	Disruption of cell wall
	Shape-making
Drying	Moisture removal
	Oxidation of tea catechins
	Transformation of aroma

Tea shoots must be suitably mature for processing Oolong tea. Dormant shoots with one bud and 3–4 leaves are plucked for Oolong tea. If the leaves are too tender, all the leaves will turn red after withering and rotating, but if the leaves are too mature, the processed Oolong tea is loose in appearance and lacks aroma and taste.

The quality of Pouchong tea produced in lowland Taiwan was increased if the roasting temperature was not higher than 120°C.

An investigation on the effect of microelement fertilizers spraying on the quality of Oolong Tea showed that Oolong tea quality was improved by spraying of Cu, Zn, Mn and B solutions. The effect on quality was in the order B > Mn > Zn > Cu, and the optimum concentrations were 50 mg B/liter, 300 mg Mn/liter, 150–450 mg Zn/liter and 400 mg Cu/liter. All the elements tested showed an obvious influence on the taste of tea infusion but only a slight influence on tea aroma (Guo, *et al.* 1992).

Some Taiwan produced Oolong teas were made from mechanically plucked leaves. However, tea made from mechanically plucked leaves is loose, has a poorer appearance, and weaker taste and liquor color than hand-plucked teas (Chen and Chang, 1989).

REFERENCES

Arvind, G., Ashu, G., Ravindranath, S.D., Satyanarayana, G., Chakrabarty, D.N. (1993) Effect of blister blight on infusion quality in orthodox tea. *Indian Phytopathology*, 46, 155–159.

Chen, Y.K., Chang, C.K (1989) A study on the technical evaluation of conversion from hand plucking to mechanical plucking in eastern Taiwan. *Taiwan Tea Research Bulletin*, 8, 57–59.

Cheng, Q.K. (ed). *Brief Introduction to Tea Biochemistry*. Tea Research Institute, Chinese Academy of Agricultural Sciences. Hangzhou, China, 1982.

Chiu, W.T.F. (1990) Factors affecting the production and quality of partially fermented tea in Taiwan. *Acta Horticulturae*, 275, 57–63.

Gogoi, A.K., Choudhury, M.N.D., Gogoi, N. (1993) Effect of phosphorus on the quality of made teas. *Two and a Bud*, 40, 15–21.

Guo, W.F., Ogawa, K., Yamauchi, K., Watanabe, N., Usui, T., Luo, S. (1996) Isolation and characterization of a β-primeverosidase concerned with alcoholic aroma formation in tea leaves. *Biosci. Biotech. Biochem*, 60, 1810–1814.

Guo, Z. and Lin, X.J. (1992) Effect of spraying microelement fertilizers on the yield and quality of Oolong Tea. *Tea Science and Technology Bulletin*, 2, 21–29.

Hazarika, M., Mahanta, P.K., Takeo, T. (1984) Studies on some volatile flavour constitutents in orthodox black teas of various clones and flushes in North-east India, *J. Sci. Fd. Agric.*, 35, 1201–1207.

ITC (1997) *Annual Bulletin of Statistics*. In: International Tea Committee (ed), Colombo, Sri Lanka

Kawakami, M., Yamanishi, T., Kobayashi, A. (1995) Aroma composition of original Chinese black tea, Zheng Shan Xiao Zhong and other black teas. In: *Proceedings of '95 International Tea-Quality-Human Health Symposium*. November 7–10, 1995, Shanghai, China, pp. 164–169.

Lelyveld, L.J. van, Smith, B.L., Frazer, C. (1990) Nitrogen fertilization of tea: effect on chlorophyll and quality parameters of processed black tea. *Acta Horticulturae*, 275, 483–488.

Lin, H.S. (1962) Some questions about the changes of soluble sugars during the processing of *Tunlu* green tea. In: Tea Research Institute, Chinese Academy of Agricultural Sciences (Ed) *Collection of National Tea Research Projects*, pp. 172–176.

Lu, H.S., Zhang, T.H., Chen, H.C., Shi, Z.P. (Eds) *Tea Sensory Evaluation and Test* (Chinese): Agricultural Publish House, 96–97. Beijing, China, 1979.

Mahanta, P.K., Hazarika, M., Baruah, S. (1990) Influence of plucking and processing on cellwall and soluble components in black tea. *Two and a Bud*, 37, 17–19

Mahanta, P.K. and Baruah, S. (1992) Changes in pigments and phenolics and their relationship with black tea quality. *Journal of the Science of Food and Agriculture*, 59, 21–26.

Moon, J.H., Watanabe, N., Ijima, Y. (1996) Cis- and trans-linalool 3,7-oxides and methyl salicylate glycosides and (Z)-3-hexenyl β-D-glucopyranoside as aroma precursors from tea leaves for oolong tea *Biosci. Biotech. Biochem*, 6, 1815–1819.

Musalam, Y., Suhartika, T., Yamanishi, T. (1991) Seasonal effect on the aroma of Indonesian black tea. *Proceedings of the International symposium on Tea Science*, August 26–29, Shizuoka, Japan, pp. 47–51.

Obanda, M., Owuor, P.O., Njuguna, C.K. (1992) The impact of clonal variation of total polyphenols content and polyphenol oxidase activity of fresh tea shoots on plain black tea quality parameters. *Tea*, 13, 129–133.

Obanda, M. and Owuor, P.O. (1994) Effects of wither and plucking methods on the biochemical and chemical parameters of selected Kenyan tea. *Discovery and Innovation*, 6, 190–197.

Ogawa, K., Ijima Y., Guo, W.F. (1997) Purification of a β-primeverosidase concerned with alcoholic aroma formation in tea leaves (cv. Shuixian) to be processed to oolong tea. *J. Agric. Food Chem*, **45**, 877–882.

Owuor, P.O. and Reeves, S.G. (1986) Optimising fermentation time in black tea manufacture. *Food. Chem*, **21**, 195–203.

Owuor, P.O. and Reeves, S.G. (1988) Comparative methods of optimising fermentaion time in black tea processing. *Acta Horticulturea*, **218**, 385–396.

Owuor, P.O. and Orchard, J.E. (1989) Changes in the biochemical constituents of green leaf and black tea to withering: A review. *Tea*, **10**, 53–59.

Owuor, P.O. and Orchard, J.E. (1990) Changes in the quality and chemical composition of black tea due to degree of physical wither, condition and duration of fermentation. *Tea*, **11**, 109–117.

Owuor, P.O., Munavu, R.M. and Muritu, J.W. (1990) Effect of pruning and altitute on the fatty acids compositon of tea (*Camellia sinensis* (L.)) shoots. *Trop. Sci.*, **30**, 211–219.

Owuor, P.O. and Obanda, M. (1992) Influence of fermentation conditions and duration on the quality of plain black tea. *Tea*, **13**, 120–128.

Owuor, P.O. (1994) Effects of lung pruning and nitrogen fertilizer on black tea quality. *Tea*, **15**, 4–7.

Owuor, P.O. and McDowell, I. (1994) Changes in theaflavins composition and astringency during black tea fermentation. *Food Chem*, **51**, 251–254.

Owuor, P.O. and Obanda, M. (1995) Clonal variation in the individual theaflavins and their impact on astringency and sensory evaluation. *Food Chem*, **54**, 293–277.

Owuor, P.O. (1996) Development of reliable quality parameters of black tea and their application to quality improvement in Kenya. *Tea*, **17**, 82–90.

Sanderson, G.W. (1972) The chemistry of tea and tea manufacture. In Runeckles, T.T.C. (Ed.). *Recent advance in Phytochemistry*, Vol. 5, pp. 247–316.

Sugha, S.K., Singh, B.M. and Sharma, D.K., Sharma K.L. (1991) Effect of blister blight on tea quality. *Journal of Plantation Crops*, **19**, 58–60.

Takeo, T. (1981a) Production of linalool and geraniol by hydrolytic breakdown of bound forms in disrupted tea shoots. *Phytochemistry*, **20**, 2149–2151.

Takeo, T. (1981b) Variation in amounts of linalool and geranial produced in tea shoots by mechanical injury. *Phytochemistry*, **20**, 2149–2151.

Takeo, T. (1983a) Effect of clonal specificity of the monoterpene alcohol composition of tea shoots on black tea aroma profile. *JARQ*, **17**, 1210–1247.

Takeo, T. and Mahanta, P.K. (1983b) Comparison of black tea aromas of orthodox and CTC tea and of black teas made from different varieties. *J. Sci. Fd. Agric*, **34**, 307–310.

Takeo, T., and Mahanta, P.K. (1983c) Why CTC tea is less fragrant? *Two and a bud*, **30**, 76–77.

Takeo, T. (1996) Effect of withering process on volatile compound formation during black tea manufacture. *J. Sci. Fd. Agric*, **35**, 84–87.

Ullah, M.R. and Roy, P.C. (1982) Effect of withering on polyphenol oxidase level in the tea leaf. *J. Sci. Fd. Agric*, **23**, 492–492.

Xiao, W.X. (1963) Changes of Chlorophyll during Green Tea Processing. In: Anahui Agricultural University (ed) *Collection of Tea Research Achievements* (3). (Chinese), pp. 105–108.

Yu, F.L. and Lin, S.Q. (1987) Original place and centre of tea plant. In: *Proceeding of International Tea-Quality-Human Health Symposium*. November 4th–9th, 1987, Hangzhou, China, pp. 7–11.

Zhang, Q. (ed) 1992, *Cha Tan (About Tea)*. Taiwan: Gaoshang Printing Co. Ltd., p. 53.

Zhou, J.S. (1976) *Biochemistry of Green Tea Processing*. In: Tea Science Group, Zhejiang Agricultural University (ed), Hangzhou, Zhejiang, China.

4. THE CHEMISTRY OF TEA NON-VOLATILES

ZONG-MAO CHEN, HUA-FU WANG, XIAO-QING YOU AND NING XU

Tea Research Institute, Chinese Academy of Agricultural Sciences, 1 Yunqi Road, Hangzhou, Zhejiang 310008, China

1. INTRODUCTION

Tea is the most widely consumed beverage in the world. The daily consumption is around 3 billion cups per day. The manufacturing process causes the fresh green tea leaves to be converted to different commercial made tea including green tea (not fermented), oolong tea (semi-fermented) and black tea (fermented tea). Green tea is preferred in China, Japan and the Middle East countries, the oolong tea is mainly consumed in the eastern part of China, China Taiwan and Japan, while 80% of the rest of the consumers prefer black tea. The manufacturing process in each type of tea has a pronounced impact on the formative and degradative reaction pathways, thus influencing the color, flavor and aroma of the end product. The color, flavor and aroma of various types of tea are dependent on different components in tea. If we say that the tea aroma is mainly dependent on the volatile compounds it contained, then the color and the taste of tea are mainly dependent on the non-volatile compounds it contained. This chapter is intended to discuss the non-volatile components in tea and the roles of these components in determining the color and taste of various type of tea.

2. CHEMICAL COMPOSITION OF TEA FLUSH

The tea flush is generally a reference to the apical shoots which consist of the terminal bud and two adjacent leaves. Overall composition of tea flush is listed in Table 4.1. Consideration is given to those chemicals in the fresh flush which are important in determining the quality and hence the value of the made tea product.

2.1. Polyphenols

2.1.1. *Polyphenols in tea flush and green tea*

The total polyphenols in tea flush ranges between 20–35% and 11–20% in green tea on a dry weight basis. The contents in macrophyll cultivars are higher than those in microphyll cultivars. It is a group of phenolic compounds in which the catechins are the main components. They are the main constituents in determining the color and taste of tea infusion and the key material basis in forming the tea quality. The contents of polyphenolic compounds are the material constituents of tea infusion; however, they are

57

Table 4.1 Composition of Fresh Tea Flush (% Dry Weight).

Components	Dry weight (%)
Soluble in water	
Flavonols	18–32
(–) – EGCG	9–14
(–) – EGC	4–7
(–) – ECG	2–4
(–) – EC	1–3
(+) – GC	1–2
(+) – C	0.5–1
Minor catechins	0.4–1
Flavonol glucosides	3–4
Proanthocyanidins	2–3
Caffeine	3–4
Amino acids	2–4
Carbohydrates	3–5
Organic acids	0.5–2
Saponins	0.04–0.07
Pigments	0.5–0.8
Vitamins	0.6–1.0
Soluble minerals	2–4
Insoluble or Slightly soluble in water	
Cellulose	6–8
Lignin	4–6
Polysaccharides	4–10
Lipids	2–4
Insoluble pigments	0.5
Insoluble minerals	1.5–3.0
Volatiles	0.01–0.02

not the sole determinative factor of tea quality. The polyphenols in tea mainly include the following six groups of compounds: flavanols, hydroxy-4-flavonols, anthocyanins, flavones, flavonols and phenolic acids. Among these, the flavonols (mainly the catechins) are most important and occupy 60–80% of the total amount of polyphenols in tea. About 90–95% of the flavonols undergo enzymatic oxidation to products which are closely responsible for the characteristic color of tea infusion and its taste. They are generally water soluble, colorless compounds.

Catechins, the major constituents of total polyphenols, are the substances responsible for the bitterness and astringency of green tea as well as the precursors of theaflavins in black tea. Tsujimura (1927–1935) first isolated three catechins from tea, i.e., (–)-EC, (–)-ECG and (–)-EGC. Bradfield (1944,1948) further isolated the (–)-EGCG. These four catechins constitute around 90% of the total catechins, and the (+)-C and (+)-GC occupy around 6% of the total catechins. The largest part of the catechins present in tea flushes is esterified with gallic acid in the 3-positon. In addition, some minor catechins (less than 2% of the total catechins) have been reported (Coxon, 1972; Saijo, 1982; Nonaka *et al.* 1983; Seihi, 1984; Hashimoto, 1987). The ether type catechins [(–)-ECG and (–)-EGCG] are stronger in bitterness and more astringent than (–)-EC and (–)-EGC. The structure of catechins in tea is listed in Table 4.2. The shape of crystalline and the physical properties of major catechins are shown in Figure 4.1 and Table 4.3. The relative amounts of various catechins and their gallates are genetically controlled and therefore a clonal characteristic. It also depends on the various seasons and other environmental factors. Usually, the contents of catechins are higher in the summer season and in the macrophyll cultivars than those in the spring season and in the microphyll cultivars. The catechin contents decrease with the increase of fiber in the shoot components. This is why high quality teas arise from the more tender shoots, which possess higher contents of catechins and lower fiber contents. So, the high catechin-fiber ratio represents a better quality of tea. The biosynthesis of catechins has been investigated tentatively in the former USSR and published as a monograph (Zaprometov, 1958, 1987, 1989).

2.1.2. Polyphenols in black tea

The major difference between the manufacture of green tea and black tea is that black tea processing undergoes a fermentation stage that involves an enzyme-catalysed oxidative reaction to the colored phenolic compounds. These brown coloured pigments formed during the fermentation process are the products of catechin oxidation. The major catechin oxidative products are theaflavin and thearubigin. Roberts and his coworkers (1950, 1957) were the first to use the paper chromatography to isolate the ethyl acetate fraction and named theaflavin and theaflavin gallate.

Theaflavin (TFs) has been shown to be a bis-flavan substituted 1′, 2′-dihydroxy-3, 4-benzotropolone, orange-red in colour, constituting about 0.3–1.8% of black tea dry

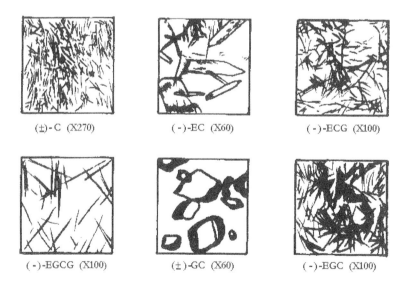

(±)-C (X270)   (-)-EC (X60)   (-)-ECG (X100)

(-)-EGCG (X100)   (±)-GC (X60)   (-)-EGC (X100)

Figure 4.1   Crystalline shape of various catechins

Table 4.2 Tea Catechins and Their Structure.

Catechins in tea	Abbreviation	R_1	R_2	R_3	R_4	R_5	% of the total catechins
(−) Epicatechin	(−) − EC	OH	OH	H	H	H	14
(−) Epicatechin gallate	(−) − ECG	OH	OH	H	G	H	9
(−) Epigallocatechin	(−) − EGC	OH	OH	OH	H	H	44
(−)- Epigallocatechin gallate	(−) − EGCG	OH	OH	OH	G	H	23
(+) Catechin	(+) − C	OH	OH	H	H	H	3
(+) Gallocatechin	(+) − GC	OH	OH	OH	H	H	6
(−) Epiafzelechin		H	H	H	H	H	
(−) Epiafzelechin-3-O-gallate		H	G	H	H	H	
(−) EC-3-O-(3″-O-methyl)gallate		OH	3″-MeG	H	H	H	
(−)-EC-3-O-(4″-O-methyl)gallate		OH	4″-MeG	H	H	H	
(−)-EGC-3-O-(3″-O-methyl)gallate		OH	3″-MeG	OH	H	H	
(−)-EC-3,5-di-O-gallate		OH	G	H	H	G	
(−)-EGC-3,5-di-O-gallate		OH	G	OH	H	G	
(−)-EGC-3,3-di-O-gallate		OH	G	OG	H	H	
(−)-EGC-3,4-di-O-gallate		OH	G	OH	G	H	
(−)-EC-3-O-p-hydroxy benzoate		OH	p-OH ben.	H	H	H	
(−)-EGC-3-O-p-coumarate		OH	p-cou.	OH	H	H	
(−)-EGC-3-O-cinnamate		OH	cin.	OH	H	H	

* Adapted from Bradfield (1948), Coxon *et al.* (1972), Saijo (1982), Nonaka *et al.* (1983) Seihi *et al.* (1984), Hashimoto *et al.*. (1987), Nagabayashi *et al.* (1992)

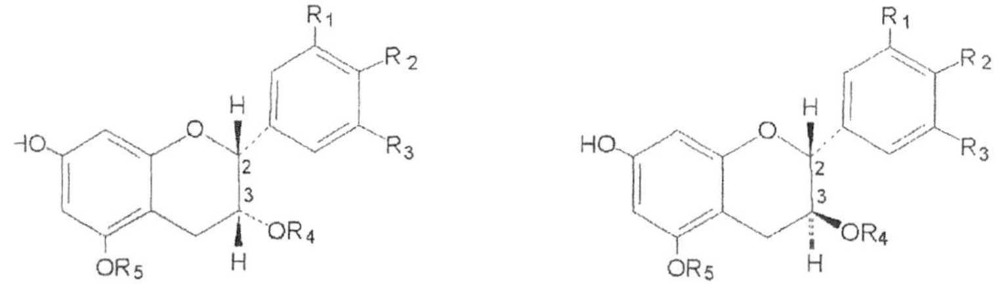

(-) epi type
(2R, 3R)

(+) type
(2R, 3S)

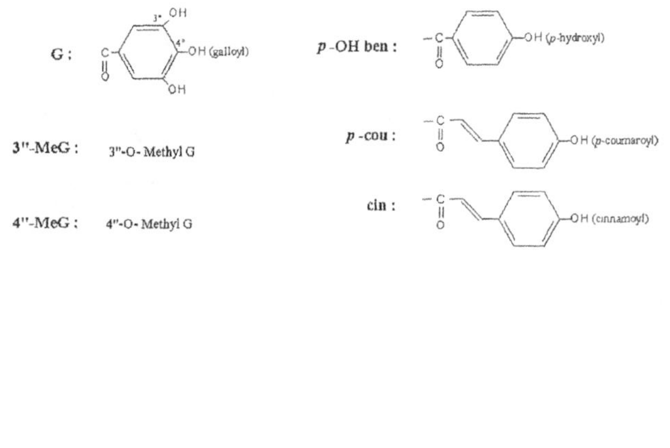

G : (galloyl)

3″-MeG : 3″-O- Methyl G

4″-MeG : 4″-O- Methyl G

p -OH ben : (p-hydroxyl)

p -cou : (p-coumaroyl)

cin : (cinnamoyl)

Table 4.3 Crystalline Shape and Physical Property of Tea Catechins.

Catechins	Crystalline shape	$\lambda_{max}(\varepsilon_{max})$ in UV	Melting point (°C)
(+)-Catechin	thick needle shape	280 (4100)	176
(−)-Epicatechin	long prism with pointed end	280 (3100)	242
(+)-Gallocatechin	Oblique square shape with unclear edge	275 (1290)	188
(−)-Epigallocatechin	thick needle shape	275 (1450)	218
(−)-Epicatechin gallate	thin needle shape	280 (13500)	253
(−)-Epigallocatechin gallate	long needle shape	276 (11250)	254

weight and 1–6% of the solids in tea infusion. It contributes significantly to the bright color and brisk taste of tea infusions. The theaflavins are formed by the enzymatic oxidation and condensation of catechins with di- and tri-hydroxylated B rings, referred to from now on as simple catechins [(+)-catechin, (−)-epicatechin, (−)-epicatechin-3-gallate] and gallocatechins [(−)-epigallocatechin-3-gallate, (−)-epigallocatechin and (+)-gallocatechin] (Takino *et al.* 1964). Coxon (1970b), Bryce (1970, 1972) discovered the epitheaflavic acid, epitheaflavic acid-3-gallate, theaflavin-3, 3′-digallate. Recently, a gallated version of theaflavate A and a non-gallate version of theaflavate A have been isolated from black tea (Wan *et al.* 1997). Up to now, there have been 11 theaflavins reported. Coxon *et al.* (1970a) reported that the approximate relative proportions of the theaflavins in black tea were theaflavin (18%), theaflavin-3-gallate (18%), theaflavin-3′-gallate (20%), theaflavin-3,3′-digallate (40%), isotheaflavin + theaflavic acids (4%). The name, structure, precursors and their proportion in total theaflavins are listed in Table 4.4.

Roberts (1950), Takino *et al.* (1964), Collier *et al.* (1973) and Robertson (1983) investigated the mechanism and pathways of theaflavin formation. It was considered that the oxidation of the catechins to their respective *o*-quinones was catalyzed by polyphenol oxidase and a net uptake of molecular oxygen was observed. Low oxygen conditions may inhibit this reaction, thus resulting in poor recovery of theaflavins during fermentation (Robertson, 1983). The quinones rapidly reacted with each other and other compounds to form theaflavins. The formation of theaflavin during fermentation reached a maximum and then declined. For CTC tea, this maximum usually occurred between 90–120 min. It was generally recognized that the theaflavins play a premier role in determining the characteristic cup quality of black tea infusion described by tea tasters as "brightness" and "briskness" (Roberts, 1962; Hilton & Ellis, 1972). However, the contribution made by these compounds to quality differs with individual theaflavin. The digallate is believed to contribute most, while theaflavin itself is considered to contribute least.

The preparation of theaflavins is implemented according to the following procedure. One kg of black tea is extracted with hot water, and further extracted with methyl isobutyl ketone. After concentration, the extract is washed with 2.5% aqueous sodium bicarbonate solution and concentrated again. The residue is dissolved in water and is

Table 4.4 Tea Theaflavin (Geissman, 1962; Takino *et al.* 1965, 1966; Brown, 1966; Bryce *et al.* 1970; Coxon *et al.* 1970; Sanderson, 1972; Nagahayashi *et al.* 1992).

Theaflavin	Structure	Precursors	%of the total theaflavins in black tea
Theaflavin (1964)	R=H R′=H	(–)-EC, (–)-EGC	9.0
Theaflavin-3-O-gallate (1970)	R=G R′=H	(–)-ECG, (–)-EGC	36.6
Theaflavin-3′-O-gallate (1970)	R=H R′=G	(–)-EC,(–)-EGCG	20.5
Theaflavin-3,3′-di-O-gallate (1970)	R=G R′=G	(–)ECG,(–)-EGCG	30.0
Isotheaflavin (1970)	R=H R′=H	(+)-C, (–)-EGC	0.2
Theaflavin isomer (1972)	R=H R′=	(–)-EC, (+)-GC	
Theaflavic acid (1970)	R′=H	(+)-C, Gallic acid	
Epitheaflavic acid* (1970)	R′=H	(–)-EC, Gallic acid	0.2
Epitheaflavic acid-3′-O-gallate (1972)	R′=G	(–)-ECG, Gallic acid	0.3
Theaflagallin (1986)	R′=H	(+)-GC, Gallic acid	
Epitheaflagallin (1986)	R′=H	(–)-EGC, Gallic acid	
Epitheaflagallin 3′-O-gallate (1986)	R′=G	(–)-EGCG, Gallic acid	

*G=galloyl

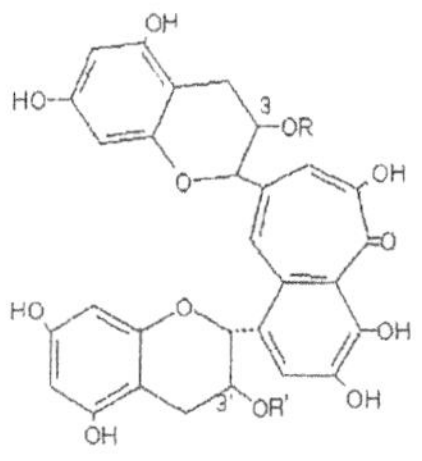

Theaflavin
Theaflavin 3-O-gallate
Theaflavin 3′-O-gallate
Isotheaflavin

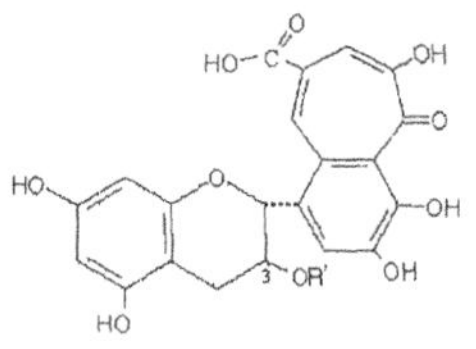

Theaflavic acid

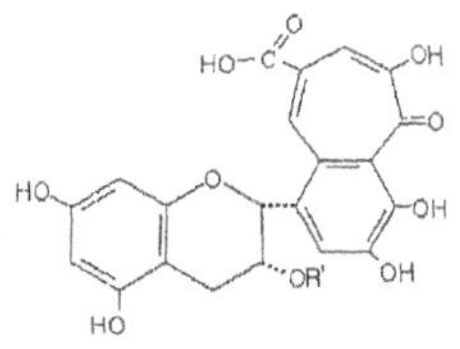

Epitheaflavic acid
Epitheaflavic acid 3′-O-gallate

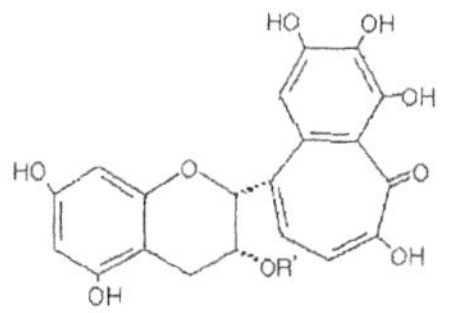

Epitheaflagallin
Epitheaflagallin 3′-O-gallate

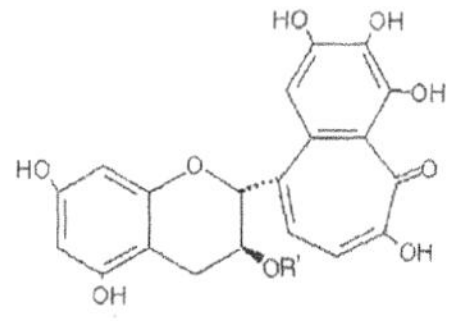

Theaflagallin

washed with chloroform to remove caffeine, and then concentrated until dry. The solid is redissolved in methanol- water (43:57 v/v) and the solution eluted through Sephadex LH-20 and cellulose in ethyl acetate, producing theaflavin (310 mg), theaflavin monogallate (300 mg) and theaflavin digallate (360 mg) (Mahanta, 1988).

Thearubigin (TRs) is the name originally assigned to a heterogeneous group of orange-brown, weakly acidic pigments formed by enzymatic oxidative transformation of flavanols during the fermentation process of black tea (Roberts, 1958). However, the withering process also showed an obvious impact on the formation of theaflavins and thearubigins (Xiao, 1987). Thearubigins comprise between 10–20% of the dry weight of black tea and between 30–60% of the solids of the black tea infusion, which is 10–20 times higher than the dry weight of theaflavins. Unlike the theaflavins, thearubigins have still not been characterized. The chemical structure of the thearubigins remains a mystery. The major difficulty is that they are diverse in their chemistry and possibly in their molecular size. Hazarika *et al.* (1984b) separated the TRs from 60% acetone CTC black tea extracts by using the Sephadex LH-20 column chromatography and divided them into three components: TR_1 (with high molecular weight), TR_2 (with moderate molecular weight) and TR_3 (with low molecular weight). The contents of low molecular weight TR was decreased and the contents of high molecular weight TR was increased with the time of fermentation. The thearubigins are formed by the oxidation of any one of the tea catechins or combination thereof. Unlike the formation of theaflavins, the thearubigins have no specific intermediate product precursors of catechin oxidation. However, the different TR was formed from different combination of catechins. Apart from the theaflavins which impact the briskness and brightness of black tea infusion, the TRs also make an important contribution to the color, strength and mouthfeel (Roberts & Smith, 1963, Millin *et al.* 1969). They are responsible for "body", "depth of color", "richness" and "fullness" of tea infusion. The amounts of TRs were increased with the decrease of TFs during the fermentation process in black tea manufacture, indicating the TFs are probably the intermediate of TRs. Dix *et al.* (1981) proved the transformation of TFs to TRs with the action of peroxidase.

The preparation of thearubigins can be implemented by successive extraction with ethyl acetate and *n*-butanol (Millin *et al.* 1969). Mahanta (1988) extracted black tea with 60% aqueous acetone. The extract was separated into six fractions over a Sephadex LH-20 column. The two methyl ethyl ketone-soluble components correspond to the above mentioned TR_1 and TR_3 fractions.

2.2. Flavonols and Flavonol Glycosides

The flavonols and flavonol glycosides occur in small quantities. There are three major flavonol aglycones in fresh tea flush: kaempherol, quercetin and myricetin

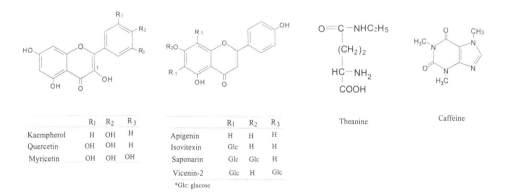

Figure 4.2   Structure of some flavonols, flavones, theanine and caffeine contained in tea

(Figure 4.2), which differ in the degree of hydroxylation on the B ring, i.e., mono-, di- and tri-hydroxy substitutions, respectively. These flavonols occur both as free flavonols and as glycosides. The glycosidic group may be glucose, rhamnose and galactose. The 3-glucosides appear to be the most significant in the macrophyll cultivars, while the rhamnodiglucosides predominate in the microphyll cultivars.

2.3. Flavones

The flavones of tea have been studied by Ul'yanova (1963) and Sakamoto (1967, 1969, 1970), who found 18 flavones in green tea infusion, some of which were identified as C-glycosyl flavones, namely, vitexin, isovitexin, 3 isomers of C-glycosyl apigenin, saponarin, vicemin-2, theiferin A and theiferin B. These flavones are water-soluble and constitute the yellow color in the infusion of green tea and black tea.

2.4. Phenolic Acids and Depsides

The major phenolic acids present in tea flush are gallic acid, chlorogenic acid and coumaryl quinic acid. The most important depside is 3-galloyl quinic acid (theogallin). It aroused attention due to its relatively high level in tea flush and its statistic correlation to black tea quality (Cartwright & Roberts, 1955; Roberts & Myers, 1958). The contents of gallic acid in green tea infusion ranged from 0.4–1.6 g/kg dry weight. The amounts of free gallic acid increase during the fermentation owing to its liberation from catechin gallates.

2.5. Amino Acids

Around 2–4% of amino acids is present in tea flush. It was considered to be important in the taste of green tea (Nakagawa, 1970, 1975) and aroma of black tea (Ekborg-Ott et al. 1997). The most abundant amino acids are theanine (5-N-ethyl glutamine) which is apparently unique to tea and found at levels of 50% of the free amino acid fraction. It was first isolated from tea leaves by Sakato (1950). It is a colorless needle shaped crystalline. The contents of theanine in tea averaged 1.37% in 100 g tea samples, with Taiwan oolong (0.6%) having the lowest concentration and Yunnan black tea (2.38%) the highest (Ekborg-Ott et al. 1997). Theanine has two enantiomers: L- and D-theanine. The average relative level of D-theanine was around 1.85% of the total theanine (Ekborg-Ott et al. 1997). The tea that had the lowest relative amounts of D-theanine (< 1%) were always of pekoe or FOP black tea grades which refer to the high quality. The relative ratio of D-theanine in tea was significantly increased under the high temperature condition of storage. Thus, it was suggested that the enantiomeric ratios of theanine might be useful as an indicator for long-term storage, or possibly for shipping and handling of tea and may be a useful tool in the grading of tea (Ekborg-Ott et al. 1997). The precursors in the biosynthesis of theanine in tea plant were identified as glutamic acid and ethylamine (Sasaoka & Kito, 1964; Takeo, 1974). The site of biosynthesis of theanine is the root from where translocation occurred to younger leaves. Thus, this amino acid is present in highest concentration in the roots of the tea plant, but is absent from tea seeds (Konishi, 1969; Wickremasinghe & Perera, 1972; Feldheit et al. 1986). It has been shown that theanine plays a role in protecting enzymes from inactivation by polyphenolic products (Wickremasinghe & Perera, 1973). Studies have also shown reduction effects on blood pressure, neurotransmitter action in the brain as well as reduction of the levels of 5-hydroxyindoles in hypertensive rats at high dose (1500–2000 mg/kg) (Yokogoshi *et al.* 1995). These pharmacological and physiological effects of theanine showed a relaxation-causing effect in human volunteers. Thus it may become a functional additive to foods to make stressed people relax.

Twenty-six other amino acids usually associated with proteins have been reported in tea flush. There are glutamic acid, arginine, glutamine, aspartic acid, glycine, serine, asparagine, lysine, threonine, histidine, α-alanine, β-alanine, tyrosine, proline, hydroxyproline, valine, S-methylmethionine, tryptophan, leucine, isoleucine, phenylalanine, γ-glutanyl methylamide, γ-aminobutylic acid, γ-N-ethylasparagine, cysteic acid and pipecolic acid (Krishnamurthy *et al.* 1952; Konishi & Takahasii, 1966; Wickremasinghe, 1978; Wang, 1982). Among these, the contents of the first four amino acids are around 200–280 mg/100 g in made tea. High levels of these free amino acids are related to the quality of green tea. However, it was regarded that the high level of free amino acids, especially leucine and isoleucine, in fresh tea flush are detrimental to the quality of black tea (Tirimanna, 1967; Wickremasinghe, 1978). The amino acids have also been established as important aroma processors. Also, the γ-aminobutylic acid was known to act as a neurotransmitter and proved to depress the blood-pressure of humans *in vitro* and *in vivo* (Omori, 1991).

2.6. Chlorophyll, Carotenoids and Other Pigments

The main pigments present in fresh tea flush are chlorophylls and carotenoids. The amounts of chlorophyll in tea leaves is reported about 0.2–0.6% dry basis. The proportion of chlorophyll A to chlorophyll B is around 2:1. The contents of chlorophyll were decreased during the black tea manufacturing process and some degradative products of chlorophyll, such as pheophytin A, pheophytin B and pheophorbide, were produced (Wickremasinghe & Perera, 1966). These products cause the blackness or brownness of black tea infusion due to the brown color of pheophorbide and black color of pheophytins.

There were 15 carotenoid compounds reported in tea flush: phytofluene, α-carotene, β-carotene, β-zeacarotene, aurochrome, mutatochrome, cryptoxanthin, cryptoflavin, cryptoxanthin-5, 8-diepoxide, lutein, zeaxanthin, lutein-5, 6-epoxide, violaxanthin, leuteoxanthin and neoxanthin (Tirimanna & Wickremasinghe, 1965; Sanderson *et al.* 1971; Venkatakrishna *et al.* 1976; Hazarika & Mahanta, 1984a). Among those, the β-carotene, lutein, zeaxanthin are the most important components. The total contents of carotenoids in tea flush were reported as 0.03–0.06% dry weight. The mature leaves contain more carotenes than younger leaves. There was a marked increase of β-carotene and lutein + zeaxanthin during leaf maturation (Venkatakrishna *et al.* 1977). The carotenes play an important role in the formation of black tea quality and act as the precursors of black tea aroma. Many investigations showed that the carotenes present in tea flush undergo appreciable degradation around 47–70% (Tirimanna & Wickremasinghe, 1965; Sanderson *et al.* 1971; Mahanta & Hazarika, 1985) during the manufacturing process of black tea. Several aromatic compounds, such as β-ionone, dihydroactinidiole, theaspirone, are formed from carotenoids via thermal degradation during tea manufacturing (Tirimanna & Wickremasinghe, 1965; Sanderson *et al.* 1971; Mahanta & Hazarika, 1985).

2.7.   Carbohydrates

Sanderson & Perera (1965) investigated the carbohydrates in tea flush. The free sugar contents in tea flush were reported as 3–5% in dry weight. They consisted of glucose, fructose, sucrose, raffinose and stachyose. The free sugar contents in tea flush change under natural and shaded conditions. Sucrose is the major primary product of photosynthesis under natural conditions, and it increases with the growth of tea shoots, occupying more than 50% of the total free sugar content. The free sugar contents in tea flush under natural conditions were 40–50% higher than those under shade (Anan *et al.* 1985). The monosaccharides and disaccharides present in tea flush are one of the sweet taste components in tea infusion. The polysaccharides were separated into hemicellulose, cellulose (6–8% dry weight basis) and other extractable polysaccharide fraction (1–3%) composed of different sugar residues, i.e., glucose, galactose, mannose, arabinose, xylose, ribose and rhamnose. The cellulose and hemicellulose contents were negatively correlated with the tenderness of the tea shoots. The lower the cellulose and hemicellulose content in tea flush, the more tender the raw material, and also the higher the quality of made tea product. Starch was found mostly in the root system of the plant with only a small amount in tea flush. As early as 1933, it was shown that tea polysaccharides are good blood-glucose depressing agents and are recommended for use in the treatment of diabetes (Konayagi & Minowada, 1933). Some recent investigations carried out in China and Japan also extracted the polysaccharides from made tea and showed they could decrease blood-glucose level, thus potentially useful in the treatments of diabetes (Mori *et al.* 1988; Wang *et al.* 1996). The composition of tea polysaccharides differs with the tenderness of the raw material and season. Mori *et al.* (1988) identified a mixture of arabinose, ribose and glucose (5.1: 4.7: 1.7). Takeo (1991) identified glucose, galactose, arabinose and fructose (12: 3: 3: 1), and Wang *et al.* (1996) reported arabinose, xylose, fructose, glucose and galactose (5.52: 2.21: 6.08: 44.2: 41.99).

2.8.   Organic Acids

Several organic acids are contained in the tea flush. They include the dicarboxylic acids and tricarboxylic acids, such as succinic acid, oxalic acid, quinic acid, malic acid, citric acid etc. and the fatty acids, such as linoleic acid, palmic acid, hexanoic acid, pentoic acid etc. Some of these organic acids are aroma components. Some of them are not aromatic themselves, however, they probably transfer to aromatic components via oxidation or other reactions. The total amounts of organic acids were reported at 0.5–2.0% dry weight in fresh tea flushes. The content of quinic acid is the highest. Oxalic acid is the next highest, which is present as crystals in the vacuoles of leaf cells.

2.9.   Caffeine and Other Alkaloids

Tea contains 2–5% of caffeine in fresh flush. Caffeine is a trimethyl derivative of purine 2,6-diol. It was first isolated from tea by Runge and named theine in 1820. It was found that tea theine was the same as the caffeine from coffee, so the "theine" name was later abolished. Daily per capita caffeine intake from all sources including tea has been estimated at 200 mg on the basis of total US consumption (Barone & Roberts, 1984). According to a survey by Phelps & Phelps (1988) of 44 countries, the caffeine intake was under 100 mg per day in 23 countries, 100–200 mg per day in 11 countries, 200–300 mg in 7 countries and over 300 mg in 3 countries. The average intake by US consumers is 1.9 mg/kg and by UK consumers is 4.0 mg/kg (Barone & Grice, 1994). That is around 110 mg and 240 mg per day, respectively (calculated by the average body weight 60 kg). Tea is one of the intake sources of caffeine. It was estimated that in the USA and Canada, 15–30% of dietary caffeine is obtained from tea; in Britain around 44% comes from tea (Marks, 1992). In the case of tea, the type of tea and the method of preparation affect the caffeine content. Caffeine content is not significantly reduced during tea processing, although it may decrease during the firing process. It has been reported that during tea processing, caffeine reacts with theaflavins to form a compound, which imparts "briskness" to the tea infusion. High caffeine levels are associated with good "cream" formation in black tea infusion. The pure caffeine is a white, light, spun–silk like crystalline with a m.p. of 238°C and sublimation from 120°C completed at 178°C. It is soluble in hot water with a bitter taste (bitter taste threshold 0.0007M). The caffeine contained in tea flush is higher in spring and gradually decreased with the growth of leaves. The caffeine contents in 1st and 2nd leaf (3.4% in dry weight) are higher than that in the mature leaf (around 1.5% in dry weight). The caffeine can be separated from polyphenols by applying the tea infusion through a column of β-cyclodextrin polymer, the caffeine passes through the column while the tea polyphenols are adsorbed on the column and eluted with 30–50% ethanol (Chu & Juneja, 1997). The caffeine is rapidly and completely absorbed after oral intake. The peak blood caffeine level is reached within 30–60 min, when tea is drunk. The half-life of caffeine in blood plasma was reported as 3–7 hr, and follows first order kinetics. The half-life of caffeine is decreased to 3 hr or less in smokers due to the activity of hepatic aryl hydrocarbon hydroxylase (Parsons & Neins, 1978).

Caffeine is pharmacologically classified as a central nervous system stimulant and a diuretic. Caffeine possesses the ability to improve the wall elasticity of blood vessels, promoting blood circulation, increasing the efficient diameter of vessel, and stimulating the urination and autooxidative activity (Dews, 1982; Elias, 1986; Shi et al. 1991). The reaction rate constant of caffeine in scavenging the hydroxyl radicals was around 5.9×10^9 M^{-1} sec^{-1}, which is comparable with that of other efficient antioxidants (Shi et al. 1991). The merits and harmfulness of caffeine have been disputed over a long period of time, and are still uncertain. Many epidemiological investigations have shown that no positive correlation can be established between the intake of caffeine and the incidence of myocardial infarction and cancer in humans (Garattini, 1993). The no-effect dosage of caffeine was set as 40 mg/kg/day by the FDA in USA (Elias, 1986). This dosage is 8–10 times higher than the normal intake dosage. It was pointed out that even if overdosed, the effect is transient and can be removed (Dews, 1982). In conclusion, the quantity of caffeine ingested by average or moderate consumers of tea, 3–6 cups per day, is highly unlikely to induce undesirable effects unless significant additional amounts are obtained from other sources.

Besides caffeine, the other methyl-xanthines are theobromine and theophylline, xanthine, hypoxanthine and guanine. They are found in very small quantities in tea.

2.10. Minerals

The average total ash content of tea is around 5% of the dry matter. There are 28 elements reported in tea flush. Besides molybdenum, iodine and lead are situated in the V-VI period in the periodic table, the remaining 25 elements contained in tea flush are situated in the I-IV period. Elements contained in fresh tea leaves can be classified into the following 4 groups.

1) Contents > 2000 mg/kg: carbon, hydrogen, oxygen, nitrogen, phosphorus, and potassium.
2) Contents 500–2000 mg/kg: magnesium, manganese, fluorine, aluminium, calcium, sodium, and sulfur.
(3) Contents 5–500 mg/kg: iron, arsenic, copper, nickel, silicon, zinc, and boron.
(4) Contents < 5 mg/kg: molybdenum, lead, cadmium, cobalt, selenium, bromine, iodide, and chromium.

Compared to other plants, the following eight elements including fluorine, potassium, aluminium, iodine, selenium, nickel, arsenic and manganese are present in higher than average levels (Figure 4.3). Some elements are accumulated in the mature leaf, such as fluorine, aluminium, selenium, calcium, iron, silicon, manganese,

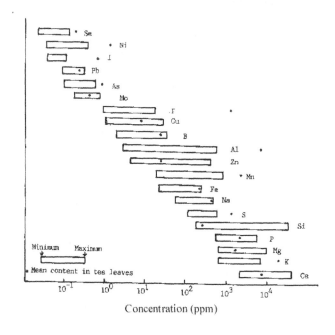

Figure 4.3   Comparison to the contents of various elements in the leaves of tea plant and other

Table 4.5 Contribution of Tea Drinking to the Requirements of Various Elements for Humans (Chen, 1990).

Element	Average daily requirements per capita (mg)	Extraction from made tea during infusion (%)	Average daily amount ingested from tea infusion of 10 g tea (mg)	As the required amounts (%)
F	0.5	50–60	0.3–0.4	60–80
Cu	2–5	70–80	0.5–0.6	10–30
Mn	3–9	33–36	4–8	60–100
Ni	0.3–0.5	50	0.03–0.6	10–20
Al	10–50	16–18	1.3–1.9	5–20
K	1500–3000	90–97	140–200	6–10
As	0.1–0.3	32	0.003–0.007	1–7
Zn	10–15	36–56	0.2–0.4	1.3–4.0
Se	0.05–0.20	8–24	0.002–0.004	1.0–8.0
Mg	220–400	46–53	6–10	1.8–5.0
Ca	400–1500	5–7	2.5–5.2	0.3–1.3
Fe	12–15	<10	0.1–0.2	<1–1.6
B	10–20	24–31	0.04–0.10	0.3–1.0
S	420–3000	50–60	5–8	0.2–1.9
P	1200–2700	25–35	0.4–5.0	0.1–0.5
Na	1600–2700	10–20	2.5–4.0	0.15–0.25
Pb	0.3–0.4	<10	0.0004–0.0005	<0.1

boron etc, and some elements are accumulated in the tender leaves, such as the zinc, magnesium, potassium, arsenic, nickel etc.

The extraction rate of various elements during the infusion process is different. Some elements could be almost completely extracted into the tea infusion, such as bromine, potassium, some elements could be well extracted into the infusion, such as copper, fluorine, nickel, zinc, chromium, manganese, magnesium, sulfur, cobalt, while some elements could be only partly extracted (Table 4.5).

Tea plant is a fluorine concentrating plant. The fluorine content in tea leaves is unusually high, being about 100–200 mg/kg dry weight in tea shoot, around 300–400 mg/kg in mature leaf, and as high as 1000 mg/kg in old leaf (Yamada, 1980). The role of fluorine in tea plant is uncertain; however, the fluorine in tea is beneficial to human health, especially in the prevention of dental caries. It was shown that the hydroxyl group in the hydroxyphosphorite of dental enamel can be replaced by fluorine in tea to form fluorophosphorite, thus increasing the hardness and antacid activity of teeth (Kakuda et al. 1994). However, it was reported that the dental fluorosis among children in Gansu province in China was caused by the consumption of milk tea made from brick tea in water containing high concentrations of fluorine (2.5–3.69 mg/kg).

Tea plant is also an aluminium concentrating plant. The aluminium content in tea flush is much higher than those in other crops. Ordinarily, the aluminium content in fresh tea leaves ranges from 200–1500 mg/kg (Pennington, 1987). However, only some aluminium is extracted into the infusion due to its low extraction rate (Takeo, 1983). Most of the Al species found in tea leaves were in the catechin complex. Other parts of tea plant had two distinctive forms – catechin and fluorine complexes, however, the Al-F forms were not found in the leaves, indicating that the Al-F complex is the translocated form which is rapidly converted into another form in the leaves. The concentration of Al in tea infusion ranged around only 2–6 mg/l. The chemical forms of Al in tea infusion mainly include a complex of one mole of Al and 3 moles of oxalate as well as a complex of Al, oxalate and fluorine (Horie *et al.* 1994). The role of Al in tea quality is obscure, however, it was reported that Al is closely related to the metabolism and storage of tea flavonol (Jurd, 1963). Some patents and papers (Edmonds *et al.* 1979; Chang *et al.* 1982) recommended adding Al to commercial black tea for the purpose of improving the infusion colour, however, it is not acceptable from the viewpoint of human health. Overdose of Al from food may induce neurotoxicity, and is possibly related to Alzheimer's disorder (Alfrey, 1986; Flaten & Degard, 1988).

Copper also deserves close attention and is of particular importance in tea biochemistry, because the polyphenol oxidase enzyme (PPO) contains this mineral. The Cu content in tea leaves is average compared to other plants. The average Cu content in tea leaves amounted to 12–18 mg/kg (Child, 1955). However, the Cu contents may be increased to several hundred mg/kg level in made tea in some areas due to the Cu residue caused by the application of copper fungicides for the purpose of controlling blister blight disease of the tea plant. Manganese is an important element which participates and catalyses the activity of many enzymes, such as DNAase, choline esterase, phosphatase, phosphohexokinase, adenosine kinase, pectin kinase, *trans*-glutaminase, polymerase etc. in humans. Tea plant is a Mn-rich plant. The Mn content in fresh tea leaves ranges from 200–1200 mg/kg. The content in old leaves is higher than that in tender leaves. The extraction rate of Mn in tea infusion is around 35%. The Mn ingested from daily tea drinking is over 60% of the daily Mn requirement of humans.

2.11. Vitamins

Tea is rich in vitamins, especially vitamin C. It was reported that the vitamin C contents in 100 g green tea is as high as 150–300 mg. However, the vitamin C content in oolong tea and black tea is less than that in green tea due to decomposition in the fermentation process. The vitamin B group content in green tea and black tea are similar, around 1–2 mg per 100 g made tea. They are water-soluble and hence are extracted into infusion during the process of brewing. Vitamin E has shown anti-senescence and anticarcinogenic activity. It mainly exists in the lipid fraction of made tea, and the content is around 24–80 mg per 100 g made tea. The vitamin K content is around 300–500 i.u. per gram of made tea, so 5 cups of tea per day may satisfy the requirements of vitamin K for humans.

2.12. Enzymes

A most important fraction in tea flush is the proteins, because this fraction includes the enzymes required for the metabolism of tea plant and tea quality. Jinech & Takeo (1984) presented an excellent review on the enzyme system in tea plant. Table 6 lists important enzymes that have been found in tea plant and their function in tea biochemistry.

The polyphenol oxidase (PPO) is the most important enzyme in tea flush, because it has a key role in black tea manufacture. It has a high degree of specificity, only attacking the B rings of tea polyphenols. It was shown to be a Cu-containing protein (Steerangacher, 1943) and found to consist of at least four isoenzymes (Gregory & Bendall, 1966) of which the major component had a molecular weight of 144,000 ± 16,000 and contained 0.32% (w/w) of Cu. The PPO activity in tea flush is around 3 times higher than that in mature leaf (Takeo & Baker, 1973) and also shows seasonal variation and processing variation (Takeo, 1966). The location of PPO in tea leaves has been reported in various sites. Early works suggested that the enzyme bound in the chloroplast (Roberts, 1941; Oparin & Schubert, 1950). Bokuchava *et al.* (1970) also found the activity of enzyme in mitochondria. Wickremasinghe *et al.* (1967) reported that the PPO was located in the epidermis around the vascular bundles of the tea leaf tissue. The activity of PPO in tea leaves is inactivated during the fixing process in green tea manufacture and increases during the withering and fermenting processes in black tea manufacture. The changes of PPO activity during the manufacturing process of black tea have attracted the attention of several authors. Most authors report that the PPO activity is increased during the withering process in black tea manufacture (Bokuchava *et al.* 1950; Takeo, 1973), however, the decreasing PPO activity in the withering process is also reported (Ullah, 1982, Xiao, 1986; Wu, 1990). An investigation from Liu *et al.* (1989) reported that the PPO activity increased in the first 16 hr period. After this period, the PPO activity decreased, illustrating that over-withering prevents the polymerization of polyphenols, and also showing that the decrease or increase of PPO activity is significantly correlated with the time of determination. The reason for decreasing PPO activity in the withering process was attributed to the dehydration (Ullah, 1982) and the changes of the active sites of PPO (Wu, 1990). After withering, the PPO activity decreased (Liu & Shi, 1989) or increased (Takeo, 1966) in the rolling process and continuously decreased through the fermentation and drying process. Before drying, the PPO activity was only 25–30% of the original activity, and after drying, only slight PPO activity remained. Other enzymes in tea flush and their function are listed in Table 4.6.

3.   CHEMISTRY OF COLOR AND TASTE OF MADE TEA

A cup of infusion of made tea is completely different from the infusion of fresh tea flushes in color, taste and flavor. These characteristics are developed during the manufacturing process after the harvesting of tea flushes. The differences in color, taste and flavor of various teas are caused by the manufacturing process.

Table 4.6 Enzymes Detected in Tea Flush.

Enzyme	Function	Reference
Adenosine nucleosidase	Accumulation of 2′-nucleotide	Imagawa *et al.*, 1979
Alcohol dehydrogenase	Catalyze the conversion of *cis*-3-hexenal and *trans*-3-hexenal to their respective alcohols	Hatanaka & Harada, 1973
Chlorophyllase	Determine the proportion of pheophytin and pheophorbide	Wickremasinghe, 1978
3-Dehydroquinic acid synthetase	Formation of B-ring in catechins	Saijo & Takeo, 1979
5-Dehydroshikimate reductase	Biosynthesis of polyphenols	Sanderson, 1966
3-Dehydroshikimic acid reductase	Formation of B-ring in catechins	Saijo & Takeo, 1979
3-Deoxy-D-Arabinose-heptulose-7-phosphosynthetase	Formation of B-ring in catechins	Saijo & Takeo, 1979
Glumatic acid dehydrogenase	Important enzyme in nitrogen metabolism	Jinesh & Takeo 1984
Glumatic acid synthetase	Important enzyme in nitrogen metabolism	Jinesh & Takeo, 1984
Glutaminase	Important enzyme in nitrogen metabolism	Jinesh & Takeo, 1984
Leucine-α-ketoglutarate transaminase	Biosynthesis of volatile compounds responsible for tea flavor	Wickremasinghe, 1978
Lipoxidase	Formation of C_6-aldehyde compounds	Jinesh & Takeo, 1984
Malate dehydrogenase	Unknown	Morchiladze, 1972
Nitrate reductase	Regulate the reduction of nitrate	Jinesh & Takeo, 1984
Nucleases	Liberation of 5′-nucleotide from nucleic acid	Imagawa *et al.*, 1982
Peptidase	Influence on the free amino acid content in tea shoot	Jinesh & Takeo, 1984; Sanderson & Robert, 1964
Pectin methylesterase	Regulate the fermentation rate in black tea manufacture	Ramaswamy & Lamb, 1958
Peroxidase	Activate the oxidation of polyphenols by PPO	Bokuchava, 1950; Roberts, 1952
Phenylalanine ammonia lyase	Biosynthesis of polyphenols	Sanderson, 1960; Stafford, 1974
Polyphenol oxidase (Catechol oxidase)	Oxidation of tea catechins in tea fermentation	Steerangacher, 1943
Tannase	hydrolyze the complexes of gallate polyphenols and caffeine	Tsushida & Takeo, 1985
L-Theanine lylase	hydrolysis of theanine	Morchiladze, 1972; Tsushida & Takeo, 1985
Theanine synthetase	Biosynthesis of theanine	Takeo, 1978

3.1. Color

Shade of color in made tea and the infusion color are two attributes besides the aroma and taste in the evaluation of various kinds of tea. The green color is the main shade of color in the infused leaf and the infusion of green tea. It is mainly determined by the chlorophyll content and the ratio of chlorophyll A which is dark green in color and chlorophyll B which is yellowish-green in color. The yellow color is the auxiliary color in constituting the shade of color of made tea and infused leaf. The TFs and the flavonols as well as their glycosides are the contributors of yellow color. The yellowish brown color usually appears in tea infusion which has brewed for longer or brewed with green tea stored under poor conditions. Chemically, it is induced by the xanthophyll, carotenoids and the primary oxidative products of phenolic compounds. The reaction between catechins and amino acids as well as sugars in water at high temperature produces yellowish brown colored substances. The red color is the main shade of color in black tea. The colored TFs and TRs are produced by the enzymatic oxidation and condensation of catechins in green leaf during the fermentation process. The different ratio of TFs and TRs constitute the different shade of red colour in black tea infusion. The black colour is the basic shade of black tea. It is produced by the decomposed products of chlorophyll (pheophytin and pheophorbide), protein, pectin, sugar and phenolic compounds which form in the manufacturing process of black tea and accumulate on the surface of made tea. Sometimes the infused leaf shows a dull appearance, caused by the inappropriate polymerization between the phenolic compounds and catechins as well as the excessive amounts of theafulvin (theabrown).

3.1.1. Color of green tea

Green tea infusion, unlike that of black tea, contains no highly colored products formed by the oxidation of polyphenolic compounds, and the desired color is greenish or yellowish green without any trace of red or brown color. The liquid should remain clear on cooling without turbidity, and the infused leaf should be green with no sign of discoloration due to damage. Greenish color of made tea and tea infusion is the first choice in selecting the green tea. The shade of green color in made tea is mainly determined by the chlorophyll content. Due to the different color in the two components of chlorophyll (chlorophyll A and chlorophyll B), the ratio of chlorophyll A to B creates various shades of made green tea. The degradative products of chlorophyll (pheophytin and pheophorbide), may cause the made tea color to become darker. The degradation is activated by the chlorophyllase enzyme, high temperature and high humidity. So, the greenish color of green tea may become dark brown under higher temperature and humid storage conditions (Nakagawa, 1970, 1975).

The color of green tea infusion is mainly yellowish green and green. In fact, the green infusion color is not produced by the soluble amounts of chlorophyll. It is because the chlorophyll is not soluble in water. However, the protein within the chloroplast plastid is degraded by the high temperature during the manufacturing process and exposed chlorophyll, thus causing the greenish colour in the infusion. In addition, the hydrolyzing products of chlorophyll (chlorophyllin, phytol) also contribute to the greenish color of tea infusion. In practice, the infusion of green tea is yellowish-green and the yellow color is more prominent. The yellow color in tea infusion is mainly determined by the water-soluble flavonols (1.3–1.5% of the tea leaves in dry weight), which include kaempherol, quercetin, isoquercetin, myricetin, myricitrin, rutin, kaempferitrin etc., and flavones (0.02% of the tea leaves in dry weight), which include apigenin, isovitexin, vitexin, saponarin, vicenin-2 etc. as well as their glycosides. In 1867 Hlasiwetz separated the first flavone compound (quercetin) from green tea infusion (Nagabayashi *et al.* 1992). Then, Roberts *et al.* (1952), Roberts *et al.* (1956), Sakamoto (1967, 1969, 1970) carried out investigations on the yellow pigments from green tea infusion. Besides the compounds of flavonols and flavones, the water-soluble anthocyanins also contribute color to green tea infusion. The yellow-greenish color of green tea infusion is generally unstable due to the autooxidation of catechins, and several browning compounds are formed via the reaction of amino acids, catechins and sugars. These compounds are yellow-brown to brown, reddish brown in color, thus making the tea infusion turn reddish brown in color after several hours.

3.1.2. *Color of black tea*

The fresh green leaves turn to yellowish brown to a reddish brown color after the rolling process. This is because the leaf cell is disrupted during this process and the mixing of cellular catechins and PPO occurs. The colored TFs and TRs are produced by the enzymatic oxidation and condensation of catechins in green leaf during the fermentation process. The TFs are generally yellow in color and TRs are generally red in color, so the different ratio of TFs and TRs constitute the different shades of red color in black tea. Brightness and attractiveness are desirable characteristics of black tea infusion. Roberts and his co-workers (1963) studied the relationship between the color and strength of black tea infusion in terms of chemical compounds. The bright reddish brown color of the black tea infusion is also mainly determined by the TFs and TRs in made tea (Roberts, 1958). Roberts *et al.* (1963) firstly reported that the TFs are closely related to the color of black tea infusion. The lower the TFs content, the more inferior the black tea infusion. The higher the TFs content, the brighter the tea infusion, with a golden tint. Theaflavins are also responsible for the development of the briskness of taste of the liquor as well as the coppery color of infused leaf. This is responsible for what tasters term "body", "depth of color", "richness" and "fullness" of black tea infusion. Too much of them and the tea becomes "soft" (McDowell & Owuor, 1992). This group of compounds contributes about 30% of the total color (Milliam & Swaine, 1981). Takeo (1976) also proved the above relationship and suggested that among the TFs components, the TF and TF-monogallate were highly related to the tea infusion evaluation, and the relative coefficient was as high as 0.89 and 0.91, respectively. TF-digallate was also positively related to the tea infusion, however, the correlative coefficient was lower (0.60).

TRs are the other important group related to the color of black tea infusion. The formation of TRs begins after plucking and continues up to the drying stage of black tea processing. The higher the content of the TRs is, the browner the color the tea infusion appears. So the different ratios of TF-digallate and TFs constitutes the different shades of color of tea infusion (Nakagawa, 1969). The CTC method of manufacture can enhance the oxidation of polyphenolic compounds to produce higher amounts of TF and TR type soluble pigments in black tea (Mahanta & Hazarika, 1985). In orthodox teas as much as a quarter of the oxidizable polyphenols remains unchanged, but in CTC teas practically all of them are converted into TF and TR. This is due to less disruption of the leaf in the orthodox method than that in CTC method. The quick and severe maceration of the leaf cells in CTC manufacture increases enzymatic oxidation of green leaf catechins thus forming more TF and TR. Liu *et al.* (1990) pointed out that the pheophytins/TR ratio value could be used to reflect the colour of made black tea. The higher the ratio value, the more black bloom the made tea has. Mahanta (1988) divided the made black tea into 3 groups according to the pheophytins/TR ratio in orthodox teas. The brownish tea with the ratio value between 0.04–0.05, the exclusively black teas with the ratio value higher than 0.05 and in brown teas < 0.04. As mentioned above the pheophytin and pheophorbide are the degradative products of chlorophyll. It was noted that the low pheophorbide contents related to the higher formation of TFs and TRs. The orthodox tea showed a 30% higher degradation of chlorophyll than that in CTC tea. So the orthodox teas have a greater tendency to blackness, while the CTC teas have a greater tendency to brown color (Mahanta, 1988). The degradative rate of chlorophyll was higher in the congou black tea than that in broken black tea. However, the chlorophyll/carotenoids ratio was higher in broken black tea than that in congou black tea. The higher degradative rate of chlorophyll in congou black tea determines the more black bloom color of made tea (Liu *et al.* 1990). Pheophorbide is one of the factors determining the blackness color of tea infusion. Besides the TF and TR, the following phenolic compounds also participate in the color formation of tea infusion: isotheaflavin, neotheaflavin, TF-3-gallate, TF-3'-gallate, TF-3,3'-digallate, theaflavic acid and epitheaflavic acid (Mahantan, 1988). As early as 1971, Berkowitz *et al.* (1971) reported that when the oxidation of mixtures of epicatechin and gallic acid, epicatechin gallate and gallic acid in a model tea fermentation system, bright red phenolic compounds, identified as epitheaflavic acid and epitheaflavic acid gallate were formed. It was proposed that the TRs were oxidation products of these epitheaflavic acids through coupled oxidation by oxidizing catechins. Wickremasinghe (1967) reported from Sri Lanka that the most deeply colored reaction was found to contain corilagin, theanine and protein. The presence of theafulvin is also considered to be one of the contributing factors to the brownish color of tea infusion, especially in low quality teas. Another factor that could be of importance in determining the color of black tea infusion is aluminium (Reeves *et al.* 1987), because it has been shown that Al can complex with TFs, resulting in a more red colored tea infusion (Reeves *et al.* 1987). On the other hand, unexpected color may possibly be a sign of poor-quality tea. A greyish appearance may possibly be due to poor processing or even an indication of spoilage or adulteration.

3.1.3.   Color of oolong tea

Oolong tea is a semi-fermented tea. It is a large group which includes the light-fermented oolong (10–29% fermentation degree, such as Pouchong), moderate-fermentation oolong (30–59% fermentation degree, such as Tie-guan-ying, Fu-shou) and heavy-fermentation oolong (60–70% fermentation degree, such as Taiwan oolong). Due to the various degrees of fermentation, the degree of oxidation of catechins differs accordingly. The catechins are entirely non-oxidized in green tea, and about 25% oxidized in light-fermented oolong; however, those in heavy-fermented oolong are around 40–45% oxidized and those in black tea are more than 50% oxidized. Accordingly, the amount of TFs and TRs are highest in black tea, almost absent in light-fermented oolong tea and in small amounts in heavy-fermented oolong tea. So, the shade of color of made oolong tea ranges from dark green to light brown black to brown black from light-fermented oolong, moderate-fermented oolong to heavy-fermented oolong. The partially oxidized catechins and the partially degraded chlorophyll cause the shade of color of light-fermented oolong of dark green. The pheophorbide A and B contents are especially high in oolong tea, around 2–5 times higher than that in black tea (Liu *et al.* 1990). Furthermore, the content of pheophorbide A is higher than that of pheophorbide B. The former is blackish brown in color and the latter is yellowish brown in color. These two chlorophyll-degradative products together with the non-degraded chlorophyll and chlorophyllin constitute the black bloom color of oolong tea. The higher contents of carotenoids and the higher pheophytin/TR ratio are also particular characteristics of oolong tea (Liu *et al.* 1990). The contents of various fat-soluble pigments in black tea and oolong tea are listed in Table 4.7.

The infusion color of oolong tea is generally reddish-brown in moderate- to heavy-fermented oolong and dark greenish color in light-fermented oolong. The color-determining compounds in light-fermented tea are composed of the flavonols and flavones in green tea and small amounts of TFs and TRs in black tea. In the moderate- and heavy-fermented oolong tea, the major color-determining compounds are TRs

Table 4.7   Various Pigments (%) in Black Tea and Oolong Tea (Liu *et al.* 1990).

Pigments	Yunnan Broken Black Tea	Keemen Black Tea	Tie-Quan-Yin Oolong Tea	Fu-Shou Oolong Tea
Chlorophylls	0.75	3.72	8.07	2.83
Carotenoids	13.99	21.82	23.27	30.25
Pheophytins +Pheophorbide	11.95	16.76	29.18	23.52
Total Amount	26.70	42.30	63.90	60.99
Chls*/Cars†	0.0534	0.1704	0.348	0.09
Chls/Phys‡	0.1731	0.2219	0.276	0.12
Phys/TR$^\Psi$	2.5383	3.1765	7.495	5.36

*Chls: Chlorophyll a,b, and Chlorophyllide a, b;
†Cars: β-carotene and α-carotene;
‡Phys: Pheophytin a, b, and Pheophorbide a, b;
$^\Psi$TR: Thearubigin

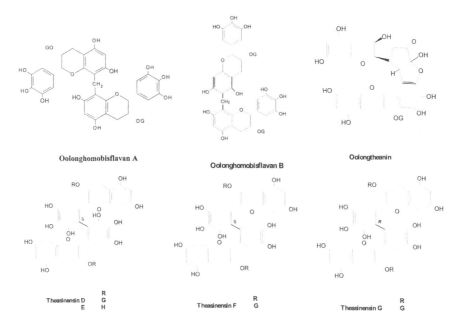

Figure 4.4 Some colored compounds in Oolong tea

and their oxidized polymers. The amount of TFs in heavy-fermented oolong tea is only one-tenth of TFs in black tea (Nagabayashi *et al.* 1992). In addition, some homobisflavan compounds such as Oolonghomobisflavan A, B (Nagabayashi *et al.* 1992), Theasinensin D, E, F, G (Nonaka *et al.* 1983; Nagabayashi *et al.* 1992) and Oolongtheanin (Nagabayashi *et al.* 1992) are related to the color of oolong tea infusion. The structure of these colored compounds is shown in Figure 4.4.

The color of made tea leaf in oolong tea often shows the character of "red-rimmed green leaf". This is because the fresh leaves are rotated after the withering process. The rotating process causes friction between the leaves, disrupts the cellular tissue at the edge of the leaves, and causes a limited degree of fermentation.

3.2. Taste

Taste of food is mainly composed of five basic sensations, i.e., sweetness, astringency, sourness, bitterness and umami (Tamura *et al.* 1969). A delicious cup of tea infusion is an ingenious balance of various taste sensations. Astringency is a drying, puckering sensation in the mouth that affects the whole of the tongue more or less uniformly (Lea & Arnold, 1978). Bitterness is usually unpleasant, but sometimes desirable in moderate amounts, and is perceived predominantly at the back of, and sometimes along the sides of, the tongue (Moncrieff, 1967). The umami is a Japanese term, it is similar to the "meaty" taste (Shallenberger, 1993) or "brothy" taste.

Table 4.8   Taste and Threshold Value of Tea Components (Yamanishi, 1993).

Compounds	Taste	Threshold (mg/100 ml)	
Amino acids			
Theanine	Brothy, Sweet	150	
Glutamic acid	Brothy, Sour	5	
Asparagic acid	Brothy, Sour	3	
Arginine	Bitter	10	
Alanine	Sweet, Bitter	60	
Serine	Sweet, Brothy	150	
Sugars			
Sucrose	Sweet	100	
Glucose	Sweet	135–200	
Fructose	Sweet	56–167	
		Astringent	Bitter
Catechins			
(+)-C	Bitter with sweet after-taste	–	60
(–)-EC	Bitter with sweet after-taste	–	60
(–)-EGC	Bitter with sweet after-taste	–	35
(–)-ECG	Bitter and astringent	50	20
(–)-EGCG	Bitter and astringent	60	30
Caffeine	Bitter	–	20
Theaflavins			
Crude theaflavin		60	70
Theaflavin		80	75–100
TF monogallate		36	30–50
TF digallate		12.5	–
Gallic acid		not astr.	not bit.
Tannic acid		20	80

3.2.1.   Taste of green tea

Its strong astringency and bitterness, median umami and sweetness, as well as slight sourness characterizes green tea. Nakagawa *et al.* (1990) suggested the relative importance of these five taste sensations in green tea as follows: astringency 4.17, bitterness 3.44, umami 1.42, sweetness 0.53, saltiness and sourness < 0.3. The astringency and bitterness are the major sensations in the taste of green tea. Nakagawa (1975) studied the correlation between chemical composition and organoleptic properties of various grades of green tea. His results indicated that the multiple correlation coefficient between the sensory evaluation and catechins, amino acids, caffeine and other soluble substances was highly significant. The astringency and bitterness of green tea infusion was mainly determined by the contents of catechins and other phenolic compounds. The taste and their threshold values of some important taste-determining compounds are listed in Table 4.8.

It can be seen from Table 4.8 that the degree of bitterness and astringency of free type catechins are less than that of gallate type catechins. Among the catechins in green tea, 60–70% are the gallate type catechins (ECG, EGCG), thus constituting the bitterness and astringency taste. The threshold value of catechins is around 12–17 $\times$ 10^{-4} M [(+)-EC, (–)-EGC] in free type catechins, and 4 $\times$ 10^{-4} M [(–)-ECG, (–)-EGCG] in gallate type catechins (Yamanishi, 1992). (–)-EC has a significantly higher bitterness and astringency than (–)-C. The (–)-EC shows an astringency response at 0.9 g/l, and the high intensity of bitterness of (–)-EC relates to the greater lipophilicity (Thorngate III, 1995). The gallic acid is regarded as the important taste-constituting component and is also the precursor in the biosynthesis of gallate type catechins. Investigation showed that the content of gallate type catechins in green tea was positively related to the quality of green tea (Guo *et al.* 1990). Caffeine is the major contributor to bitter taste of tea. Caffeine concentration ranges between 2.5–5.5% dry weight basis in fresh harvested tea flushes. In tea beverage, caffeine concentration ranges from 2–4%. The threshold value of caffeine is approximately 20 mg/100 ml in water.

Besides the catechins and caffeine, some amino acids (such as arginine, alanine etc.) also contribute to the bitterness of green tea infusion. The umami taste of green tea infusion was shown to be due to some amino acids (such as theanine, serine etc.) and the sweetness to sugars (Nakagawa, 1975). It was estimated that 70–75% of bitterness and astringency of green tea was caused by flavanols and 70% of the umami taste of green tea was caused by amino acids (Nakagawa, 1975, 1976). The saponin compounds from tea leaf also related to the bitterness of green tea. Hashizume (1973) isolated two major saponins, i.e., barringtogenol C and R$_1$-barrigenol as well as two minor saponins, i.e., camelliagenin A and A$_1$-barrigenol. They do not cause extremely bitter taste because of their very low solubility in water. However, when combined with some components in tea leaf, solubility is increased (Hashizume, 1973).

3.2.2. Taste of black tea

A cup of good quality black tea infusion is characterized by the bright reddish brown color, brisk, strong taste and rich flavour. Astringency in black tea is divided into two types: tangy and non-tangy, by Sanderson *et al.* (1976). The tangy astringency with a sharp and puckering action and little after taste effect, and the non-tangy astringency which is tasteless, mouth-drying and mouth-coating, with a lingering (more than 60 sec.) after taste effect. The tangy astringency of black tea infusion is possibly related to the "bitterness" (Wood *et al.* 1964; Millin *et al.* 1969). It was noted that the compounds with a galloyl group, TFs and caffeine are closely related to the tangy astringency (Wood *et al.* 1964; Sanderson *et al.* 1976). So, the caffeine together with the black tea polyphenols was necessary for the expression of reasonable amounts of tangy astringency. However, the caffeine does not contribute to the non-tangy astringency. Decaffeination may change the nature of the astringency in black tea infusion from the tangy type to a non-tangy type (Sanderson *et al.* 1976). Experiments have demonstrated the relationship between individual polyphenolic compounds and the taste of black tea infusion. The gallated tea flavonols are related to the astringency and also to the bitterness taste; the non-gallated tea flavonols are related to the bitterness, however, are not related or only slightly related to the astringent taste of tea infusion. Among the TFs, theaflavine is less astringent. The contribution of TF-digallate and TF-monogallate to astringency is 6.4 and 2.2 times higher than that of theaflavine. TFs in solution are normally very astringent, but in tea the astringency is reduced due to an interaction with bitter caffeine. TRs are also closely related to the taste of black tea. The TR_1 contributes significantly to the "dull" colour of infusion and is negatively related to the "briskness", while TR_3 is positively related to "briskness". Deb and Ullah (1968) pointed out that caffeine contributes towards the briskness of black tea.

There were inconclusive conclusions on the individual chemical components which have been evaluated as contributing to the total quality of black tea, only total TF contents showed some degree of correlation (Roberts *et al.* 1963; Deb *et al.* 1968; Hilton *et al.* 1972). However, the total TF content is not the sole determinant factor of market price. Analysis of Kenyan tea samples showed that the correlation between total TF contents and tasters' evaluation was sometimes low (Owuor *et al.* 1986), suggesting that there are other factors which affect the valuation of black tea. In tea with lower TF levels, such as those produced in central Africa, Sri Lanka, and India, the correlation between TF contents and market price is generally positive but with statistically non-significant correlation coefficients (Owuor *et al.* 1986). In contrast, a very good correlation between tasters' evaluation and TF contents was established in Malawi (Cloughley *et al.* 1980, 1981) and also showed a good correlation between the prices and the contents of TFs or TFs + TRs in northeast India (Biswas *et al.* 1973). Obandu & Owuor (1997) also suggested the green leaf (–)-ECG, (–)-EGCG and caffeine as potential quality indicators before the processing of tea leaves.

3.2.3. *Taste of oolong tea*

Mellowness and sweet taste as well as its special aroma characterizes a cup of good quality oolong tea infusion. The taste of oolong tea infusion is quite unusual and depends on the various fermentation degrees. The decreasing rate of catechins during the manufacturing process was shown to be lower than that in black tea and higher than that in green tea (Nagabayashi *et al.* 1992). The content of TFs was very low or entirely absent. In the light-fermented oolong tea, TFs were entirely absent. Even in heavy-fermented oolong tea, TFs content was also only one tenth of that in black tea due to the low cell breakage rate (around 30%). However, TRs content formed via the oxidation of EGC and its gallate (Takayanagi *et al.* 1984). In addition, some of the secondary polyphenolic compounds, such as oolonghomobis-flavane, theasinensin, and oolongtheanine, were formed and related to the infusion taste (Nonaka *et al.* 1983; Nagabayashi *et al.* 1992). Thus, the mellowness and sweetness of oolong tea infusion are the integrated taste of non-oxidized catechins, TRs, some secondary oxidative polyphenolic compounds, caffeine, free amino acids and related sugars. The astringency taste of oolong tea is lower and the sweetness taste is stronger than those of green tea. The compounds responsible still need clarification.

3.3. **Tea Cream and Tea Scum**

Tea cream is used to define the precipitation formed when the black infusion cools. Chemically, tea cream is the complex of theaflavin, TF-gallate, EGCG, ECG, TRs, caffeine, caffeic acid, gallic acid, ellagic acid, chlorophyll, and bisflavanol A, B

(Wickremasinghe & Perera, 1960: Millin *et al.* 1969). Roberts (1963) reported that the components of tea cream are as follows: TRs 66:TFs 17:caffeine 17. It was soluble in hot water; however, it showed only a low solubility in cold water. The cream properties are accepted as one of the quality attributes of black tea infusion. The amount of tea cream is a measure of the strength and briskness of black tea. In other words, the value of teas increases as the tendency of their infusion to form cream increases (Sanderson, 1976). This phenomenon of "cream-down" in black tea infusion may not occur in some black teas, especially those black teas manufactured from microphyll cultivars. Investigations showed that the "cream-down" of black tea infusion occurred in those black teas in which the TFs content is more than 0.4% and these black teas can produce an infusion with bright reddish brown color. The color of "cream-down" teas is determined by the ratio of TFs/TRs; the lower the ratio, the brighter the infusion color. The higher the ratio, the duller the tea infusion. It is because the complex of TF and caffeine is orange-yellow in color, and the complex of TR and caffeine is dark reddish brown in color.

In hard water areas, an unsightly film called "tea scum" or "tea scale" forms on the surface of tea infusions. Kooijmans (1940) was the first author to discuss tea scum. He divided the tea scum into two types. The kind of scum formed due to the spontaneous extraction of tea organic compounds, which is independent of the water hardness and occurs even in distilled water. The second type is produced within teapots and mugs. It depends on water quality. Spiro & Jaganyi (1994b) reported that the tea scum is amorphous, high molecular weight organic material together with calcium carbonate. The empirical formula of the organic scum matrix contained approximately 45 carbon atoms. The scum also possessed the hydroxyl groups, carbonyls and some unsaturated groups. Chemical microanalysis showed that the percentage of calcium ranged from 3–7%. Besides the calcium, a small amount of magnesium, manganese, potassium, sodium and aluminium also occurred. Virtually all the calcium and sodium originated in the water, the potassium, manganese and aluminium came almost entirely from the tea leaf, while magnesium was provided in comparable amounts by both sources. The tea scum is only formed in water that contains both calcium (or magnesium) ions and bicarbonate ions. Boiling the water reduced but did not eliminate scum. More scum formed if the period of scum development took place at a high temperature. The activation energy of scum formation was found to be 34 KJ/M. This relatively high value showed that the overall process is chemically controlled and not diffusion controlled. The mechanism of the formation of tea scum involves the aerial oxidation of tea solubles, probably the polyphenolic components such as catechins, theaflavins and thearubigins (Spiro & Juganyi, 1994b).

4. FUNCTIONAL EFFECTS OF TEA NON-VOLATILES

Tea aroma is mainly determined by the tea volatiles and the impact on human health is mainly determined by the tea non-volatiles. As above mentioned, tea is not only a nutritional, flavored beverage, but is also a physiological function-modulating healthy beverage. According to medicinal research on tea over the last 10 years, these functional components possess multiple activities which include anticaries, blood-

Table 4.9   Relative Biological Activity of Catechins and Theaflavins in Tea (Chen, 1998).

Biological activity	(+)-C	(-)-EC	(+)-GC	(-)-EGC	(-)-ECG	(-)-EGCG	TF	TF-3-G	TF-3'-G	TF-3,3'-G
Cariogenic bacteria	+	+	+++	+++	++	++	?	?	?	?
Clostrium spp.	?	+	?	+	++	++	++	++	++	+++
Staphylococcus aurens toxin	+	+	+	+	++	+++	?	?	?	?
Vibrio cholea toxin	+	++	?	–	+++	+++	?	?	?	?
Phytopathogenic bacteria	?	++	?	+++	++	+++	?	?	?	?
AIDS-HI virus	?	?	?	+	++	+++	?	?	?	?
Insoluble glucan formation inhibiting activity	?	+	?	+	++	+++	+++	+++	+++	+++
Angiotensin I transferase inhibiting activity	+	+	+	+	+++	+++	+++	++	++	++
a-amylase inhibiting activity	?	?	?	?	+	++	++	+++	+++	?
Glucose S-transferase inhibiting activity	+	+	?	+	++	+	+++	?	?	?
Blood-platelet aggregation inhibiting activity	++	++	+++	+	+	+	+	++	++	+
Liver protecting activity	+++	+	?	?	?	?	?	?	?	?
Antioxidative activity	++	++	+	++	+++	+++	?	?	?	?
Halitosis inhibiting	?	++	?	+++	++	+++	?	?	?	?
Anticarcinogenic activity	?	+	?	++	++	+++	?	?	?	?
Histamine release inhibiting activity	+	+	?	++	++	+++	++	+++	+++	+++
Antiulceric activity	+	+	?	?	?	++	?	?	?	?

pressure lowering, blood-glucose depressing, blood-lipid reducing, prevention of atherosclerosis and thrombus formation, corpulance reducing, antisenescent, anti-radiation damage, antioxidative, immune function improving, deodorizing, germicidial and antiviral, diuretic, liver-protecting, antiulceric, antimutagenic and anticarcinogenic activities (Chen, 1991, 1998; Blot *et al.* 1996; Dou, *et al.* 1997). Table 4.9 lists the specific functions of tea catechins and theaflavins reported in last 5 years. These functions are looked at afresh by medical circles and will be discussed separately in this monograph.

REFERENCES

Alfrey, A.C. (1986) Aluminium. In Mertz, W. (Ed.), *Trace elements in human and animal nutrition*. Acad. Press, USA, pp. 281–317.

Anan, T., Takayanagi, H. and Ikegawa, K. (1985) Changes in the contents of free sugars of tea shoots during development and by shade culture, *J. Jp. Soc. Fd. Sci. & Technol*, **32**, 43–50.

Barone, J.J. and Grice, H.C. (1994) Seventh international caffeine workship, *Santorini, Greece*, June 13–17, 1993, *Fd. Chem. Toxicol,* **32**, 65–77.

Barone, J.J. and Roberts, H. (1984) Human consumption of caffeine. In Dews, P.B. (Ed.), *Caffeine: Perspectives from recent research*, Springer, Germany, pp. 59–73.

Berkowitz, J.E., Coggan, P. & Sanderson, G.W. (1971) Formation of epitheaflavic acid and its transformation to thearubigins during tea fermentation. *Phytochemistry*, **10**, 2271–2278.

Biswas, A.K., Sankar, A.R. and Biswas A.K. (1973) Biological and chemical factors affecting the valuation of North East Indian plains teas. III. Statistical evaluation of the biochemical constituents and their effects on colour, brightness and strength of black teas. *J. Sci. Fd. Agricul,* **24**, 1457–1477.

Blot, W.J., Chow, W.H. and McLanghlin, J.K. (1996) Tea and cancer: A review of the epidemiological evidence. *Europ. J. Cancer Preven,* **5**, 425–438.

Bockuchava, M.A. (1950) Transformation of tannins under the action of polyphenol oxidase (Review) (Russian). *Biochem,* **6**, 100–109.

Bockuchava, M.A., Shalamheridze, T.K. and Soboleva, G.A. (1970) Localization of polyphenol oxidase in a tea leaf (Russian). *Dokl. Akad. Nauk SSSR,* **192**, 1374–1375.

Bradfield, A.E. and Penney, M. (1944) The chemical composition of tea. The proximate composition of an infusion of black tea and its relation to quality. *J. Soc. Chem. Ind. (London),* **63**, 306–310.

Bradfield, A.E., Penny, M. and Wright W.B. (1948) The catechins of green tea, Part. I. *J. Chem. Soc,* **32**, 2249.

Bryce, T., Collier, P.D., Fowlis, I. *et al.* (1970) The structures of the theaflavins of black tea. *Tetrahedron Lett,* **26**, 2789–2792.

Bryce, T., Collier, P.D., Mallows, R. *et al.* (1972) Three new theaflavins from black tea. *Tetrahedron Lett,* **28**, 463–466.

Cartwright, R.A. and Roberts, E.A.H. (1955) Theogallin as a galloyl ester of quinic acid. *J. Soc. Chem. Ind. (London),* **74**, 230–231.

Chang, S.S. and Guapason, G.V. (1982) Effect of addition of aluminium salts on the quality of black tea. *J. Agricul. Fd. Chem,* **30**, 940–943.

Chen, Z.M. (1990) Tea – microelements – human health (Chinese). *Tea Abstracts,* 4, 1–10.

Chen, Z.M. (1991) Contribution of tea to human health. In T. Yamanishi (ed.), World tea, *International Symposium on Tea Science*, Aug. 26–29, Shizuoka, Japan, pp. 12–20.

Chen, Z.M. (1999) Pharmacological function of tea. In Jain, N.K. (Ed.) *Global advances on tea sciences*, 333–358.

Child, R. (1955) Copper: its occurrence and role in tea leaf. *Trop. Agricul.* (London), **32**, 100–106.

Chu, D.C. & Juneja, L.R. (1997) General chemical composition of green tea and its infusion, In Yamamoto, T. *et al.* (Eds.) *Chemistry and applications of green tea.*, CRC Press LLC., pp. 13–24.

Cloughley, J.B. and Ellis, R.T. (1980) The effect of pH modification during fermentation on the quality parameters of Central African black teas. *J. Sci. Fd. Agricul,* **31**, 924–934.

Cloughley, J.B. (1981) Storage deterioration in Central African tea: Changes in chemical composition, sensory characteristics and price evaluation. *J. Sci. Fd. Agricul,* **32**, 1213–1223.

Collier, P.D., Bryce, T., Mallows, R. *et al.* (1973) The theaflavins of black tea. *Tetrahedron Lett,* **29**, 125–142.

Coxon, D.T., Holmes, A. and Ollis, W.D. (1970a) Isotheaflavin, A new black tea pigment. *Tetrahedron Lett,* **26**, 5241–5246.

Coxon, D.T., Holmes, A. and Ollis, W.D. (1970b) Theaflavic acid and epitheaflavic acid. *Tetrahedron Lett,* **26**, 5247–5250.

Coxon, D.T., Holmes, A., Ollis, W.D. *et al.* (1972) Flavanol digallates in green tea leaf. *Tetrahedron Lett,* **28**, 2819–2826.

Deb, S.B. and Ulah, M.R. (1968) The role of theaflavins (TF) and thearubigins (TR) in the evaluation of black tea. *Two and a bud,* 15, 101–102.

Dews, P.B. (1982) Caffeine. *Annu. Rev. Nutrition,* 2, 323–341.

Dix, M.A., Fairley, C.J., Millin, D.J. *et al.* (1981) Fermentation of tea in aqueous suspension. Influence of tea peroxidase. *J. Sci. Fd. Agricul,* 32, 920–932.

Dou, J.H. Steele, L. and Pillai (1997) Chemoprevention: A review of the potential therapeutic antioxidant properties of green tea *(Camellia sinensis)* and certain of its constituents. *Medicinal Res. Rev,* 17, 327–365.

Edmonds, C.R. *et al.* (1979) United States Patent 413500.

Ekborg-Ott, K.H., Taylor, A., Armstrong, D.W. (1997) Varietal differences in the total and enantiomeric composition of theanine in tea. *J. Fd. Agricul. Chem,* 45, 353–363.

Elias, P.S. (1986) Current biological problems with coffee and caffeine. *Cafe Cacao The,* 30, 121–138.

Feldheit, W., Yongvanit, P. and Cummings, P.H. (1986) Investigation of the presence and significance of theanine in the tea plant. *J. Sci. Fd. Agricul,* 37, 527–534.

Flaten, T.P. and Degard, M. (1988) Tea, Al and Alzheimer's disease. *Fd. Chem. & Toxicol,* 26, 595–596.

Garattini, S. (1993) *Overview.* In Garattini, S. (ed.), *Caffeine: coffee and health,* New York, Raven Press, pp. 402–403.

Gregory, R.P.F. and Bendall, D.S. (1966) The purification and some properties of the polyphenol oxidase from tea *(Camellia sinensis). Biochem. J,* 37, 661–667.

Guo, B.Y., Ruan, Y.C. and Cheng, Q.K. (1992) Relationship between the quality of green tea and the content of gallic acid (Chinese). *J. Tea Sci,* 10, 41–43.

Hashizume, A. (1973) Saponin from the leaf of *Thea sinensis.* III. Component Sapogenins and sugars of the saponin from the leaf of *Thea sinensis. J. Jp. Agricul. Chem. Soc,* 47, 237–240.

Hatanaka, A. and Harada, T. (1973) Formation of *cis*-3-hexenal, *trans*-2-hexenal and *cis*-3-hexenol in macreated *Thea sinesis* leaves. *Phytochemistry,* 12, 2341–2346.

Hazarika, M. and Mahanta, P.K. (1984a) Compositional changes in chlorophylls and carotenoids during the four flushes of tea in NE India. *J. Sci. Fd. Agricul,* 35, 298–303.

Hazarika, M., Chakravarty, S.K. and Mahanta, P.K. (1984b) Studies on the thearubigin pigments in black tea manufacturing systems. *J. Sci. Fd. Agricul,* 35, 1208–1218.

Hilton, P.L. and Ellis, R.T. (1972) Estimation on the market value of Central Africa tea by theaflavin analysis. *J. Sci. Fd. Agricul,* 25, 227–232.

Horie, H, Mukai, T. *et al.* (1994) Analysis of chemical forms of alumiinium in tea infusions by using ^{27}Al-NMR., *Jp. Fd. Sci. J,* 41, 120–122,

Imagawa, H., Yamano, H, Inouem, K. *et al.* (1979) Purification and properties of adenosine nucleosidaoes from tea leaves. *Agricul. Biol. Chem,* 43, 2337–2342.

Imagawa, H., Torya, H., Ozawa, T. *et al.* (1982) Purifucation and characterization of nuclease from tea leaves. *Agricul. Biol. Chem,* 46, 1261–1269.

Jinesh, C.J. and Takeo, T. (1984) Enzyme system of tea and the role in the tea manufacture. *J. Fd. Chem,* 8, 213–280.

Jurd, L. (1963) Spectral properties of flavonol compounds. In Geissman, T.A. *et al.* (ed.), *Chemistry of flavonoid compounds.* Pergamon, USA, pp. 101–155.

Kakuda, T., Takihara, Sekaul, I. *et al.* (1994) Antimicrobial activity of tea extracts against periodontopathic bacteria. *J. Jp. Agricul. Chem. Soc,* 68, 241–243.

Konayagi, S. and Minowada, M. (1933) *Study of physiology, Kyoto Univ,* 10, 449–454.

Konishi, S. and Takahashi, E. (1966) Existence and synthesis of *L*-glutamic acid γ-methylamide in tea plants. *Plant Cell Physil,* 7, 171–175.

Konishi, S. and Takahashi, E. (1969) Metabolism of N-ethyl-^{14}C and its metabolic distribution in the tea plant. VI. Metabolism and regulation of theanine and related compounds in the tea plant., *J. Jp. Agricul. Chem. Soc*, **40**, 479–484.

Krishnamurthy, K. *et al.* (1952) Circular paper chromatographic analysis of the amino acids of tea and coffee infusions. *Current Sci*, **21**, 133–134.

Lea, A.G.H. & Arnold, G.M. (1978) The phenolics of cider: bitterness and astringency. *J. Sci. Fd. Agricul*, **20**, 478–483.

Liu, Z.H. and Shi, Z.P. (1989) Variation in isoenzymes and activities of polyphenol oxidase during black tea manufacturing (Chinese). *J. Tea Sci*, **9**, 141–158.

Liu, Z.H., Huang, X.Y. and Shi, Z. P. (1990) Relationship between pigments and the colours of black tea and oolong tea (Chinese). *J. Tea Sci*, **10**, 59–64.

Mahanta, P.K. and Hazarika, M. (1985) Chlorophylls and degradation products in orthodox and CTC black teas and their influence on shade of colour and sensory quality in relation to thearubigins. *J. Sci. Fd. Agricul,* **36**, 1133–1139.

Mahanta, P.K. (1988) Colour and flavour characteristics of made tea. In Linskens, H.F. *et al.* (eds.) *Modern Methods of Plant Analysis*. New series volume 8. Springer, Berlin, pp. 220–295.

Marks, V. (1992) Physiological and clinical effects of tea. In Wilson, K.C. and Clifford, M.N. (eds.) *Tea: Cultivation to Consumption*, Chapman & Hall, pp. 708–739.

McDowell, I. and Owuor, P. (1992) The taste of tea. *New Scientist*, **11**, 24–27.

Millin, D.J. and Rustidge, D.W. (1967) Tea manufacture. *Process Biochem*, **2**, 9–13.

Millin, D.J., Swaine, D. and Dix, P.L. (1969) Separation and classifacation of the brown pigments of aqueous infusions of black tea. *J. Sci. Fd. Agricul,* **20**, 296–307.

Millin, D.J. and Swaine, D. (1981) Fermentation of tea in aqueous suspension. *J. Sci. Fd. Agricul*, **32**, 905–919.

Moncrieff, R.W. (1967) *The chemical senses, 3rd Ed.,* Leonard Hill Press, London, pp. 58–59.

Morchiladze, Z.N., Tkemaladze, G.S., Soseliya, M.F. *et al.* (1972) Isolation and purification of malate dehydrogenase from the tea plants (Russian). *Soobshch Akad. Gruz. SSR*, **65**, 181–184.

Mori, M., Morita, N. and Ikegaya, K. (1988) Polysaccharides from tea for manufacture of hypoglycemics, antidiabetics and health foods. *Japan Patent 63308001.*

Nagata, T., Hayatsu, M and Kosuge, N. (1993) Aluminium kenetics in the tea plant using ^{27}Al- and ^{19}F-NMR. *Phytochmistry*, **32**, 771–775.

Nagabayashi, T. *et al.* (1992) *Chemistry and function of green tea, black tea and oolong tea.* Hong-Xie Publisher, Japan.

Nakagawa, M. (1969) Correlation of theaflavin and thearubigin contents with tea tasters' evaluation (Japanese). *J. Jp. Soc. Fd. Sci. & Technol*, **16**, 266–271.

Nagakawa, M. (1970) Constituents in tea leaf and their contribution to the taste of green tea liquors. *Jp. Agricul. Res. Quart*, **5**, 43–47.

Nakagawa, M. (1975) Chemical components and taste of green tea. *Jp. Agricul. Res. Quart*, **9**, 156–160.

Nagakawa, M. (1975) Contribution of green tea constituents to the intensity of taste elements of brew., *J. Jp. Soc. Fd. Sci. & Technol*, **22**, 59–64

Nonaka, G., Kawahara, O., Nishioka, I. *et al.* (1983) Tannins and related compounds. XV. A new class of dimeric flavan-3-ol gallates, theasinensins A and B, and proanthocyanidin gallate from green tea leaf. (I). *Chem. Pharmac. Bull*, **31**, 3906.

Obanda, M. and Owuor, P.O. (1997) Flavanol composition and caffeine content of green tea, quality potential indicators of Kenyan black teas. *J. Sci. Fd. Agricul,* **74**, 209–215.

Omori, M. (1991) Effect of anaerobically treated tea (Gabaron tea) on the blood pressure of spontaneously hypertensive rates loaded common salts, *Proceedings of the International Symposium on Tea Science, Aug. 26–29, 1991,* Shizuoka, Japan, p. 230–235.

Oparin, A.I. and Shubert, T.A. (1950) On oxidative respiratory systems in the tea leaf (Russians). *Biochem,* **6,** 82–89.

Owuor, P.O., Reeves, S.G., Wanyoko, J.K. (1986) Correlation of theaflavins contents and evaluation of Kenyan black teas. *J. Sci. Fd. Agricul,* **37,** 507–513.

Parsons, W.D. and Neims, A.D. (1978) Effect of smoking on caffeine clearance. *Clin, Pharmac. and Therap,* **24,** 40–45.

Pennington, J.A.T. (1987) Aluminium contents of foods and diets, *Fd. Addit. and Contam,* **5,** 161–232.

Phelps, H.M. and Phelps, C.E. (1988) Caffeine ingestion and breast cancer: A negative correlation. *Cancer,* **61,** 1051–1054.

Ramaswamy, M.S. and lamb, J. (1958) Fermentation of Ceylon tea. X. Pectic enzymes in tea leaf. *J. Sci. Fd. Agricul,* **9,** 46–51.

Reeves, C.G., Owuor, P.O. and Gore, F. (1987) *Fd. Chem,* **30,** 940–944.

Roberts, E.A.H. (1941) The fermentation process in tea manufacture – the influence of external factors on fermentation rate. *Biochem. J,* **35,** 909–919.

Roberts, E.A.H. (1950) The phenolic substances of manufactured tea. II. Their origin as enzymatic oxide products in fermentation. *J. Sci. Fd. Agricul,* **9,** 212–216.

Roberts, E.A.H. (1952) The chemistry of tea manufacture. *J. Sci. Fd. Agricul,* **3,** 193–198.

Roberts, E.A.H., Cartwtight, R.A., Wood, D.J. *et al.* (1956) The flavonols of tea, *J. Sci. Fd. Agricul,* **7,** 637–646.

Roberts, E.A.H., Cartright, R.A. and Oldschool, M. (1957) Fractionation and paper chromatography of water soluble substances from manufactured tea. *J. Sci. Fd. Agricul,* **8,** 72–80.

Roberts, E.A.H. and Myers, M. (1958) Theogallin, a polyphenol occurring in tea. II. Identification as a galloyl quinic acid. *J. Sci. Fd. Agricul,* **9,** 701–705.

Roberts, E.A.H. (1962) Economic importance of food substances: Tea fermentation. In Geissmann, T.A. (ed.), *The chemistry of flavonoid compounds.* Pergamon, Oxford. pp. 468–512.

Roberts, E.A.H. and Smith, R.F. (1963) The phenolics of manufactured teas – spectrophotometric evaluation of tea liquors. *J. Sci. Fd. Agricul,* **14,** 687–700.

Roberts, E.A.H. (1963) The phenolic substances of manufactured tea. X. The creaming down of tea liquors. *J. Sci. Fd. Agricul,* **14,** 701–705.

Robertson, A. (1983) Effect of physical and chemical conditions on the *in vitro* oxidation of tea leaf catechins. *Phytochemistry,* **22,** 889–896.

Saijo, R. (1982) Isolation and chemical structures of two new catwchins from fresh tea leaves. *Agricul. Biol. Chem,* **46,** 1969–1970.

Saijo, R. and Takeo, T. (1979) Some properties of the initial form enzymes involved in shikimic acid biosynthesis in tea plant. *Agricul. Biol. Chem,* **43,** 1427–1432.

Sakamoto, Y. (1967) Flavones in green tea. Part I. Isolation and structure of flavones occurring in green tea infusion. *Agricul. Biol. Chem,* **31,**102–1034.

Sakamoto, Y. (1969) Flavones in green tea, part. II. Identification of isovitexin and saponarin. *Agricul. Biol. Chem,* **33,** 959–961.

Sakamoto, Y. (1970) Flavones in green tea, Part III. Structure of pigments IIIa and IIIb. *Agricul. Biol. Chem,* **34,** 919–925.

Sakato, Y. (1950) Studies on the chemical constituents of tea. III. On a new amide – theanine. *J. Jp. Agricul. Chem. Soc,* **23,** 262–264.

Sanderson, G.W. and Roberts, G.R. (1964) Peptidase activity in shoot tips of the tea plant (*Camellia sinensis*). *Biochem. J,* **93,** 419–423.

Sanderson, G.W. and Perera, B.P.H. (1965) Carbohydrates in tea plants.I. The carbohydrate on tea shoot tips. *Tea Q,* **36,** 6–13.

Sanderson, G.W. (1966) γ-Dehydroshikimate reductase in the tea plant (*Camellia sinensis*). *Biochem. J*, **98**, 248–252.

Sanderson, G.W., Co, H. and Gonzalez, J.G. (1971) Biochemistry of tea fermentation: the role of carotenes in black tea aroma formation. *J. Fd. Sci*, **36**, 231–236.

Sanderson, G.W. (1972) The chemistry of tea and tea manufacture. In Runeckles, T.T.C. (ed.). *Recent advance in Phytochemistry*, Vol. 5, pp. 247–316.

Sanderson, G.W., Kanadive, A.S., Eisenburg, L.S. *et al.* (1976) Contribution of polyphenolic compounds to the taste of tea. In Charala, M. and Katz, I., (eds.), *Phenolic, Sulfur and Nitrogen Compounds in Food Flavour*. ACS Symposium Series, No. 26, American Chemical Society, Washington, pp. 14–46.

Sasaoka, K. and Kito, M. (1964) Synthesis of theanine by tea seedling homogenerate. *Agricul. Biol, Chem,* **28**, 313–317.

Setho, V.K., Taneja, S.C., Dhar, K.L. *et al.*, (1984) (–)-Epiafzelechin-5-O-b=B-D-glucoside from *Crataeve religiosa*. *Phytochemistry*, **23**, 2402–2403.

Shallenberger, R.S. (1993) *Taste Chemistry,* Chapman & Hall, London, UK, pp. 5–20.

Shi, X. and Dalal, N.S. (1991) Antioxidant behaviour of caffeine: Efficient scavenging of hydroxyl radicals. *Fd. Chem. Toxicol,* **29**, 1–6.

Spiro, M. and Juganyi, D. (1994a) Kinetics and equilibria of tea infusion. Part 10. The composition and structure of tea scum. *Fd. Chem,* **49**, 351–357.

Spiro, M. and Juganyi, D. (1994b) Kinetics and equilibria of tea infusion. Part 11. The kenetics of the formation of tea scum. *Fd. Chem,* **49**, 359–365.

Stagg, G.V. and Millin, D.J. (1975) The nutritional and therapeutic value of tea – A review, *J. Sci. Fd. Agricul,* **26**, 1439–1459.

Steeranguachar, H.B. (1943) Studies on the fermentation of Ceylon tea. 6. The nature of the tea oxidase system. *Biochem. J,* **37**, 661–667.

Takayanagi, H., Anan, T., Ikegaya, K. *et al.* (1984) Chemical composition of oolong tea and pouchung tea. *Tea Res. J,* **60**, 54–68.

Takeo, T. (1966) Tea leaf polyoxydase. Part. III. Studies of the changes of polyphenol oxidase during black tea manufacture. *Agricul. Biol. Chem,* **30**, 529–535.

Takeo, T. and Baker, J.E. (1973) Changes in multiple forms of polyphenol oxidase during maturation of tea leaves. *Phytochemistry*, **12**, 21–24.

Takeo, T. (1974) *L*-Alanine as a precursor of ethylamine in *Camellia sinensis*. *Phytochemistry*, **13**, 1401–1406.

Takeo, T. (1978) *L*-Alanine decarboxylase in *Camellia sinensis*. *Phytochemistry*, **17**, 313–314.

Takeo, T. (1983) Natural mineral contents in tea and solubilities of minerals in tea infusion, *Bull. of Nat. Res. Inst. of Tea*, **19**, 87–128.

Takeo, T. and Imamura, Y. (1984) Comparison of liquid properties and colored components found in liquids among oolong tea, black tea and green (Kamairi) tea. *Tea Res. J,* **60**, 46–53.

Takino, Y., Imagawa., Hikikawa, H. *et al.* (1964) Studies on the mechanism of the oxidation of tea leaf catechins-formation of a reddish/orange pigment and its spectral relationship to some benztropolone derivatives. *Agricul. Biol. Chem,* **28**, 64–71.

Tamura, S., Ishima, N., Saito, E. *et al.* (1969) *Proceedings of Japanese Symposium on taste and smell,* **3**, 3–5.

Thorngate III, J.H. and Noble, A.C. (1995) sensory evaluation of bitterness and astringency of 3R(–)-epicatechin and 3S(+)-catechin. *J. Sci. Fd. Agricul,* **67**, 531–535.

Tirimanna, A.S.L. and Wickremasinghe, R.L. (1965) Studies on the quality and flavour of tea. 2. The carotenoids. *Tea Q,* **36**, 115–121.

Tirimanna, A.S.L. (1967) Aroma complex with special reference to tea. *Tea Q,* 38, 293–298.

Tsujimura, M. (1929) *Sci. Pap. Inst. Phys. Chem. Res. (Tokyo)*, 10, 253; (1930) 10, 63 (1934) **24**, 149; (1935) **26**, 186.

Tsushida, T. and Takeo, T. (1985) *L*-Theanine lylase in tea leaves. *Agricul. Biol. Chem*, **49**, 2913–2917.

Ullah, M.R. (1982) Effect of withering on the polyphenol oxidase level in the tea leaf. *J. Sci. Fd. Agricul*, **33**, 492–495.

Ul'yanova, M.S. (1963) Flavones in tea leaves (Russian). *USSR Cong. Biochem*, 1st Abst., No. 31.

Venkataramani, S., Premachandra, B.R. and Cama, H.R. (1976) Comparative study of the effects of processing on the carotenoid composition of Chinese (*Thea sinensis)* and Assamese (*Thea assamica)* tea. *Agricul. Biol. Chem*, **40**, 2367–2371.

Venkataramani, S., Premachandra, B.R. and Cama, H.R. (1977) Distribution of carotenoid pigments in tea leaves. *Tea Q*, **47**, 28–31.

Wan, X.C., Nursten, H.E. *et al.* (1997) A new type of tea pigment – from the chemical oxidation of epicatechin gallate and isolated from tea *J. Sci. Fd. Agricul*, **74**, 401–408.

Wang, D.F., Xie, X.F., Wang S.L. *et al.* (1996) Composition and the physical and chemical characteristics of tea polysaccharide (Chinese). *J. Tea Sci*, *16*, 1–8.

Wang, Z.N. *et al.* (1982) *Tea Biochemistry*(Chinese), Agricul. Publish. pp. 279.

Wickremasinghe, R.L. and Parera, V.H. (1966) The blackness of tea and colour of tip. *Tea Q*, **37**, 75–79.

Wickremasinghe, R.L., Roberts, G.R. and Perera, K.P.W.C. (1967) The localization of the polyphenol oxidase of tea leaf. *Tea Q*, **38**, 9–10.

Wickremasinghe, R.L., Perera, B.P.M. and Desilva U.L.C. (1969) Quality and flavour of tea. IV. Biosynthetic of volatile compounds. *Tea Q*, **40**, 26–30.

Wickremasinghe, R.L. and Perera, K.P.W.C. (1972) Site of biosynthesis and translocation in the tea plant. *Tea Q*, **43**, 125–137.

Wickremasinghe, R.L. and Perera, K.P.W.C. (1973) Factors affecting quality, strength and colour of black tea liquors. *J. Natl. Ser. Counc. Sri Lanka*, **1**, 111–112.

Wickremasinghe, R.L. (1978) Tea, *Adv. in Fd. Res*, 24, 229–286.

Wood, D.J. and Roberts, E.A.H. (1944) The chemical basis of quality in tea. III. Correlation of analytical results withtea tasters' reports and valuation, *J. Sci. Fd. Agricul*, **15**, 19–25.

Wu, X.C. (1990) Change of soluble polyphenol oxidase activity during the process of withering (Chinese). *J. Tea Sci*, **10**, 44.

Xiao, W.X. (1986) Changes of polyphenol oxidase activity in the process of withering (Chinese), *Tea Communication*, **3**, 9–11.

Xiao, W.X. (1987) Progress in the investigation of thearubigin in black tea (Chinese). *Foreign Agricul.(Tea)*, **1**, 1–6.

Yamada, H. (1980) Biogeochemical studies on the absorption of fluorine by plants., *Proceed. of Kyoto Univ*, **32**, 138–170.

Yamanishi, T. (1992) *Science of tea*, Zanhua Press, pp. 31–45.

Yamanishi, T. *et al.* (1995) Tea. *Fd. Rev. Internat*, **11**, 371–545.

Yokogoshi, H. *et al.* (1995) Reduction effect of theanine on blood pressure and brain 5-hydroxyindoles in spontaneously hypertensive rats., *Biosci. Biotech. Biochem*, **59**, 615–618.

Zaprometov, M.N. and Bukhlaeva, V. Ya. (1971) The effectiveness of the use of various ^{14}C-precursors of the biosynthesis of flavonoids in the tea plants (Russian). *Biochem. J*, **36**, 272–276.

Zaprometov, M.N. (1989) The formation of phenolic compounds in plant celll and tissue cultures and the possibility of its regulation, *Adv. in Cell Culture*, 7, 2201–260.

5. THE CHEMISTRY OF TEA VOLATILES

HUA-FU WANG, XIAO-QING YOU AND ZONG-MAO CHEN

Tea Research Institute, Chinese Academy of Agricultural Sciences, 1 Yunqi Road, Hangzhou, Zhejiang 310008, China

1. INTRODUCTION

Aroma is one of the critical aspects of tea quality which can determine acceptance or rejection of a tea before it is tasted. Early research on tea aroma can be traced back 160 years (Mulder, 1838), but progress on a more scientific basis has been achieved by the development and application of modern analytical techniques since the 1960's, when gas chromatography was widely used, especially when capillary column techniques became available. Tremendous advances in gas chromatography and combined gas chromatography-mass spectrometry have greatly increased our knowledge of tea aroma. An assessment of all data known shows that more than 630 compounds have been reported in tea aroma. One of the primary goals in aroma research is to identify those constituents which are responsible for the characteristic aroma of tea. Many attempts have been made to look for the key compounds for the aroma of tea (Takei *et al.* 1976; Yamaguchi and Shibamoto, 1981; Yamanishi, 1978a), but no single compound or group of compounds has been identified as responsible for the full tea aroma. It is generally believed that the characteristics of

Table 5.1 Aroma Constituents of Tea (Yamanishi 1995).

Compounds	Number	Compounds	Number	Compounds	Number
I. Hydrocarbons, total	72	3. Aromatic	16	IX. Nitrogenous	
1. Aliphatic	14	4. Terpenoid	3	compounds, total	86
2. Aromatic	25	5. Ionone derivatives	16	1. Pyrroles	12
3. Terpenoid	33	V. Acids, total	69	2. Pyridines	17
II. Alcohols, total	89	1. Aliphatic	63	3. Pyrazines	24
1. Aliphatic	51	2. Aromatic	3	4. Others	33
2. Aromatic	5	3. Terpenoid	3	X. Oxygenated	
3. Terpenoid	33	VI. Esters, total	82	compounds, total	36
III. Aldehydes, total	68	1. Aliphatic	52	1. Furanoid	17
1. Aliphatic	45	2. Alicyclic	3	2. Aromatic	10
2. Aromatic	18	3. Aromatic	19	3. Ionone-related	5
3. Terpenoid	5	4. Terpenoid	8	4. Others	4
IV. Ketones, total	75	VII. Lactones	25	XI. Sulfur compds	14
1. Aliphatic	30	VIII. Phenolic			
2. Alicyclic	10	compounds	22	Grand total	638

various kinds of tea consist of a balance of very complicated mixtures of aroma compounds in the tea. In Table 5.1, these compounds have been arranged into chemical categories to demonstrate their distribution. Eleven selected classes were considered (Yamanishi, 1995). For details of each aroma compound, refer to the review by Yamanishi (1995).

Research on tea aroma has been well reviewed in a series of papers (Schreier, 1988; Yamanishi, 1995, 1996; Takeo, 1996; Kawakami, 1997). This review is mainly on the chemistry of tea aroma with emphasis on work of recent years, although earlier studies are included where they are relevant.

2. EXTRACTION AND ANALYTICAL METHOD FOR TEA AROMA

No analysis can be made without appropriate sample preparation. Procedures used to isolate flavour components can have profound effects on the composition of the final product. The choice of isolation method depends on many factors, such as the concentration of the analyte and the physical and chemical properties of the different volatiles. The ideal methodology for sample preparation should be fast, accurate, precise, and consume minimal solvent with little cost.

2.1. Steam Distillation

Steam distillation is the method conventionally used in the analysis of flavor compounds in tea. It can be used for isolating volatile components either at atmospheric pressure or under reduced pressure. Steam distillation was used by Yamanishi *et al.* (1972), when 57 compounds were identified.

Steam distillation at normal pressure is a popular method for the isolation of tea volatiles, but it is regarded as too harsh a treatment because of the formation of off-flavors by thermal degradation or hydrolysis of some of the components. Consequently, the final concentration of analytes in the distillate may not correspond to that in tea. For example, using steam distillation as the extraction method, (Z)-3-hexenal was not detected because the isomerization of this aldehyde to (E)-2-hexenal is inevitable under such conditions (Hatanaka and Harada, 1973).

2.2. Simultaneous Distillation Extraction (SDE)

This method was first reported by Likens and Nickerson (1964), and then modified by Maarse and Kepner (1970) and Schultz *et al.* (1977). SDE can be operated at atmospheric pressure or under vacuum. The technique has been used for the analysis of volatile compounds from various kinds of tea by many researchers (Owour *et al.* 1989; Saijo and Takeo, 1973; Yamanishi *et al.* 1988; Luo *et al.* 1990; Wang and You, 1996). With SDE, the volatile compounds of interest in tea can be highly concentrated from a dilute solution in a single operation within a short time (usually 20-min) using only small volumes of solvent (ca. 50 ml). Artifact formation was also observed with this method (Kawakami *et al.* 1987a). Kawakami (1982) has reported that some disadvantages may occur with the SDE method. These include (1) glycosides can be easily hydrolysed; (2) the lactone structures can be easily opened; and (3) ketones are formed as degradation products from carotene by heating in aqueous media. Shimoda (1995) also found that SDE caused serious decomposition of volatile compounds in green tea infusion.

2.3.   Solvent Extraction

Solvent extraction has the advantage over steam distillation that lower temperatures are used during most of the extraction. It is an effective method for isolating highly water-soluble flavor constituents which are usually poorly recovered by distillation and headspace techniques. Therefore, the volatiles extracted by this method are claimed to have a more "natural" composition that is far superior to steam-distilled concentrates, in which thermal alteration may have occurred.

Kawakami *et al.* (1995) investigated the different aroma compositions of oolong tea and other black teas by comparing the brewed extraction method with SDE. They found that the GC pattern of brewed tea extracted with dichloromethane differs greatly from that of a similar tea extract prepared by SDE. The brewed extract contained higher amounts of acids, aromatic alcohols and monoterpenediols and lower amounts of monoterpene alcohols than those in the SDE extract.

The direct brewed extraction method using solvents is most suitable for producing "real" aroma of tea under experimental conditions. Using this procedure, alteration of aroma compounds can be minimized. (Kawakami *et al.* 1993a, b). However, the main disadvantage of solvent extraction lies in the fact that non-volatile materials (i.e. fats and polyphenols) are also extracted and can interfere with the analysis. The analysis can be improved by using a Porapak Q column. In the modified method, the brewed extract is not directly extracted with solvent, but is passed through a Porapak Q column. Aroma compounds together with catechins are adsorbed onto the Porapak Q adsorbent. The aroma compounds are then eluted with solvent. The authors found that the results of analysis for aroma compounds differ according to the solvent used. Diethyl ether was found to be the most suitable extraction solvent (Kawakami, 1997).

2.4.   Headspace Analysis

In headspace analysis, there are usually two major techniques: static and dynamic. In the former, the vapor phase is in equilibrium with the solute; in the latter this equilibrium is continually altered.

2.4.1. *Static*

Direct headspace analysis can be thought of as the optimum approach to sensory analysis since the headspace contains those volatiles that are most easily released from a tea infusion and so reach the human nose. In the static method a specific volume of the headspace is directly injected on to a gas chromatographic column with or without concentration. The main advantages of this method are: (1) no sample preparation is needed; (2) simple and timesaving; (3) no solvents are used; (4) avoids changes in the volatile composition due to chemical or enzymatic reactions. Therefore, this method provides the most accurate results on the composition of the perceived odor. However, this method is limited to products that contain high concentrations of volatile materials. It can not detect odorants that are present in low concentrations in the headspace. Because of the difficulty in detecting odorants that are present in low concentrations by direct headspace analysis, a further concentration process is needed. This technique was used for identifying aroma components in dry tea leaves by Heins *et al.* (1966).

2.4.2. *Dynamic*

In the dynamic method the volatiles are swept by an inert gas into a trap containing a porous polymer which adsorbs the majority of the organic constituents. Volatiles concentrated in adsorbent traps require posterior recuperation, usually achieved through thermal desorption. Horita and Hara (1985) analysed the headspace volatile components of green tea by using the Tenax TA trapping system. They found that the headspace volatiles could be detected with this system and smelt as green tea odor. However, as the yields of the odorants are strongly dependent on the velocity of the carrier gas (Werkhoff *et al.* 1989) and on the selectivity of the adsorption and desorption processes (Jennings and Filsoof, 1977), the composition of the volatile fraction detected by the dynamic procedure is different from that in the static method.

A typical device for dynamic headspace analysis is the microprocessor-based purge and trap-concentrator. In the purge and trap method, samples contained in a gas-tight, glass vessel are purged with an inert gas, causing volatile compounds to be swept out of the sample and into the vapor phase. The organic compounds are trapped on an adsorbent that allows the purged gas and any water vapor to pass through to the GC column. With this technique, analyses of the enantiomers of volatile compounds in black tea (Werkhoff *et al.* 1991), volatile acids in tea (Clark and Bunch, 1997) and volatile components in green tea (Lee *et al.* 1997) were accomplished.

2.5. Supercritical Fluid Extraction (SFE)

This methodology is based upon the principle that a fluid (gas) above its critical point exhibits the solution properties of a liquid solvent. CO_2 is considered as a most suitable extractant because it has moderate critical parameters (7.38 MPa and 31.04°C) (Otto and Thomas. 1967). The extraction of components can also be achieved using a static or dynamic method. The first example of the application of this method in tea aroma research was carried out by Vitzthum *et al.* (1975). In their experiment, the volatile components of black tea were isolated by extraction with supercritical CO_2 under pressure followed by atmospheric steam distillation and enrichment of steam volatiles on Porapak Q.

Although SFE is able to recover the majority of the aroma compounds, it is also more selective, and therefore less effective, than solvent extraction, i.e. fewer compounds are extracted with SFE than with solvent extraction (Polesello *et al.* 1993).

2.6. Solid Phase Extraction (SPE)

Solid-phase extraction is a rapid and sensitive sample preparation technique. Its use has increased considerably in recent years. Sample preparation and concentration via SPE can be achieved in a one-step extraction, and the methodology is valuable for isolating trace amounts of chemical compounds from complex matrices. This procedure is to condition a column with an appropriate solvent and then force the sample through the packing material (solid phase). The next step is to wash out the impurities, leaving the compounds of interest (volatile and non-volatile) bound to the adsorbent. In the final step, the compounds are eluted from the solid phase using a strong solvent.

Compared to conventional liquid/liquid extraction, SPE is timesaving, more robust and reliable, and less labor intensive. In addition, it provides better recoveries and reduces sample and solvent consumption by a significant factor. A method using SPE and GC/MS analysis has been applied to determining the volatile organic compounds in tea samples (Guidotti, 1997).

2.7. Identification and Quantification

Gas-liquid chromatography (GLC) is the most frequently used method for the separation of the aroma compounds of tea. For detection, the flame ionisation detector (FID) is widely used because of its high sensitivity to the different organic compounds present in tea flavor. However, sometimes tea aroma analysis requires more specific component information. In this case specific GC detectors, for example, N-, S- and P-specific detectors, have to be used. A flame therminoic detector (FTD) can also be used to analyse the pyrazine and pyrrole compounds in green tea because of its sensitivity to these nitrogen- containing compounds (Horita and Hara, 1985a). Mass spectrometry as a GLC detector is a powerful tool for the identification and structure elucidation of aroma compounds from tea (Wang and Li, 1989; Clark and Bunch, 1997).

In the elucidation and confirmation of the structure of an unknown compound, additional information can be obtained by nuclear magnetic resonance (NMR) and UV- and IR-spectroscopic measurements. The use of capillary gas chromatography/ fourier transform-infra red spectroscopy (GC/FT-IR) to study tea aroma (Kawakami et al. 1995) has also been reported.

As tea aroma components are so complex, no real quantification has been carried out. Usually, the relative amounts of volatiles are obtained by calculating the ratio of peak areas with that of the internal standard. Ethyl decanoate has been widely used as an internal standard (Yamanishi, 1978a, b; Takeo *et al.* 1985)

3. FORMATION OF TEA AROMA

3.1. Green Odor

The green odor is described as the odor of freshly cut grass or ground leaves and green plant materials. Since 1959 Hatanaka (1993) and his colleagues have systematically studied green odor in plants including that in tea plants. They found that eight volatile compounds involved the formation of green odor in green leaves. There are (E)-2-hexenal (leaf aldehyde), (Z)-3-hexenol (leaf alcohol), (Z)-3-hexenal, (E)-3-hexenol, (E)-3-hexenal, (E)-2-hexenol, n-hexanal and n-hexanol. Those are synthesised in green leaves from α-linolenic and linoleic acids via their respective hydroperoxides or by oxidative breakdown of fatty acids during tea processing.

Polyphenol oxidase from fresh tea leaves was found to form aldehydes from amino acids (alanine, valine and leucine) in the presence of (+)-catechin or other diphenols. These aldehydes contribute to the development of the aroma in tea (Srivastava, 1986).

Baruah et al. (1986) also found that unsaturated fatty acids that are released during the processing of tea give rise to the volatile alkanals, alkanols, and alkanones which are believed to have a green note.

3.2. Floral Aroma

The floral note can be defined as the odor emitted by flowers and contains sweet, green, fruity characters. Terpenes, especially monoterpenes, are often the most important components responsible for the characteristic floral odors of tea. The monoterpene citronellol shows fresh and rose-like aroma, the geraniol with sweet, fruity and rose-like aroma, the linalool with fresh, sweet and citrus aroma, nerol with fresh, sweet, and rose-like aroma, and the α-terpineol with sweet and peach-like as well as floral aroma. These seemingly diverse compounds have the 5-carbon isoprene unit (2-methyl-1,3-butadiene) in common as their skeletal building block. These can be linked together in multiples of two (monoterpenes), three (sesquiterpenes), four, six and nine and even higher. These structures may be openchain, cyclic, saturated or unsaturated or oxidised. As compared to monoterpenes, the flavor value of sesquiterpenes is less pronounced.

In addition to the free, odor-producing forms of monoterpene alcohols, the presence of glycosidically bound monoterpene alcohols in tea was first suggested by Takeo (1981a, b). He found terpene alcohols in tea are generated by the enzymatic action of β-glucosidase and proposed the presence of β-glucosides as the precursors in tea leaves. Yano et al. (1990) also found the liberation of some aroma constituents including monoterpene alcohols when non-volatile materials from tea were treated with a crude enzyme preparation from tea leaves. Fischer et al. (1987) investigated the enzymatic hydrolysis of non-volatile compounds in fresh tea leaves during the black tea manufacturing process, especially during the fermentation procedure. They found that the main aroma components, such as (Z)-3-hexenol, benzyl alcohol, 2-phenyl ethanol, linalool and geraniol are present in their bound forms. However, they reported that the occurrence of the bound forms of such components are not naturally present but are produced during the withering or fermentation stage of the black tea production process. In recent years, besides glucosides, some primeverosides and vicianosides have also been identified in fresh tea leaves of an oolong tea cultivar (Guo et al. 1994, 1995; Moon et al. 1994). This will be reviewed later. The most common aglycones occurring as glycosidically bound volatiles are listed in Table 5.2 according to the paper by Stahl-Biskup et al. (1993).

Table 5.2   Compounds Occurring as Glycones of Glycosidically Bound Volatiles.

No	Compounds	No	Compounds	No	Compounds
1	Pentan-1-ol	8	Octan-3-ol	15	Geranylacetone
2	Pentan-2-ol	9	Nonanol	16	Linalool
3	Penten-3-ol	10	Benzyl alcohol	17	9-Hydroxy-linalool
4	(Z)-2-Penten-1-ol	11	Methyl salicylate	18	Nerol
5	Hexanol	12	Phenol	19	α-Terpineol
6	(Z)-3-Hexenol	13	2-Phenylethanol	20	α-ionone
7	(E)-3-Hexenol	14	Geraniol	22	Dihydroaclinldoline

Wang and You (1996a) investigated the varietal and seasonal changes in both free and bound monoterpene alcohols in the clones cultured in the Qimen (Keemun) area in China. The content of the flavor components released from the bound forms increased greatly after addition of crude enzyme extracted from fresh tea leaves after incubation at 37°C for 10 hrs. Most of the b/f values for monoterpene alcohols, the ratio of bound to free forms, were greater than 1, indicating that the bound forms were in higher concentrations in fresh tea leaves than the corresponding aglycones. Geraniol was the highest among the monoterpene alcohols both in the amount released from the bound form and the b/f value. This indicated that the crude enzyme used in this experiment was very effective in releasing geraniol. The sequence for the content of both bound and free flavours in different seasons was, generally: spring > summer ≥ autumn. Both forms of monoterpene alcohols exhibited a marked decrease in concentration in summer and autumn compared with those in spring. Ogawa *et al.* (1995) measured amounts of alcoholic aroma precursor and glycosidase activity in each part of the tea shoot (*Camellia sinensis var. sinensis* cv Yabukita and a hybrid of *var. Assamica & var. sinensis* cv Izumi) by means of a crude enzyme assay. They found that the aroma precursors were abundant in young leaves and decreased as the leaf aged. Glycosidase activity also decreased as leaves aged, but was high in stems. Wang and You (1995) investigated the β-glucosidase activity of crude enzyme preparations and the liberation of monoterpene alcohols in made teas. They found that the free forms of monoterpene alcohols in green made tea are not only connected with the content of the bound forms of monoterpene alcohols, but also with β-glucosidase activity in the corresponding fresh leaves of cultivars. In contrast, in black tea, the liberation of free monoterpene alcohols in made tea mainly depended on the corresponding content of bound forms of monoterpene alcohols. This is possibly due to the full fermentation process during manufacture of black tea. It is known that in various tea cultivars, the amount of non-volatile terpenyl-glycosides can be more abundant than the amount of free aromatic terpenols. Consequently, it could be of value to increase the varietal character by hydrolysing and liberating these aromatic compounds.

Xia *et al.* (1996) investigated the changes of β-glucosidase activity during the withering and fermentation of black tea. Enzyme activity in the leaves increased gradually during the withering process and reached about 2–2.5 times that in the fresh leaves, and then decreased during the fermentation process. The higher the fermentation temperature, the faster the enzyme activity decreased. Therefore, the authors recommended that withering and fermentation under low temperature were beneficial for increasing the enzyme activity and the formation of aroma.

The formation of various terpene alcohols via the mevalonate pathway was shown by Erman (1985). Withopf *et al.* (1997) screened benzyl, 2-phenyl ethyl, geranyl, citronellyl, and 2,5-dimethyl-4-hydroxy-3-(2H)-furanone 6′-O-malonyl β-D-glucopyranosides in green tea and other plant tissues using synthesised reference compounds. The results indicated that malonylation of glycoconjugates was a common pathway in plant secondary metabolism.

The hydrophilic nature of bound monoterpenes means they do not contribute to aroma of tea. Therefore, many researchers are interested in the hydrolysis of these potential aroma precursors to release the free floral terpenes and enhance tea aroma. You *et al.* (1994) added crude acetone powder extracts into autumn tea leaves at the rolling stage of green tea manufacturing, and they found that the contents of linalool and geraniol increased markedly, and the aroma quality was found to be improved. Guo *et al.* (1992) tried to produce congou black tea from stale green tea of low commercial value. The experiment showed that the main constituents of the black tea such as geraniol, linalool and its oxides, methyl salicylate, benzyl alcohol, 2-phenyl ethanol, and β-ionone increased markedly. Geraniol, which is present in very much lower amounts in green tea, became the most abundant peak in the black tea product.

Ionones and the ionone isomers are cyclic terpenoid compounds. It is reported that these ionone and ionone derivatives in tea have a floral aroma, like violet and rose, and contribute to fresh taste (Nose *et al.* 1971). Sanderson *et al.* (1973) suggested that the carotenoid derived compounds illustrated in Figure 5.1 were largely responsible for tea flavor, particularly in black tea. These aroma compounds are thought to be formed from carotenoids by thermal degradation during tea manufacture. This supposition was confirmed by Hazarika and Mahanta's experiment (1983), in which a decrease of the four major carotenoids (β-carotene, lutein, violaxanthin and neoxanthin) was found during black tea manufacture.

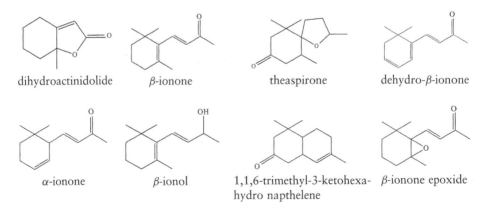

dihydroactinidolide   β-ionone   theaspirone   dehydro-β-ionone

α-ionone   β-ionol   1,1,6-trimethyl-3-ketohexahydro napthelene   β-ionone epoxide

Figure 5.1 Carotenoid derivatives found in black tea as flavor compounds

4-Decanolide
(γ-Decalactone)

5-Decanolide
(δ-Decalactone)

Figure 5.2 Common lactone structures

Lactones are widely distributed in black tea (Yamanishi *et al.* 1973a), pouchong tea (Nobumoto *et al.* 1993) and green tea (Kawakami and Yamanishi, 1983). They have potent, generally pleasing, aroma and flavor properties (Maga, 1976), and are associated with odor characters such as fruity, coconut-like, buttery, sweet or nutty. The naturally occurring lactones generally have γ- and δ-lactone structure, while a few have a macrocyclic lactone structure. Common lactone structures are shown in Figure 5.2.

Generally, lactones are formed via β-oxidation of saturated and unsaturated hydroxy fatty acids or their lipid precursors (Okui *et al.* 1963; Mizugaki *et al.* 1965). Contrary to glycosidically bound monoterpenols (e.g. conjugates of linalool) or phenols (e.g. conjugates of raspberry ketone) which after enzymatic hydrolysis liberate an "attractive" aglycon, most of the known C13-glycosides liberate an aglycon (polyol) which is flavorless. In this case, further modifications, e.g., acid-catalysed conversion at elevated temperatures, are required to finally generate the odor-active form (Winterhalter, 1992).

3.3. Roasted and Nutty Aroma

The roasted and nutty aroma is mostly described as a slightly burnt odor of nuts. Pyrazines are primarily responsible for the roasted and nutty flavours that are produced in tea. These arise through a process known as non-enzymatic browning (Maillard reaction). This browning reaction occurs when the carbonyl group of a reducing sugar condenses with a protein amino group at temperatures of 100°C or higher (Shallenberger and Birch, 1975). Pyrazines are also biosynthesised by plants and microorganisms, and appear to be ubiquitous in fermented foods, beverages and vegetables (Maga, 1982). The basic pyrazine structure is shown in Figure 5.3. These nitrogen-containing heterocyclic compounds may be, and usually are, highly substituted, thus yielding a large family of

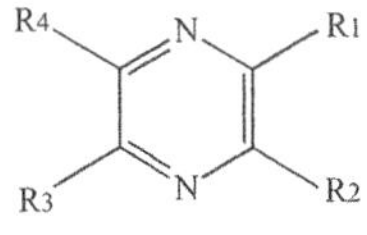

Figure 5.3 Basic pyrazine structure

compounds which produce unique organoleptic characteristics at extremely low concentrations. The flavour character of any given pyrazine depends strongly upon its concentration and chemical structure. Increasing the chain length of the 3-alkyl substituent of methoxyalkyl pyrazines results in a succession of aromas ranging from a nutty-earthy-green to a bell pepper aroma and, when the chain length increases above C-6, to an earthy aroma (Parliment and Epstein, 1973).

Beside pyrazines, the Maillard reaction leads to many other important classes of flavor compounds including furans, pyrroles, thiophenes and other heterocyclic compounds, giving rise to an extremely complex array of volatile components in tea. These compounds are also regarded as contributors to the roasted and nutty aroma of tea (Yamanishi et al. 1973b).

L-theanine is an important and the most abundant amino acid in tea. When L-theanine is heated to 180°C, a large amount of N-ethyl formamide, as well as ethyl amine, propyl amine, 2-pyrrolidone, N-ethyl succinimide, and 1-ethyl-3, 4-dehydro-pyrrolidone are found. But if it is heated with glucose above 150°C, the main product is 1-ethyl-3, 4-dehydropyrrolidone together with small amounts of pyrazines and furans (Yamanishi et al. 1989b).

Hara (1985) investigated the effect of the firing temperature on the production of tea aroma. The experiment showed that pyrazines can be produced only when the firing temperature is over 100°C. The production of pyrazines depends not only on firing temperature, but also on heating time and the amount of tea leaves (Sawamura and Masuzawa, 1982).

3.4. Off-flavor

Off-flavor in food is defined as any flavor that is not normally associated with the food in question. This flavor may be of a type that is quite unacceptable in food, or it may simply be a flavor which is not associated with that particular food product (Saxby, 1982). Although tea can be roughly divided into three categories, i.e. non-fermented (green), partially fermented (oolong) and fermented (black) teas, there are a number of subtypes under each of these categories, including some special types of tea. They all have their unique flavors and tastes. Therefore, sometimes an acceptable flavor in one kind of tea may be an off-flavor in another kind of tea. The off-flavor in tea differs from tea to tea because the characteristic flavor of different teas varies quite extensively.

3.4.1. Stale flavor

Tea can be, but usually is not consumed immediately after it is produced. Therefore, a period of storage is inevitable. In the early stage of storage, maturation of tea may occur and helps the tea to reach its best quality. However, in the later stage of storage, many chemical changes due to enzymatic and non-enzymatic or other oxidative reactions may occur and cause the quality, especially aroma quality, to fall sharply. The related flavor is called stale or stored flavor. This problem has attracted much attention from many researchers (Nose, 1971; Saijo, 1972; Dougan *et al.* 1978; Horita, 1987; Hara *et al.* 1987; Hara, 1989). Saijo (1972) investigated the changes of aroma compounds in some types of black tea after they had been stored for four years. The results showed that old black tea had high contents of n-hexanal, β-myrcene, (Z)-β-ocimene, acetic acid, and benzyldehyde; while the freshly made black tea had high contents of 1-penten-3-ol, (E)-2-hexenal, (Z)-2-penten-ol, (Z)-3-hexenol, (E)-2-hexenol, linalool and linalool oxides, phenyl acetaldehyde, methyl salicylate, geraniol, and β-phenyl ethanol. He indicated that those compounds present in stale black tea were closely related to the old, aged flavor of the black tea. Stagg (1974) reported that the volatile fraction resulting from stale tea extraction showed an overall decline which was accelerated by the uptake of moisture and, to some extent, in the case of black tea by storing at elevated temperature. In contrast, for green tea, an increase in its volatiles content was found during storage (Anan, 1983). This is due to the auto-oxidation of fatty acids in the green tea. Linolenic and linoleic acids are abundant in tea. The former can be degraded to form hexanal and hexanol, and the latter can be degraded to form unsaturated C_6 aldehydes and alcohols, such as (Z)-3-hexenal, (Z)-3-hexenol, (E)-2-hexenal and (E)-2-hexenol (Muritu *et al.* 1988). Hara and Kubota (1982) found that 1-penten-3-ol, (Z)-2-penten-1-ol, (E,Z)-2, 4-heptadienal and (E,E)-2,4-heptadienal increased with length of storage. By sensory evaluation, they found that 1-penten-3-ol had a strong stinging green odor, (Z)-2-penten-1-ol had a weak stinging green odor and 2,4-heptadienals had an oily green smell. It was thought that these four compounds were the major off-flavor compounds responsible for the aged flavor of green tea during storage. To avoid or reduce the aged flavor, it was recommended that tea should be suitably dried and stored in moisture-proof packing in an inert atmosphere of nitrogen (Horita, 1987; Hara *et al.* 1987). In addition, it was found that the aged flavor of green tea could be decreased significantly using a refiring process (Hara, 1989).

3.4.2. *Photo-induced flavor*

Photo-induced flavor is an off-flavor produced when tea is exposed to light. This off-flavor could be detected in green tea after just one-day of exposure to light, and after four days a strong off-flavor was noted (Horita and Hara, 1986). Masuzawa (1974) observed the effect of light on the quality of green tea. He found that after green tea was kept under light of 2000 lux for one month, the color of the tea showed little change, but the infusion had a very strong off-flavor, unacceptable for this drink. The author reported that light had only a small influence on the appearance of the tea or the color of the infusion, but had a marked deleterious effect on the aroma and taste of the infusion. Hara (1989) investigated the changes in volatile components in green tea after exposure to light. His results showed that the levels of pentanal, 1-pente-3-ol, butanol, 2-methyl propanal, propanal, bovolide and dibovolide increased significantly in the light exposed tea, while dimethyl sulfide, which is believed to be one of the major compounds closely related to the fresh flavor of green tea, decreased significantly. Meanwhile, the content of linalool and nerolidol, which contribute the most to green tea flavour, changed little. Among those compounds showing an increase, bovolide was the most evident. It could be formed even under light of 50 lux. Therefore, the author concluded that bovolide could be used as an indicator to assess whether or not tea had been exposed to light.

3.4.3. Retort smell

The flavor deterioration of canned tea drink at retort-sterilization was examined by Kinugasa and Takeo (1989). The contents of 2,4-heptadienals, linalool, geraniol, benzyl alcohol, β-ionone, 4-vinylphenol and indole were increased in sterilized Sencha drink. 4-Vinylphenol has been reported as the major compound of the retort smell. Kinugasa and Yakeo (1990) investigated the changes of aroma compounds during the manufacturing of green tea canned drinks. They found that the retort smell was produced after sterilization by hydrolysis of precursors of volatile compounds. To avoid the retort smell, adding a suitable amount of β-cyclodextrin during manufacturing is recommended (Kinugasa and Takeo, 1989). β-cyclodextrin can also be used for tea aromatization in order to improve the aroma stability and sensory quality (Szente et al. 1988). Kinugasa et al. (1997) also found that the content of precursors in tea leaves is higher in young leaves than in matured leaves. Furthermore, the precursors in tea leaves are easily hydrolyzed during the firing process of Kamairicha (a type of roasted green tea) at 200–250°C. Therefore, the content of precursors of volatile compounds in Kamairicha is low, and hence the off-flavor becomes weaker in the drinks of Sencha made from matured leaves and the drink of Kamairicha.

3.4.4. Smoky-burnt odor

Smoky and burnt odors are major problems that exist widely in roasted green tea. Wang and Li (1989) reported that pyrazines and pyrroles formed the chemical basis of burnt tea, and 2,5-dimethyl pyrazine could be used as an indicator for burnt odor in tea, while guaiacol, naphthalene and indene were the compounds responsible for the smoky odor of tea. The experiments showed that the burnt odor of tea mainly occurred during the drying process, but the smoky odor could occur in all steps of heat treatment. Tsutomu et al. (1985) found some naphthalenes and indenes in the headspace of Rooibos tea infusion. They presumed that these compounds came from the absorption of contaminants.

The contributions of non-volatile compounds, such as polyphenols, caffeine, fatty acids and chlorophylls to the off-flavors of tea should not be forgotten. But as it is beyond the range of this chapter, we are not going to discuss this in detail.

4. AROMA CHARACTERISTICS OF VARIOUS TEAS

4.1. Green Tea

For the production of green tea, the enzyme activities in freshly plucked tea leaves need to be inactivated. There are two ways to deactivate the enzymes in the fresh leaves in the first step of green tea making, i.e. pan-firing and steaming, resulting in pan-fired or steamed teas.

4.1.1. Pan-fired green tea

Kosuge *et al.* (1981) investigated the difference between the aroma of Chinese and Japanese pan-fired green teas. They found the former contained large quantities of geraniol, 2-phenylethanol, benzyl alcohol and phenol. Linalool oxides (pyranoid), (Z)-3-hexenyl hexanoate and 2,6,6-trimethyl-2-hydroxy cyclo-hexan-1-one were much lower in Chinese pan-fired green teas than in Japanese pan-fired green teas. Both Chinese and Japanese pan-fired green teas contained a small quantity of linalool, which was found in high concentrations in Sencha, and characteristically contained pyrazines at relatively high concentration.

Longjing tea is one of the most famous green teas in China produced by manual pan-firing. Kawakami and Yamanishi (1983) analyzed the aroma components of Longjing tea. Higher concentrations of linalool, linalool oxides, geraniol, 2-phenyl ethanol, lactones and pyrazines were found in Longjing tea. They deduced that these compounds contribute to the characteristic floral sweet note and pleasant roasted aroma of Longjing.

You *et al.* (1992a) investigated the effect of the "tan-fang" treatment on the aroma formation of Longjing tea. In the so-called "tan-fang" fresh tea leaves are thinly spread in a cool shaded place as the first step of Lonjing tea making. The result showed that the amounts of volatile compounds in Longjing tea were increased in tea leaves during the progression of the "tan-fang" period. The contents of 1-penten-3-ol, (E)-2-hexenal, hexanal and (Z)-3-hexenol which are induced from tea leaf lipid (Hatanaka *et al.* 1982), and linalool, its oxides, geraniol 2-phenylethanol and methyl salicylate which are liberated from glucosides (Takeo, 1981a, b) were significantly increased during the "tan-fang" process. Therefore, "tan-fang" is an important process for developing the characteristic aroma of Longjing tea.

4.1.2. Steamed green tea

The quality characteristics of steamed green tea are freshness, naturalness and briskness. The components that are responsible for the briskness and freshness were found to be (Z)-3-hexenol and its esters, i.e. hexanate and (E)-2-hexenoate. In addition, an adequate contribution of indole and dimethylsulfide seems to make the aroma of spring green tea more attractive (Yamanishi, 1978a). High contents of β-cyclocitral, 2,6,6,-trimethyl cyclohexanone, 2-pentenol, 2,4-heptadienal and 5-octadien-2-one were found in the best Japanese green tea in Gyokuro (Yamaguchi and Shibamoto, 1981).

In Japan, a roasted flavor, mainly caused by pyrazines, pyrroles and furans, is not regarded as a good flavor for ordinary steamed green tea (Omori, 1997). Kubota *et al.* (1996) compared the difference in aroma compounds of different green teas after microwave heating and hot air heating which is a traditional method for tea refining in Japan. They found that the microwave treatment substantially suppressed the formation of pyrazines, pyrroles and furan compounds, well known as typical thermally generated compounds and responsible for a roasted aroma. Their results indicated that microwave treatment is very effective for preparing green tea that has a refreshing aroma and low astringency.

4.1.3. Scented tea

Scented tea is mainly produced from steamed green tea by scenting with fresh flowers. Jasmine tea is the most popular. In jasmine tea the most abundant aroma components were found to be (Z)-3-hexenyl benzoate, benzyl acetate, linalool and methyl anthranylate (Yamanishi, 1988). Other kinds of scented teas are also produced. Youzi tea is one of the scented teas produced in China that is scented with fresh flowers of Youzi (*Citrus grandis* L.). The main aromatic components of Youzi tea were linalool, nerolidol, (E,E)-farnesol, methyl anthranilate, phytol, etc. (Luo *et al.* 1990). Fu *et al.* (1991) also investigated the aroma compounds of tea scented with Zhulan flowers (*Chloranthus spicatus* Mak.). They found the main aroma components were (Z)-methyl jasmonate, (E)-methyl jasmonate, linalool, nerolidol, methyl N-methyl anthranilate, cedrol and phytol.

4.2. Semi-fermented Tea

Semi-fermented tea can be classified as pouchong tea and oolong tea according to the degree of fermentation during their manufacture. Pouchong tea belongs to the lightly fermented tea mainly produced in Taiwan, while oolong tea is a relatively heavier fermented tea mainly produced in Fujian and Taiwan. The characteristic sensory qualities of oolong and pouchong teas are their unique floral flavor and pleasant taste. In comparison with black tea or green tea, the flavor quality of oolong and pouchong tea is much higher, especially the superior elegant floral flavour that solely decides their market price.

4.2.1. Pouchong tea

Yamanishi *et al.* (1980) identified 48 compounds from Taiwan pouchong tea. It was found that the aroma compounds, such as (Z)-jasmone, jasmine lactone, nerolidol, methyl jasmonate etc. were present in high concentration. In addition, esters of (Z)-3-hexen-1-ol, linalool oxides, benzyl alcohol, 2-phenylethanol, α-farnesene and benzyl cyanides were increased during the special processing of pouchong tea manufacture, including solar withering followed by indoor withering with a shake-turnover treatment. These authors regard jasmine lactone as probably the key substance contributing to the characteristic aroma of pouchong tea. Takeo (1981c) and Nobumoto *et al.* (1993) also found that pouchong tea contains large amounts of a specific jasmine lactone and other lactones.

4.2.2. Oolong tea

Takeo *et al.* (1985) published a long report on the flavor of oolong tea. Compared to light-fermented pouchong, the oolong teas produced in Fujian, such as Ti Kuan Yin, Shui Hsien, Se Zhong and Huang Chin Gui contain large amounts of nerolidol, jasmine lactone and indole. Kobayashi *et al.* (1988) found the levels of methyl epijasmonate, which has an aroma 400 times stronger than methyl jasmonate, in tea samples ranging from 4.4 to 16.8 mg%. This compound was shown to contribute to the aroma of tea, particularly to that of Chinese semi-fermented tea. Nobumoto *et al.* (1990) reported that terpenes, such as linalool and α-farnesene, especially nerolidol were the main oolong tea aroma compounds. However, Kawakami *et al.* (1995) compared the aroma composition of oolong tea after SDE and brewed extraction methods. What they found was that there were only small amounts of the above-mentioned compounds in the brewed extract.

4.3. Black Tea

Darjeeling in India, Uva in Sri Lanka and Keemun in China are well known black tea producing areas. These areas produce very famous teas with their own characteristic aroma and color. The aroma of Keemun tea has a rosy and woody note, while that of Uva tea contains a sweet flower-like fragrance with a refreshing green odor, and that of Darjeeling tea is between these two types of teas. The pleasant flowery notes of linalool oxides have been especially observed in high-quality Darjeeling black tea (Schreier. 1988). In Keemun black tea, a higher content of geraniol, benzyl alcohol, 2-phenyl ethanol and a lower content of linalool and linalool oxides have been observed compared to Uva tea. In addition there are larger amounts of aroma concentrate in Keemun black tea than that in Sri Lanka tea. (Aisaki *et al.* 1978).

Yamanishi *et al.* (1973a) and Wickremasinghe *et al.* (1973) found that compounds with high boiling points, such as jasmine lactone, 2,3-dimethyl-2-nonen-4-olide, 4-octanolide, 4-nonanolide, 5-decanolide etc. are important components of black tea. Of these, methyl jasmonate is regarded as the most important. Wang *et al.* (1993) studied the characteristic aroma components of Qimen (Kemmun) black tea using different varieties of Zhuye population and Anhui No. 7, both cultured in Qimen area. They found the flavour pattern similarity between Zhuye population and Anhui No 7 was low (0.414), but after geraniol was taken out, the reduced flavor pattern similarity increased up to 0.959, indicating that geraniol plays an important role in the characteristic aroma of Qimen black tea.

Takeo and Mahanta (1983a) compared the different aroma components in orthodox and CTC black teas and found that orthodox tea usually contains double the amount of volatiles than CTC tea. The amount of (Z)-3-hexenal, linalool, linalool oxides, methyl salicylate and geraniol is found to be higher in orthodox teas, but the (E)-2-hexenal and phenylacetaldehyde contents are higher in CTC teas. The less fragrant nature of CTC teas is probably due to the lower amount of essential volatile compounds, especially linalool and its oxides together with methyl salicylate (Takeo and Mahanta, 1983b).

Black and semi-fermented tea were made from the same variety of tea leaves and their aroma components analysed by Takeo (1983c). He found that black tea had more (Z)-3-hexenol, (E)-2-hexenyl formate, monoterpene alcohols and methyl salicylate than semi-fermented tea, while semi-fermented tea had more (E)-jasmone + β-ionone, nerolidol, jasmine lactone, methyl jasmonate and indole.

4.4. Speciality Tea

4.4.1. Post fermented tea

In China some famous teas such as Pu-er cha, Liu-bow cha, Fu-chuang cha, etc. are made through microbial fermentation. The typical characteristic of these teas is a "moldy" or "aged" flavour, and the more aged the better.

Pu-er tea: The main compounds in Pu-er tea were found to be (E)-2-hexenal, (E)-2-pentenal, (E,Z)-2,4-heptadienal, ionones and their oxides, and (Z)-jasmone. These compounds were reported as being produced during the microbial-fermentation and solar drying processes of Pu-er tea manufacture (Liu *et al.* 1989).

Brick tea: This is a kind of post-fermented tea and is often called "dark green tea". Wang *et al.* (1991) studied the changes of the aroma compounds of Fuzhuan brick tea during the fungal growing process, and found that almost all aldehydes, ketones, 2,5-dimethyl pyrazine, and 2,6-dimethyl pyrazine increased during fungal growth, especially (E,Z)-2,4-heptadienal, furfural, (E,E)-2,4-heptadienal and (E,E)-2,4-octadienal which increased greatly. It was presumed that these increased compounds greatly contributed to the formation of the typical fungus flower flavor of Fuzhan brick tea. The aroma compounds in dark green teas manufactured by orthodox pile-fermenting (OPF) and sterile pile-fermenting (SPF) respectively were analyzed and compared by Wang *et al.* (1992). The experiment showed that the OPF sample contained more terpenols and phenols, while the SPF sample contained more aldehydes and ketones.

Goishi-cha and Awa-cha are kinds of pickled tea in Japan. Their major aroma compounds were identified as (Z)-3-hexenol, linalool and its oxides, methyl salicylate, benzyl alcohol, 2-phenylethanol, acetic acid and 4-ethylphenol (Kawakami *et al.* 1987a). Toyama kurocha is another kind of pickled tea produced in Japan. Kawakami and Shibamoto (1991a) found that the amounts of terpene alcohols increased and methylether phenolic compounds were produced by natural fungal fermentation during Toyama kurocha manufacture. Aliphatic alcohols, aldehydes, and ketones were greatly increased by solar drying. 125 volatile components were identified in the stored sample. The formation of volatile phenolic compounds produced from ferulic acid and *p*-coumaric acid by *Aspergillus niger* and *Leuconostoc mesenteroides* was also investigated. Methylation of phenolic hydroxy groups by *Aspergillus niger* was observed, similar to that of the fungal fermentation process.

4.4.2. Smoked tea

Smoked tea is mostly consumed in the west part of China. Their main aroma compounds were identified as linalool, guaiacol, β-ionone, geraniol, cedrol and 6-10-14-trimethyl-2-pentadecanone, which were considered to consist of the original compounds in tea leaves, the secondary products produced during the manufacturing processes and the fumigating materials adsorbed by the tea (Wang *et al.* 1991). Shen and Yang (1989) investigated the aroma compounds of smoked black tea made with small leaves. They found that the characteristic smoky flavor compounds of this tea were mainly 5-methyl furfural, phenol, o-cresol, p-cresol, guaiacol, 2,6-dimethyl phenol, 4-ethyl guaiacol, 2,6-dimethyl guaiacol and anthracene.

4.5. Herbal Tea

Kawakami and Kobayashi (1991) analyzed the volatile components of green mate and roasted mate produced in South America. Of 250 components separated from the concentrate, a total of 196 compounds were newly identified as mate volatiles. The volatile components of mate were similar to those of *Camellia sinensis* tea, including 144 of the common components such as terpene alcohol, linalool, α-terpineol, geraniol, and nerolidol and ionone-related compounds α-ionone, β-ionone, and 2,6,6-trimethyl-2-hydroxycyclohexanone. A high level of 2-butoxyethanol was characteristic, and 3,3,5-trimethylcyclohexanone-related compounds, although present at low levels, were specific to the mate flavor. Roasted mate contained more furans, pyrazines, and pyrroles formed by the Maillard reaction during the roasting process.

Rooibos tea is produced and consumed as a beverage similar to tea in South Africa. Habu *et al.* (1985) analyzed the volatile components of Rooibos tea. They found that the major components of the extract were guaiacol, 6-methyl-3,5-heptadien-2-one, demascenone, geranylacetone, β-phenyl ethyl alcohol and 6-methyl-5-hepten-2-one. With the brewed extraction method, 50 components such as 2-phenylethanol, 2-methoxy-2-buten-4-olide, guaiacol, dihydro-actinidiolide, 4-butanolide, methyl-ethylmaleimide, and hexanoic acid were found, and with the SDE method, 123 components such as guaiacol, acetic acid, 2-phenylethanol, geranylacetone, β-damascenone, hexanoic acid, 3-methylbutanoic acid, and 6,10,14-trimethylpentadecanone were found. A total of 42 components were newly identified as Rooibos tea volatiles, including 5 hydrocarbons, 7 alcohols, 7 aldehydes, 5 ketones, 4 acids, 1 ester, 8 lactones, 2 anhydrides, 1 imide, 1 phenol, and 1 furan by Kawakami *et al.* (1993a).

Tengcha (*Ampelopsis grossedentata*) is produced by a method similar to the procedure of green tea processing in China. It is consumed as a medicine beneficial for the common cold. It has a special flavor when it is brewed. Its main aroma compounds were newly identified with GC and GC/MS as (E)-2-hexenal, (Z)-3-hexenyl hexanote, benzyle acetaldehyde, α-ionone, (Z)-jasmone, cedrol, and 6-10-14-trimethyl-2-pentadecanone (Wang and You, 1996b).

5. DEVELOPMENT AND APPLICATIONS OF RESEARCH ON TEA AROMA

5.1. Separation of Chiral Compounds

A significant proportion of the identified aroma compounds in tea is known to be chiral. Chiral discrimination has been recognized as one of the most important principles in biological activity and also odor perception (Mosandl *et al.* 1986; Mosandl and Gunther, 1989). The tremendous progress in chirospecific analysis of natural flavors in recent years was mainly due to the commercial introduction of modified cyclodextrin columns as well as multidimensional gas chromatographic instrumentation.

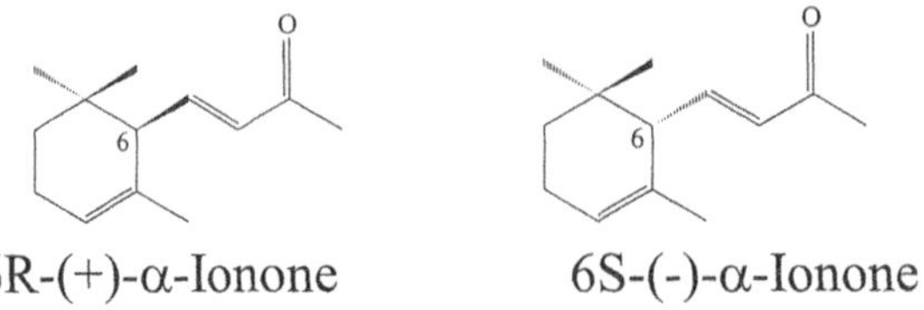

6R-(+)-α-Ionone   6S-(-)-α-Ionone

Figure 5.4   Structures of α-Ionone

There is an asymmetric carbon at C6 of α-ionone, so two types of enantiomeric compositions can be obtained as shown in Figure 5.4. The direct capillary gas chromatographic separation of *trans-α*-ionone and *trans-α*-damascone enantiomers was reported by Werkhoff *et al.* (1991) using heptakis-(2,3,6-tri-O-methyl)-*β*-cyclodextrin in polysiloxane as a suitable chiral stationary phase. The method described has been applied to determine the naturally occurring enantiomeric composition of *trans-α*-ionone in tea and the optical purity of *trans-α*-ionone in solvent extracts of black tea. *Trans-α*-ionone was either isolated by headspace stripping *in vacuo* or by organic solvent extraction; subsequently, enriched by multidimensional preparative chromatographic techniques. *Trans-α*-damascone was isolated from black tea aroma and was enriched for direct chirospecific analysis by medium pressure liquid chromatography followed by multidimensional preparative capillary gas chromatography.

Enantiomeric purity and enantiomeric excess (ee) are the usual terms used to describe the results of analyzing enantiomers. Enantiomeric purity is defined as the measured ratio (expressed as a percentage) of the detected enantiomers, whereas ee values describe the relative difference of the separated enantiomers (expressed as a percentage). Wang *et al.* (1994) enantioselectively isolated the main aroma components of oolong and black tea, linalool and four diastereomers of linalool oxides (LOs), by capillary gas chromatography, using a column coated with an optically active liquid phase, permethylated *β*-cyclodextrin. They found that the R/S ratio varied among linalool and LOs, and among the different types of tea, the ratio for a particular compound also being different. However, the complete patterns of R/S ratio were similar in the semi-fermented and fermented teas, respectively. Using a specific cultivar of black tea, the R/S ratio for each of the five compounds was compared in the free state in black tea with that of an aglycone of the glycoside in fresh tea leaves or in black tea. While the ee values of the compounds varied, those for a specific compound were similar, except for linalool, regardless of their free or combined state. Their results showed that LOs are not directly transformed from linalool, but are formed enzymatically from glycoside precursors. The configuration and odor of linalool oxide enantiomers are shown in Table 5.3.

It is well known that the stereochemistry of a flavor compound can determine its sensory properties as well as its aroma intensity. Considerable differences in sensory properties have been found for the enantiomers of methyl jasmonate and epijasmonate. The former has an odor threshold 400 times lower than that of the

Table 5.3 Configuration and Odor of Linalool Oxide Enantiomers (Wang *et al.* 1994).

Compound	Configuration	Odour
Linalool oxide I (furanoid)	2R, 5R	leaf, earthy
	2S, 5S	sweet, floral, creamy
Linalool oxide II (furanoid)	2R, 5S	leaf, earthy
	2S, 5R	sweet, floral, creamy
Linalool oxide III (pyranoid)	2S, 5R	sweet, floral, creamy
	2R, 5S	earthy
Linalool oxide IV (pyranoid)	2S, 5S	sweet, floral, creamy
	2R, 5R	earthy

latter (Acree *et al.* 1985). Different sensory properties of enantiomeric compositions of theaspiranes have also been found in a green tea sample. The 2R, 5R enantimeric structure showed a camphor and minty odor, 2S, 5S structure showed a camphor and woody odor, while the 2S, 5R and 2R, 5S structures showed the camphor and naphthalene-like as well as intense fruit and black currant odor, rspectively (Schmidt *et al.* 1992).

Subsequently Wang *et al.* (1996) resolved two epimers of methyl jasmonate with capillary gas chromatography, using heptakis (2,3,6-tri-O-methyl)-β-cyclodextrin as the chiral stationary phase. In the tea volatile concentrates, both of these epimers were present as only one enantiomer, their absolute configurations being ascertained to be (−)-(1R,2R)-methyl jasmonate and (+)-(1R,2S)-methyl epijasmonate. The thermal isomerization of methyl epijasmonate to methyl jasmonate was also clarified by optically resolved gas chromatography, and was shown to occur at the asymmetric carbon of the C-2 position that is connected to the carbonyl group.

5.2. Separation and Identification of Glycosides

Generally because glycosides are water soluble, they can be analyzed by HPLC directly. However, for structure determination, the glycosides in tea extract have to be isolated and purified. The crude glycosides can be extracted by water first, and then separated on a Sephadex LH-20 column, and finally, the purified glycoside can be analyzed by IR, MS, and NMR. For the analysis of related aglycone, the isolation is achieved with acid or enzymatic hydrolysis, and subsequent GC and GC-MS analyses. With this method, Guo *et al.* (1992) first identified the structure of linalyl glycoside as (S)-linalyl β-primeveroside in plants. At the same time, three glycosides, 6-O-β-D-xylopyranosyl-β-D-glucopyranosides (β-primeverosides) of the aroma constituents, linalool, 2-phenylethanol, and benzyl alcohol, were isolated as aroma precursors from tea leaves (*Camellia sinensis* var. *sinensis* cv. Shuixian and Maoxie, cultivars for oolong tea).

Matsumura *et al.* (1997) investigated the role of diglycosides as tea aroma precursors. They compared the hydrolysis rate of each of the 12 synthesised glycosides by a crude tea enzyme, and found that among a series of diglycosides with the same aglycon, the hydrolysis rates of primeveroside are high without exception. They found the primeverosidase in tea leaves was specific to the glycoside and to the aglycon. This specificity of the tea glycosidases explains the authors' previous experimental results in which the composition of volatiles freed from the glycoside fraction of tea leaves by crude tea glycosidase differed greatly from that obtained with commercial β-glucosidase or a nonspecific glycosidase (Morita *et al.* 1994). Halder and Bhaduri (1997) reported another result on glycosidases from tea leaf (*Camellia sinensis*). A wide range of glycosidase activities could be detected in the acetone powder extract of tea leaves of Assam tea cultivars. They found that β-galactosidase (EC 3.2.1.23) was the most active among these enzymes at the fermentation temperature needed for black tea processing and at an acidic pH. But when assayed at the optimum conditions for glucosidase at pH 6.1 and at 37°C, the enzyme showed activity that is comparable to the activity of β-galactosidase at pH 4.0. Different β-glucosidase activities in various acetone powders prepared from different tea varieties have been found by Wang and You (1995). Based on the above analysis, it seems that tea leaves contain a battery of glycosidases that in combination are probably fully capable of releasing volatile flavoring compounds from complex glycosides. To summarize the above research results, 13 kinds of glycosides were presented as the precursors of glycosidic aroma in tea leaves (Watanabe and Singh, 1997).

5.3. Quality Assessment

In practice, to detect the potent odorants in tea, gas chromatography olfactometry (GCO) is often used. This is a method of examination of the effluent from a capillary gas chromatography column by smell. In this technique, a splitter is used at the end of the GC column, where one part of the elute flows towards the detector (normally FID) and the other part of the elute to the nose. There is a quantitative GCO procedure for determining the potency of odorants in food extracts known as aroma extract dilution analysis (AEDA). Hall and Anderson (1983) found a good correlation between the sensory ranking by taste panels and odor intensities by using the dilution olfactometric technique. An application of the AEDA technique to the characterisation of the most odor-active compounds in tea has been reported (Guth and Grosch, 1993). Comparing the potency of aroma components by using AEDA technique, they found that the abundance of (Z)-3-hexenal and (Z)-1,5-octendien-3-one and a low level of linalool in green tea were key differences between green tea and black tea aromas.

Several sensory properties in tea aromas such as fresh floral, sweet floral, citrus, sweet fruity, fresh green and roasted were statistically correlated to the GC profiles of volatile flavor components by multivariate calibration methods (Togari *et al.* 1995a). Stepwise multiple linear regression analysis (MLR), principal component regression (PCR) and partial least squares (PLS) regression analysis were comparatively applied to the sensory scores and the 77 GC peaks. The result showed that MLR, PCR and PLS succeeded in calculating highly predictable regression models for most aroma properties. Among 77 volatile components, linalool and jasmine lactone, 2-phenylethanol and jasmine lactone, and 2-phenylethanol contributed highly to the prediction of fresh floral, sweet floral and sweet fruity smells respectively. Linalool oxide

(II) heavily contributed to citrus but the sign for linalool oxide (I) became negative due to multicolinearity existing between the two isomers. The positive contribution of two pyrrole derivatives for roasted tea agreed well with expectations.

Genetic algorithms (GA) incorporating a randomisation process before starting the optimisation process is regarded as a promising method to find true optimum combination of variables easily and steadily (Leardi *et al.* 1992). Aishima *et al.* (1996) applied this method to optimising sets of aroma components in black teas for calculating MLR models to predict sensory scores. The result showed that comparing coefficients calculated from stepwise MLR and GA-MLR, the predictability of GA-MLR was generally superior to those from stepwise MLR.

Unsupervised and supervised pattern recognition techniques were applied to differentiate the three tea categories based on volatile components of green tea, oolong tea and black tea. Three distinct clusters each corresponding to green tea, oolong tea and black tea were observed in the dendrogram and the principal component (PC) score plot. However, a subcluster of oolong tea was observed in the vicinity of the black tea cluster in both the dendrogram and the PC plot. The first and second PC corresponded to the fermentation products and aroma components originally contained in tea leaves, respectively. Both the PLS analysis and linear discriminant analysis correctly differentiated tea samples into the three categories. (E)-2-Hexenal, the major fermentation product from unsaturated fatty acids, was the most efficient for discrimination (Togari *et al.* 1995b).

Horie *et al.* (1993) introduced chemical sensors to evaluate the quality of Japanese green tea (Sencha) objectively. They found that the aroma of green tea sensitively reflects quality deterioration. Teas with bad aroma, judged from sensory tests, show higher odor values than teas of normal aroma with this sensor. Shimoda *et al.* (1995) compared the volatile compounds among different grades of green tea. The result showed that *D*-nerolidol, 6-methyl-α-ionone, methyl jasmonate, coumaran, indole, and coumarin were possible contributors to a typical green tea odor.

Although it is generally believed that there is no single key compound for the characteristic aroma of tea, people are always interested in finding out some simple relationships from complicated tea aroma components in order to represent the tea quality. Below are some examples.

Wickremasinghe-Yamanishi Ratio: Yamanishi *et al.* (1968) compared the compositions of volatiles in steam distillates of black tea from different geographical locations. They found variations in the relative amounts of linalool plus linalool oxides, and in the ratios of the sum of the peak areas of components having retention times lower than that of linalool to the sum of the peak areas of components appearing after linalool. Wickremasinghe *et al.* (1973) and Yamanishi *et al.* (1978b) used this ratio to evaluate the quality of black tea. The smaller the ratio, the better the quality of black tea.

Mahanta Ratio: Biochemical correlations between volatile flavour constituents of CTC black teas manufactured from different fresh leaves plucked with different standards and at different altitudes were made by Mahanta *et al.* (1988). They found that in general, fine plucking produced more monoterpenoids and less non-terpenoids than coarse plucking. Thus the ratio between terpenoid and non-terpenoid may be crucial for identifying high quality flavor in tea. In Assam teas the ratio of terpenoids and non-terpenoids lies in the range of 0.12–0.18, whereas in Darjeeling teas, which are more flavorsome, the range is 0.62–0.90.

Owuor Flavor Index: This index is based on the concept of Wickremasinghe-Yamanishi ratio. In this index, volatile flavor compounds (VFC) can be broadly classified into two groups: VFC (I) and VFC (II). VFC (I) is dominated by C6 aldehydes and alcohols that are lipid degradation products imparting a green grassy smell to black tea and correlating negatively with price evaluation (Owour *et al.* 1988). VFC (II) constituting mainly of terpenoid compounds and products from amino acids contributes principally to the desirable sweet flowery aroma quality (Saijo and Takeo, 1970, 1973). The ratio of VFC (II) to VFC (I) is called the "flavor index" (Owour *et al.* 1989). Clones with higher levels of this index will have a better aroma quality.

Yamanishi-Botheju ratio: This ratio is based on the gas chromatographic peak areas of linalool and (E)-2-hexenal, ignoring all the other VFC. This ratio was shown to have a relationship with prices at the auction of some Sri Lankan orthodox black teas (Yamanishi *et al.* 1989a).

5.4.   Chemotaxonomic Method for Classification of Tea Clones

Tea (*Camellia sinensis* (L.) O. Kuntze) has many varieties. These can be classified into *assamica* and *sinensis* varieties by observation of the physical appearance of the plants. However, in order to improve both yield and quality, many clones have been developed based on the same variety or crossed variety by various breeding techniques, and because of this some clones look very similar and are occasionally confused. Takeo (1981a, b, 1983a) suggested a chemotaxonomic method, called the terpene index, for classifying tea clones. The terpene index (TI ratio) was defined as the ratio of the gas chromatographic peak areas of linalool to those of linalool plus geraniol. These studies showed that the content of monoterpene alcohols differs greatly according to the kind of tea plant. In the Assamese tea family, the major ingredients of terpene alcohols are linalool and its oxides, while in the Chinese tea family geraniol is the major component. The Darjeeling tea of India contains both linalool and geraniol. The TI ratio for each tea, therefore, is closely related to the place of cultivation, i.e., the propagation route of tea types. The terpene index has been improved by the inclusion of (E)-geranic acid by Owuor *et al.* (1987). The modified TI ratio has been shown to be unique to certain clones irrespective of variation in agronomic or manufacturing practices except for the plucking standard (Owuor *et al.* 1987). Therefore the TI index is believed to be a reliable chemotaxonomic statistic for differentiating clonal teas. A tea with a high TI ratio has a bright and brilliant aroma, while the TI ratio of Darjeeing tea that is low, has a solemn, rosy and strong aroma (Takeo, 1996).

You *et al.* (1990) analysed the volatile flavour constituents of the tea flower and the major aroma compounds were identified as 2-pentanol, 2-heptanol, bezaldehyde, linalool and its oxides, acetophenone, geraniol, nerol and 2-phenylethanol. Baik *et al.* (1996) analysed the volatile constituents in the green tea flower. 56 compounds in the sample were separated and identified as 22 hydrocarbons, 14 alcohols, 6 aldehydes, 5 esters and 9 others. Substances in higher concentrations were heneicosyl formate,

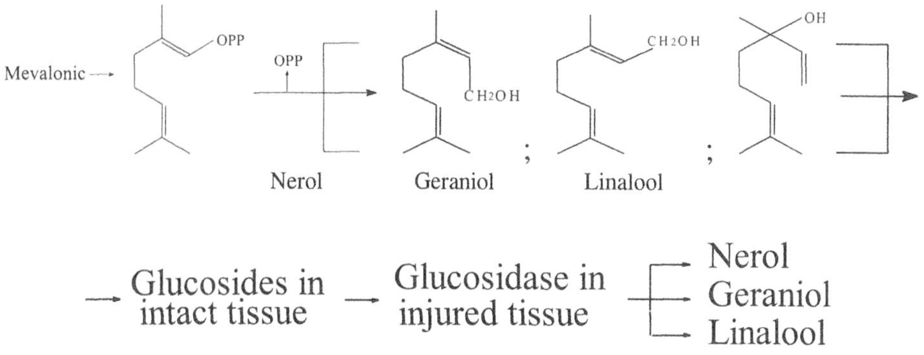

Nerol   Geraniol   Linalool

→ Glucosides in → Glucosidase in ┌→ Nerol
  intact tissue   injured tissue ├→ Geraniol
                                    └→ Linalool

Figure 5.5 Metabolic pathways of monoterpene alcohols

α-phenyl etheyl alcohol and acetophenone. Based on the investigation of the changes of aroma compounds in tea flowers during the blossoming process, You *et al.* (1990) put forward the concept of the terpene index ratio of tea flowers. They found that the total amount of the terpenes varied remarkably during the blossoming process of tea flower, but the TI ratio remained more constant in tea flowers than that in tea leaves. Because the content of nerol is much higher in tea flowers than that in tea leaves and it shares the same precursor with linalool and geraniol (Takeo *et al.* 1985) as shown in Fig. 5.5, the TI ratio of tea flower has been modified slightly as follows: TI = (Linalool + Linalool oxides)/(Linalool + Linalool oxides + Geraniol + Nerol). You *et al.* (1992b) also found that in the same clone the TI value changed greatly before the steam formed, but remained almost stable afterwards. The results of hybrid experiments showed that the TI value of the F1 generation varied within the range of the values of the parents.

By using the TI ratio, which is specific for the genetic characteristics, the genetic relationship among tea cultivars growing in China were studied by Takeo *et al.* (1992). The results showed that tea plants growing in Yunnan province which is supposed to be the original source of tea plants had a TI of near 1.0 and showed a property akin to var. *assamica*. Some clones growing in Zhejiang and Fujian provinces of China had TI values of 1.0-0.2 and it was thought that these clones might be the most remote cultivars from the original cultivars of Yunnan province in China. TI values of tea cultivars collected in China displayed the dispersion route of tea plant from the native place to various cultivation areas. Figure 5.6 shows the dispersion routes of tea plant in China and to surrounding areas.

You and Wang (1993) investigated the changes of aroma pattern of the Fuding cultivar after it was transplanted to other places in China. The result showed that the difference of the aroma pattern of related green tea depends on the distance between the original place and the site of transplantation. The longer the transplanted distance is, the larger the difference in aroma pattern will be.

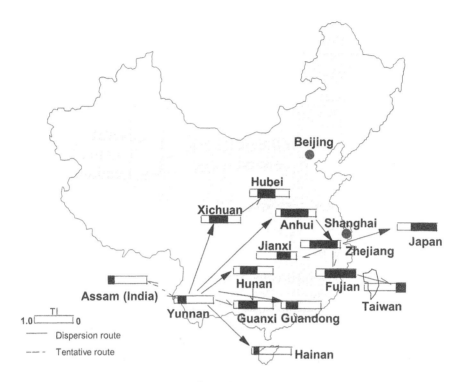

Figure 5.6 The dispersion routes of tea plant in China and to surrounding areas

5.5. Functional Effects

The antibacterial activities of green tea aroma components were first reported by Kubo *et al.* (1992). The antimicrobial activity of the 10 most abundant volatile components of green tea flavor was examined. Most of the volatiles tested inhibited the growth of one of the most important cariogenic bacteria, *Streptococcus mutans*. Among them, nerolidol was the most potent; linalool was the least effective. In addition, indole significantly enhanced the activity of δ-cadinene and caryophyllene against *S. mutans*. These two sesquiterpene hydrocarbons also showed potent activity against a dermatomycotic bacterium, *Propionibacterium acnes*. Lastly, but most importantly, indole inhibited the growth of all of the gram-negative bacteria tested: *Pseudomonas aeruginosa*, *Enterobacter aerogenes*, and *Escherichia coli*. Later on Muroi and Kubo (1993) investigated in detail the combinatorial effects of indole and other aroma compounds in green tea against *Streptococcus mutans*. They found that the bactericidal activities increased from 128-fold to 256-fold after combining sesquiterpene hydro-carbons (δ-cadinene and β-caryophyllene) with indole. Synergetic effects were also found for linalool, geraniol and nerolidol when they were combined with indole.

A series of long-chain alcohols were tested to gain new insight into their effects on *Streptococcus mutans*. Maximum activity seems to depend on the hydrophobic chain length from the hydrophilic hydroxyl group. Among the alcohols tested, 1-tridecanol was found to be the most effective for controlling this cariogenic bacterium (Kubo *et al.* 1995a).

The cytotoxicity of green tea aroma components against two human carcinoma cell lines was investigated by Kubo and Morimitsu (1995b). Among these compounds, nerolidol, β-ionone, δ-cadinene, and β-caryophyllene were found to exhibit moderate cytotoxicity *in vitro*. The nerolidol, β-ionone, β-caryophyllene and δ-cadinene showed the cytotoxicity to BT-20 breast and Hela epithelioid cervix carcinoma cells (IC50 = 2.96–3.92 *ug*/ml) (Kubo and Morimitsu, 1995).

Young *et al.* (1994) investigated the effect of flavors of tea on the growth of *Bacillus subtilis* bacterium. They found that thujone, caryophyllene and farnesol showed bactericidal effects on *Escherichia coli*, *Enterobacter aerogenes*, *Vibrio parahaemolyticus*, *Pseudomonas aeruginosa*, *B. subtilis* and *Staphylococcus aureus* using the paper disc method (8 mm diameter). The mixture of caryophyllene and farnesol was more bactericidal than the mixture of thujone, caryophyllene and farnesol or each compound separately. Caryophyllene and farnesol showed a strong bactericidal effect (diameter of inhibition zone greater than 40 mm) for *V. parahaemolyticus*, *E. aerogenes* and *B. subtilis*.

These results may have significant implications for the future development of tea essential oil as an antimicrobial agent.

5.6.   Chemical Communication among Tea Plant, Tea Geometrid, and *Apanteles sp.* Wasp

Plant volatiles play an important role in the foraging behaviour of entomophagous insects. Xu (1997) investigated the role of volatile compounds of tea in the tritrophic chemical communication among tea plants, tea geometrid, and *Apanteles sp.* wasp. The electroantennogram (EAG) response of adult tea geometrid to alcohols was higher than that to aldehydes. C6 compounds elicited the maximum EAG responses. The volatile compounds of 1-penten-3-ol, (Z)-3-hexen-1-ol, n-pentanol, (E)-2-hexenal, n-heptanol, and geraniol aroused stronger EAG responses; methylsalicylate, nerolidol weaker EAG responses. In contrast, the EAG response of *A. sp.* to aliphatic aldehydes was higher than that to alcohols. The attraction of *A. sp.* to different volatile compounds showed that C5-C6 compounds had the strongest attractiveness, terpenoid and aromatic compounds a medium attractiveness while indole and linalool had the least attractiveness. These results may be relevant to integrate pest management research in the future.

Acknowledgements: The authors would like to thank Mr Bob Howker and Dr Chris Powell (Unilever Research Colworth Laboratory, UK) for their literature support.

REFERENCES

Acree, T.E., Nishida, R. and Fukami, H. (1985) Odor thresholds of the stereoisomers of methyl jasmonate. *J Agric. Food Chem.*, 33, 425–427.

Aisaki, H., Kosuge, M. and Yamanishi, T. (1978) Comparison of the flavours of Chinese "Keemum" black tea and Ceylon black tea. *Agric. Biol. Chem.*, 42, 2157–2159.

Aishima, T., Togari, N. and Leardi, R. (1996) Selection of aroma components to predict sensory quality of Kenyan black teas using a genetic algorithm for multiple linear regression models. *Food Sci. Technol. Int.*, **2**, 124–126.

Anan, T. (1983) The lipids of tea. *Japan Agric. Res. Quarterly*, **16**, 253–257.

Baik, S.O., Bock, J.Y., Han, S.B., Cho, K.S., Bang, G.P. and Kim, II K. (1996) Analysis of volatile flavour constituents in green tea flower. *Anal. Sci. and Technol.*, **9**, 331–335.

Baruah, S., Hazarika, M., Mahanta, P.K., Horita, H. and Murai, T. (1986) Effect of plucking intervals on the chemical constituents of CTC black teas. *Agric. Biol. Chem.*, **50**, 1039–1041.

Clark, T.J. and Bunch, J.E. (1997) Determination of volatile acids in tobacco, tea, and coffee using derivatization-purge and trap gas chromatography-selected ion monitoring mass spectrometry. *J Chromatographic Sci.*, **35**, 206–208.

Dougan, J., Glossop, E.J., Howard, G.E. and Jone, B.D. (1978) A study of the changes occurring in black tea during storage. *Tropical Products Institute Bulletin*, **116**, 1–54.

Erman, W.F. (1985) Chemical of the monoterpenes. Gassman, P.G. (ed), *Studies in Organic Chemistry 11*, Dekker, New York, pp. 34–126.

Fischer, N., Nitz, S., and Drawert, F. (1987) Bound flavour compounds in plants. 2: Free and bound flavour compounds in green and black tea (*Camellia sinensis*). *Z. Lebensm. Unters Forsch.*, **185**, 195–201.

Fu, H., Guo, W. and Luo, S. (1991) Chemical composition of the aroma of Zhulan tea (Chinese). *J. Tea Sci.*, **11**, 59–61.

Guidotti, M. (1997) Determination of volatile organic compounds in tea. *Industrie delle Bevande*, **26**, 19–22.

Guo, W., Sakata, K., Yagi, A., Ina, K. and Luo, S.J. (1992) Preparation of congou black tea from stale green tea. *Biosci. Biotech. Biochem.*, **56**, 992–993.

Guo, W., Hosoi, R., Sakada, K., Watanabe, N., Yagi, A., Ina, K. *et al.* (1994) (S)-Linaly, 2-phenylethyl, and benzyl disaccharide glycosides isolated as aroma precursors from oolong tea leaves. *Biosci. Biotech. Biochem.*, **58**, 1532–1534.

Guo, W., Yamaguchi, K., Watanabe, N., Usui, T., Luo, S.J. and Sakata, K. (1995) A primeverosidase as a main glycosidase concerned with the alcoholic aroma formation in tea leaves. *Biosci. Biotech. Biochem.*, **59**, 962–964.

Guth, H. and Grosch, W. (1993) Identification of potent odourants in static headspace samples of green tea and black tea powders on the basis of aroma extract dilution analysis (AEDA). *Flavour Fragrance J*, **8**, 173–178.

Habu, T., Flath, R.A., Mon, T.R. and Morton, J.F. (1985) Volatile components of Rooibos tea (*Aspalathus linearis*). *J. Agric. Food Chem.*, **33**, 249–254.

Halder, J. and Bhaduri, A. (1997) Glycosidases from tea-leaf (Camellia sinensis) and characterisation of beta-galactosidase. *J. Nutr. Biochem.*, **8**, 378–384.

Hall, G. and Anderson, J. (1983) Volatile fat oxidation products 1. Determination of odour thresholds and odour intensity functions by dynamic olfactometry. *Lebensm. Wiss. Unters. Technol.*, **16**, 354–361.

Hara, T. and Kubota, E. (1982) Changes in aroma components of green tea after storage. *Nippon Nogeikagaku Kaishi*, **56**, 625–630.

Hara, T. (1985) Effect of various drying methods on the green tea aroma. *Tea*, **38**, 2–5.

Hara, T., Kubota, E. and Horita, H. (1987) Off-flavour components in stored packaged green tea. *Nippon Nogeikagaku Kaishi*, **61**, 471–473.

Hara, T. (1989) Studies on the firing aroma and off-flavor components of green tea. *Bull. Natl. Res. Inst. Veg. Ornam. Plants and Tea*, **3**, 9–54.

Hatanaka, A. and Harada, T. (1973) Formation of *cis*-3-hexenal, *trans*-2-hexenal and *cis*-3-hexenol in macerated *Thea sinensis* leaves. *Phytochem.*, **12**, 2341–2346.

Hatanaka, A., Kajiwara, T., Sekiya, J. and Inoue, S. (1982) Solubilization and properties of the enzyme leaving 13-L-hydroperoxylinolenic acid in tea leaves. *Phytochem.*, **21**, 13–17.

Hatanaka, A. (1993) The biogeneration of green odour by green leaves. *Phytochem*, **34**, 1201–1218.

Hazarika, M. and Mahanta, P.K. (1983) Some studies on carotenoids and their degradation in black tea manufacture. *J. Sci. Food Agric.*, **34**, 1390–1396.

Heins, J. Th., Maarse, H., ten Noever de Branw, M.C. and Weurman, C. (1966) Direct food vapour analysis and component identification by a coupled capillary GLC-MS arrangement. *J. Gas Chromatog.*, **4**, 395–397.

Horie, H., Fukatsu, S., Mukai, T., Goto, T., Kawanaka, M. and Shimohara, T. (1993) Quality evaluation on green tea. *Sensory and Actuators B-Chemical*, **13**, 451–454.

Horita, H. and Hara, T. (1985a) A quantitative analysis of heating aroma components of green tea detected by flame thermionic detector. *Study of Tea*, **67**, 44–46.

Horita, H. and Hara, T. (1985b) Analysis of headspace volatile components of tea using Tenax TA trapping system. *Study of Tea*, **68**, 17–24.

Horita, H. and Hara, T. (1986) The light-produced aroma components of green tea. *Study of Tea*, **69**, 58–67.

Horita, H. (1987) Off-flavour components of green tea guring preservation. *Jap. Agric. Res. Quarterly*, **21**, 192–197.

Janssens, L., Pooter, H.L. De., Schamp, N.M. and Vandamme, E.J. (1992) Production of flavours by microorganisms. *Process Biochem.*, **27**, 195–215.

Jennings, W.G. and Filsoof, M. (1977) Comparsion of sample preparation techniques for gas chromatographic analysis. *J. Agric. Food Chem.*, **25**, 440–445.

Kawakami, M. (1982) Ionone series compounds from carotene by the thermal degradation in aqueous medium. *Nippon Nogeikagaku Kaishi*, **56**, 917–921.

Kawakami, M. and Yamanishi, T. (1983) Flavour constituents of Longjing tea. *Agric. Biol. Chem.*, **47**, 2077–2083.

Kawakami, M., Kobayashi, A. and Yamanishi T. (1987a) Flavour constituents of the fermented teas Goishi-cha and Awa-cha. *Nippon Nogeikagaku Kaishi*, **61**, 345–352.

Kawakami, M., Chairote, G. and Kobayashi, A. (1987b) Flavour constituents of pickled tea, miang, in Thailand. *Agric. Biol. Chem.*, **51**, 1683–1687.

Kawakami, M. and Shibamoto, T. (1991a) The volatile constituents of piled tea – Toyama kurocha. *Agric. Biol. Chem.*, **55**, 1839–1847.

Kawakami, M. and Kobayashi, A. (1991b) Volatile components of green mate and roasted mate. *J. Agric. Food Chem.*, **39**, 1275–1279.

Kawakami, M., Kobayashi, A. and Kator, K. (1993a) Volatile constituents of rooibos tea (Aspalathus linearis) as affected by extraction process. *J. Agric. Food Chem.*, **41**, 633–636.

Kawakami, M., Kobayashi, A. and Yamanishi, T. (1993b) The effect on tea aroma composition by the different method. In: *Proc. Symp. on the Agricultral Chemical Society of Japan*, Sendai, p. 351.

Kawakami, M., Ganguly, S.N., Banerjee, J. and Kobayashi, A. (1995) Aroma composition of oolong tea and black tea by brewed extraction method and characterising compounds of Darjeeling tea aroma. *J. Agric. Food Chem.*, **43**, 200–207.

Kawakami, M. (1997) Comparison of extraction techniques for characterising tea aroma and analysis of tea by GC-FTIR-MS. In: *Plant volatile analysis* Linkskens H.F. and Jackson J.F. (eds), Springer, Saladruck, Berlin. pp. 211–229.

Kinugasa, H. and Takeo, T. (1989) Mechanism of retort smell developmet during sterilizattion of canned tea drink and its deodorization measure. *Nippon Nogeikagaku Kaishi*, **63**, 29–35.

Kinugasa, H. and Yakeo, T. (1990) Deterioration mechanism for tea infusion aroma by retort pasteurization. *Agric. Biol. Chem.*, **54**, 2537–2542.

Kinugasa, H., Takeo, T. and Yano, N. (1997) Differences of flavour components found in green tea canned drinks made from tea leaves plucked on different matured stage *Nippon Shokuhin Kagaku Kaishi*, **44**, 112–118.

Kobayashi, A., Kawamura, M., Yamamoto, Y., Shimizu, K., Kubota, K. and Yamanishi, T. (1988) Methyl epijasmonate in the essential oil of tea. *Agric. Biol. Chem.*, **52**, 2299–2303.

Kobayashi, A., Kubota, K., Joki, Y., Wada, E. and Wakabayashi, M. (1994) (Z)-3-hexenyl-D-glucopyranoside in fresh tea leaves as a precursor of green tea. *Biosci. Biotech. Biochem.*, **58**, 592–593.

Kosuge, M., Aisaka, H. and Yamanishi, T. (1981) Flavour constituents of Chinese and Japanese pan-fired green tea. *J. Nutr. and Food*, **34**, 545–549.

Kubo, I., Muroi, H. and Himejima, M. (1992) Antimicrobial activity of green tea components and their combination effects. *J. Agric. Food Chem.*, **40**, 245–248.

Kubo, I., Muroi, H. and Kubo, A. (1995a) Structural functions of antimicrobial long-chain alcohols and phenols. *Bioorg. and Med. Chem.*, **3**, 873–880.

Kubo, I. And Morimitsu, Y. (1995b) Cytotoxicity of green tea flavor compounds against two solid tumor cells. *J. Agric. Food Chem.*, **43**, 1626–1628.

Kubota, K., Kumeuchi, T., Kobayashi, A., Osawa, Y., Nakajima, T. and Okamoto, Y. (1996) Effect of refining treatment with microwave heating drum on aroma and taste of green tea. *Nippon Shokuhin Kagaku Kaishi*, **43**, 1197–1204.

Leardi, R., Boggia, R. and Terrile, M. (1992) Genetic algorithms as a strategy for feature selection. *J Chemeometrics*, **6**, 267–281.

Lee, J.G., Kwon, Y.J., Chang, H.J., Kwang, J.J., Kim, O.C. and Choi, Y.H. (1997) Volatile components of green tea (*Camellia sinensis* L. *var*. Yabukita) determined by purge-and-trap headspace sampler. *Han' guk Sikp' um Yongyang Hakhoechi*, **10**, 25–30.

Likens, S.T. and Nickerson, G.B. (1964) Detection of certain hop oil constituents in brewing products. *Proc. Am. Soc. Brew Chem.*, 5–13.

Liu, Q.J., Horita, H., Hara, T., Yagi, A. and Ina, K. (1989) Flavor constituents of Chinese microbial-fermented tea pu-er cha. *Nippon Shokukin Kagaku Kogaku Kaishi*, **36**, 486–489.

Luo, S., Fu, H. and Guo, W. (1990) Chemical composition of the aroma of Chinses Youzi tea. *J. Tea Sci.*, **10**, 45–48.

Luo, S.J., Guo, W. and Fu, H. (1991) Changes of aroma during process of Chinese jasmine tea. *Proc. Int. Symp. on Tea Sci.*, Aug 26–29 Shizuoka, Japan, pp. 57–61.

Maarse, H. and Kepner, R. (1970) Changes on composition of volatile trpenes in Douglas fir needls during maturation. *J. Agric. Food Chem.*, **18**, 1095–1101.

Maga, J.A. (1976) Lactones in foods. *CRC Crit. Rev. Food Sci. Nutr.*, **10**, 1–56.

Maga, J.A. (1982) Pyrazines in foods: an update. *CRC Crit. Rev. Food Sci. Nutr.*, **16**, 1–48.

Mahanta, P.K., Baruah, S., Owuor, P.O. and Murai, T. (1988) Flavour volatiles of Assam CTC black teas manufactured from different plucking standards and Orthodox teas manufactured from different altitudes of Darjeeling. *J. Sci. Food Agric.*, **45**, 317–324.

Matsumura, S., Takahashi, S., Nishikatani, M., Kubota, K. and Kobayashi, A. (1997) The role of diglycosides as tea aroma precursors: Synthesis of tea diglycosides and specificity of glycosidases in tea leaves. *J. Agric. Food Chem.*, **45**, 2674–2678.

Masuzawa, T. (1974) Effects of light on the qualities of green tea. *Tea Res. J.*, **41**, 54–58.

Mizugaki, M., Uchiyama, M. and Okui, S. (1965) Metabolism of hydroxy fatty acids. V. Metabolic conversion of homoricinoleic and homoricinelaidic acids by *Escherichia coli* K 12. *J. Biochem.*, **58**, 273–278.

Moon, J.H., Watanabe, N., Sakata, K., Yagi, A., Ina, K. and Luo, S. (1994) Studies on the aroma formation mechanism of oolong tea. 3. Trans-linalool and cis-linalool 3,6-oxide 6-O-beta-D-xylopyranosyl-beta-D-glucopyranosides isolated as aroma precursors from leaves for oolong tea. *Biosci. Biotech. Biochem.*, 58, 1742–1744.

Morita, K., Wakabayashi, M., Kubota, K., Kobayashi, A. and Herath, N.L. (1994) Glycoside precursor of tea aroma. 2. Aglycon constituents in fresh tea leaves cultivated for green and black tea. *Biosci. Biotech. Biochem.*, 58, 687–690.

Mosandl, A., Heusinger, G. and Gessner, M. (1986) Analytical and sensory differentiation of 1-octen-3-ol enantiomers. *J. Agric. Food Chem.*, 34, 119.

Mosandl, A. and Gunther, C. (1989) Stereoisomeric flavor compounds. 20. Structure and properties of *-lactone enantiomers. *J. Agric. Food Chem.*, 37, 413.

Mulder, G.J. (1838) Chemische untersuchung des chinesischen und javanischen Thees. *Annalen Physik Chemie Leipzig*, XIII (161), 161–180

Muritu, J.W., Munavu, R.M., Owuor, P.O. and Njuguna, C.K. (1988) Distribution of lipids in the young shoots of tea (*Camellia sinensis* L.). *Tea*, 9, 76–80.

Muroi, H. and Kubo, I. (1993) Combination effects of antibacterial compounds in green tea flavour against *Streptococcus mutans*. *J. Agric. Food Chem.*, 41, 1102–1105.

Nobumoto, Y., Kubota, K., Kobayashi, A. and Yamanishi, T. (1990) Structure of *-farnesene in the essential oil of oolong tea. *Agric. Biol. Chem.*, 54, 247–248.

Nobumoto, Y., Kubota, K. and Kobayashi, A. (1993) Lactones newly identified in the volatiles of pouchong-type semi-fermented tea. *Biosci. Biotech. Biochem.*, 57, 79–81.

Nose, M., Nakatani, Y. and Yamanishi, T. (1971) Studies on the flavor of green tea. Part IX. Identification and composition of intermediate and high boiling constitutents in green tea flavor. *Agric. Biol. Chem.*, 35, 261–271.

Ogawa, K., Moon, J-H., Guo, W., Yagi, A., Watanabe, N. and Sakata, K. (1995) A study on tea aroma formation mechanism: alcoholic aroma precursor amounts and glycosidase activity in parts of the tea plant. *Z. Naturforsch.*, 50c, 493–498.

Okui, S., Uchiyama, M. and Mizugaki, M. (1963) Metabolism of hydroxy fatty acids. II. Intermediates of the oxidative breakdown of methyl ricinoleate by genus *Candida*. *J. Biochem.*, 54, 536–540.

Omori, M. (1997) Characterization of tea flavor. *Koryo*, (193), 59–74.

Otto, J. and Thomas, W. (1967) Termodynamische Eigenschaften von Gasen. In: Landolt-Bornstein: *Zahlenwerte und Funktionen aus Physik, Chemie, Astronomie, Geophysik und Technik*. 6 Auflage, Teil 4, H. Hausen ed., Springer-Verlag, Berlin, pp. 172–416.

Owuor, P.O., Takeo, T. and Horita, H. (1987) Differenation of clonal tea by terpene index. *J. Sci. Agric.*, 40, 341–345.

Owour, P.O., Tsushida, T., Harita, H. and Murai, T. (1988) Effect of geographical area of production on the composition of volatile flavour compounds in Kenya clonal black CTC teas. *Experimental Agric.*, 24, 227–235.

Owour, P.O., Wanyiva, J.O., Niuru, K.E., Munavu, R.M. and Bhatia, B.M. (1989) Comparsion of the chemical composition and quality changes due to different withering methods on black tea manufacture. *Trop. Sci.*, 29, 207–213.

Parliament, T.H., Epstein, M.F. (1973) Organoleptic properties of some alkyl-substituted alkoxy- and alkylthiopyrazines. *J. Agric. Food Chem.*, 21, 714–716.

Polesello, S., Lovati, F., Rizzolo, A. and Rovida, C. (1993) Supercritical fluid extraction as a preparative tool for strawberry aroma analysis. *J. High Res. Chromatog.*, 16, 555–559.

Saijo, R. and Takeo, T. (1970) The formation of aldehyde from amino acids by tea leaves extracts. *Agric. Biol. Chem.*, 34, 227–23.

Saijo, R. (1972) *Tea Research Journal*, 44, 31–37.

Saijo, R. and Takeo, T. (1973) Volatile and non volatile forms of aroma compounds and their changes due to injury. *Agric. Biol. Chem.*, 37, 1367–1373.

Sanderson, G. and Graham, H. N. (1973) On the formation of black tea aroma. *J. Agric. Food Chem.*, **21**, 576–584.

Sawamura, S. and Masuzawa, T. (1982) Refiring of over-steamed tea (Fukamushi-cha). *Tea Res. J.*, **55**, 63–67.

Saxby, M.J. (1982) Taints and off flavors in foods. In: Morton I.D. and Macleod A.J. (eds), *Food Flavours*, Part A: Introduction. Elsevier, New York, 439–457.

Schmidt, G., Full, G., Winterhalter, P. and Schreier, P. (1992) synthesis and enantio differentiation of isomeric theaspiranes. *J. Agric. Food Chem.*, **40**, 1188–1191.

Schreier, P. (1988) Analysis of black tea volatiles. In: Linskens H.F. and Jackson J.F. (eds). *Analysis of Nonalcoholic Beverages*, Springer, Saladruck, Berlin, pp. 296–320.

Schultz, T.H., Flath, R.A., Mon, T.R., Eggling, S.B. and Teranishi, R. (1977) Isolation of volatile compounds from a model system. *J. Agric. Food Chem.*, **25**, 446–449.

Shallenberger, R.S. and Birch, G.G. (1975) Nonenzymic browning reactions. *Sugar Chemistry*, AVI Publishing Co. Inc., Westport. CT. USA, pp. 169–193.

Shen, S. and Yang, X. (1989) Study on smoked flavour of black tea made with small leaves (Chinese). *Fujian Tea*, (1), 27–30.

Shimoda, M., Shigematsu, H., Shiratsuchi, H. and Osajima, Y. (1995) Comparsion of volatile compounds among different grades of green tea and their relations to odor attributes. *J. Agric. Food Chem.*, **43**, 1621–1625.

Srivastava, R.A.K. (1986) Polyphenol oxidase activity in the development of acquired aroma in tea (*Thea sinensis var. Assmica* L.). *Current Sci.*, **55**, 284–287.

Stahl-Biskup, E., Intert, F., Holthuijzen, J., Stengele, M. and Schulz, G. (1993) Glycosidically bound volatiles – a review 1986–1991. *Flavour and Fragrance J.*, **8**, 61–80.

Stagg, G.V. (1974) Chemical changes occurring during the storage of black tea. *J. Sci. Food Agric.*, **25**, 1015–1034.

Szente, L., Gal, F.M. and Szejtli, J. (1988) Tea aromatisation with beta-cyclodextrin complexed flavours. *Acta Alimentaria*, **17**, 193–199.

Takei, Y., Ishiwata, K. and Yamanishi, T. (1976) Aroma components characteristic of spring green tea. *Agric. Biol. Chem.*, **40**, 2151–2157.

Takeo, T. (1981a) Production of linalool and geraniol by hydrolytic breakdown of bound forms in disrupted tea shoot. *Phytochem.*, **20**, 2145–2147.

Takeo, T. (1981b) Variation in amounts of linalol and geraniol produced in tea shoots by mechanical injury. *Phytochem.*, **20**, 2149–2151.

Takeo, T. (1981c) Chemical analysis of aromatic components on semi-fermented tea (oolong, pouchong tea). *Nippon Shokuhin Kogyo Gakkaishi*, **28**, 176–180.

Takeo, T. and Mahanta, P.K. (1983a) Comparsion of black tea aromas of orthodox and CTC tea and black teas made from different varieties. *J. Sci. Food Agric.*, **34**, 307–310.

Takeo, T. and Mahanta, P.K. (1983b) Why CTC tea is less fragrant. *Two and A Bud*, **30**, 76–77.

Takeo, T. (1983c) Variation in the aroma compound content of semi-fermented and black tea. *Nihon Nogei Kagakkai Shi*, **57**, 457–459.

Takeo, T., Tsushida, T., Mahanta, P.K., Tashiro, M., and Imamura, Y. (1985) Study on the food chemical aroma of Oolong tea and black tea. *J. Rep. Tea Study*, **20**, 91–180.

Takeo, T., You, X., Wang, H., Kinugasa, H., Li, M., Chen, Q. *et al.*, H. (1992) One speculation on the origin and dispersion of tea plant in China – one speculation based on the chemotaxonomy by the content-ration of terpen-alcohols found in tea aroma composition. *J. Tea Sci.*, **12**, 81–86.

Takeo, T. (1996) The relation between clonal characteristic and tea aroma. *FFI Journal*, **168**, 35–45.

Togari, N., Kobayashi, A. and Aishima, T. (1995a) Relating sensory properties of tea aroma to gas chromatographic data by chemometric calibration methods. *Food Res. Int.*, **28**, 485–493.

Togari, N., Kobayashi, A. and Aishima, T. (1995b) Pattern recognition applied to gas chromatographic profiles of volatile components in three tea categories. *Food Res. Int.*, **28**, 495–502.

Tsutomu, H., Flath, R.A., Mon, T.R. and Morton, J.F. (1985) Volatile components of rooibos tea (*Aspalathus linearis*). *J. Agric. Food Chem.*, **33**, 249–254.

Vitzthum, O.G., Werkhoff, P. and Hubert, P. (1975) new volatile constituents of black tea aroma. *J. Agric. Food Chem.*, **23**, 999–1103.

Wang, D., Ando, K., Morita, K. and Kobayashi, A. (1994) Optical isomers of linalool and linalool oxides on tea aroma. *Biosci. Biotech. Biochem.*, **58**, 2050–2053.

Wang, D., Kubota, K. and Kobayashi, A. (1996) Optical isomers of methyl jasmonate in tea aroma. *Biosci. Biotech. Biochem.*, **60**, 508–510.

Wang, H. and Li, M. (1989) Smoky-burnt odour in roasted green tea and its analytical methods (Chinese). *J. Tea Sci.*, **9**(1), 49–63.

Wang, H., Liu, Z., You, X. and Li, M. (1991) Flavour components of some smoked teas (Chinese). *J. Tea Sci.*, **11**(1), 51–58.

Wang, H., Li, M., Shi, Z., Wang, Z. and Liu, Z. (1992) Aromatic constituents in dark green tea (Chinese). *J. Tea Sci.*, **11** (Supplementary), 42–47.

Wang, H., Li, M., Liu, Z., Wang, Z. and Shi, Z. (1992) Changes of the volatile flavour constituents in Fuzhuan brick tea during the fungus growing process (Chinese). *J. Tea Sci.*, **11** (Supplementary), 81–86.

Wang, H., Takeo, T., Ina, K. and Li, M. (1993) Characteristic aroma components of Qimen black tea (Chinese). *J. Tea Sci.*, **13**(1), 61–68.

Wang, H. and You, X. (1995) Free and glycosidically bound monoterpene alcohols in tea. *Proc Int Symp on Tea-Quality-Human Health*, Nov 9–13, Shanghai, PR China, pp. 133–136.

Wang, H., You, X. (1996a) Free and glycosidically bound monoterpene alcohols in Qimen black tea. *Food Chem.*, **56**, 395–398.

Wang, H. and You, X. (1996b) Determination of aroma components of *Ampelopsis grossedentata* by gas chromatography. *Natural Product Reas. and Develop.*, 8 (4), 47–50.

Watanabe, N. and Singh, I.P. (1997) Analysis of aroma release from scented teas. In:, Linkskens H.F. and Jackson J.F. (eds), *Plant Volatile Analysis*. Springer, Saladruck, Berlin, pp. 231–258.

Werkhoff, P., Bretschneider, W., Herrmann, H.J. and Schreiber, K. (1989) Developments in the analysis of flavour compounds. II. Headspace techniques. *Labor Praxis*, 13, 426–430.

Werkhoff, P., Bretschneider, W., Guntert, M., Hopp, R. and Surburg, H. (1991) Chirospeific analysis in flavour and essential oil chemistry. B. Direct enantiomer resolution of trans-alpha-ionoe and trans-alpha-damascone by inclusion gas-chromatography. *Zeitschrift fur Lebensmitiel-Untersuchung und Forschung*, **192**, 111–115.

Werkhoff, P., Brennecke, S., Bretschneider, G.M., Hopp, R. and Surburg, H.Z. (1993) Chirospecific analysis in essential oil, fragrance and flavor research. *Lebensm. Unters. Forsch.*, **196**, 307–328.

Wickremasinghe, R.L., Wick, E.L. and Yamanishi, T. (1973) Gas chromatographic mass spectrophotometric analysis of flavoury and non-flavoury Ceylon black tea aroma concentrates prepared by different methods. *J. Chromatog.*, **79**, 75–80.

Winterhalter, P. (1992) Oxygenated C-13 norisoprenoids – important flavor precursors. In: Teranishi, R., Takeoka, G.R. and Guntert, M. (eds.), *Flavour Precursors – Thermal and Enzymatic Conversions*. ACS Symp Ser 490, American Chemical Society, Washington, DC, 1992, pp. 98–115.

Withopf, B., Richling, E., Roscher, R., Schwab, W. and Schreier, P. (1997) Sensitive and selective screening for 6'-O-malonylated glucoconjugates in plants. *J. Agric. Food Chem.*, **45**, 907–911.

Xia, T., Tong, Q., Dong, S. and Luo, Y. (1996) Studies on the change of glucosidase activity during the withering and fermentation of black tea (Chinese). *J. Tea Sci.*, **16** (1), 63–66.

Xu, N. (1997) Role of volatiles in the tritrophic chemical communication among tea plant, tea geometrid, and *Apanteles* sp. wasp. PhD thesis, Zhejang Agricultural University.

Yamaguchi, K. and Shibamoto, T. (1981) Volatile constituents of green tea Gyokuro (*Camellia sinensis* L. *var*. Yabukita). *J. Agric. Food Chem.*, **29**, 366–370.

Yamanishi, T., Kobayashi, A., Nakamura, H., Uchida, A., Mori, S., Ohsawa, K. and Sakamura, S. (1968) Flavor of black tea. Part V. Comparison of aroma of various types of black tea. *Agric. Biol. Chem.*, **17**, 379–386.

Yamanishi,T., Kita, Y., Watanabe, K. and Nakatani, Y. (1972) Constituents and composition of steam volatile aroma from Ceylon tea. *Agric. Biol. Chem.*, **36**, 1153–1158.

Yamanishi, T., Kawatsu, M., Yokoyama, T. and Nakatani, Y. (1973a) Methyl jasmonate and lactones including jasmine lactone in Ceylon tea. *Agric. Biol. Chem.*, **37**, 1075–1078.

Yamanishi, T., Shimojo, S. and Ukita, M. (1973b) Aroma of roasted green tea (Hoji-cha). *Agric. Biol. Chem.*, **37**, 2147–2153.

Yamanishi, T. (1978a) Flavour of green tea. *Japan Agric. Res. Quarterly*, **12**, 205–210.

Yamanishi, T., Wickremasinghe, R.L. and Perrera, K.P.L.W. (1978b) Studies on the flavour and quality of tea. 3. Gas chromatographic analysis of aroma complex. *Tea Quart.*, **39**, 81–86.

Yamanishi, T., Kosuge, M., Tokitomo, Y. and Maeda, R. (1980) Partially fermented tea: Pouchong tea. *Agric. Biol. Chem.*, **44**, 2139–2142.

Yamanishi, T., Luo, S.J., Guo, W. and Kobayashi, A. (1988) Aroma of Chinese scented green tea. *Frontiers of Flavor. Proc 5th Int Flavor Conf.*, 1–3 July 1987, Porto Karras, Chalkidiki, Greece, pp. 181–190.

Yamanishi, T., Botheju, W.S. and De Silva, J.M. (1989a) An index for assessing the quality of Uva seasonal black tea. *Sri Lanka J. Tea Sci.*, **58**, 40–49.

Yamanishi, T., Kawakami, M., Kobayashi, A., Hamada, T. and Musalam, Y. (1989b) In: Parliament T.H., McGorrin R.J. and Ho C.T. (eds) *Thermal Generation of Aroma*, ACS Symposium series 409, American Chemical Society, Washigton, DC, p. 310.

Yamanishi, T. (1995) Flavour of tea. *Food Rev. Int.*, **11**, 477–525.

Yamanishi, T. (1996) Tea Aroma *FFI Journal*, **168**, 23–34.

Yano, M., Okada, K., Kubota, K. and Kobayashi, A. (1990) studies on the precursors of monoterpene alcohols in tea leaves. *Agric. Biol. Chem.*, **4**, 1023–1028.

You, X., Wang, H. and Li, M. (1990) Volatile flavour constituents and terpene index of tea flower (Chinese). *J. Tea Sci.*, **10** (2), 71–75.

You, X., Li, M. and Takeo, T. (1992a) Effect of tan-fang treatment on the aroma formation of Longjing tea (Chinese). *J. Tea Sci.*, **12**(2), 161–162.

You, X., Li, M., Yang, Y. and Wang, H. (1992b) Tea varietal expression through terpene index of tea flower (Chinese). *J. Tea Sci.*, **12**(1), 27–31.

You, X. and Wang, H. (1993) Aromatic style and expression of Fuding Dabai – a national level fine variety – in the places there it is induced to (Chinese). *China Tea*, **15**(5), 6–7.

You, X., Wang, H. and Yang, Y. (1994) Progress in the study on linalool and geraniol and their glycosides in tea (Chinese). *China Tea*, **16**, 5.

Young, S.K., Mu, N.K., Jeong, O.K. and Jong, H.L. (1994) The effect of hot water-extract and flavor compounds of mugwort on microbial growth. *J. Korean Soc. Food Nutr.*, **23**, 994–1000.

6. BIOCHEMICAL AND CELLULAR BASES FOR TEA ACTIVITY

SHEN-DE LI

Cancer Institute, Chinese Academy of Medical Sciences and Peking Union Medical College, Panjiayuan Beijing 100021, China

A large number of investigations on the nutritional value and pharmacological activity of tea have been carried out by scientists since the beginning of the early eighties. Chen (1991) reviewed the main advances in this field. Recently, many papers have appeared concerning the biochemical and cellular bases of tea activity on cardiovascular diseases, brain diseases, radiation prevention and cancer prevention.

Due to the differences in processing, there are three different types of tea products: green tea (unfermented tea), black tea (fermented tea), and semi-fermented tea, such as Oolong tea and Paochung tea. Both Green tea extract (GTE) and black tea extract (BTE) are used extensively by most scientists.

Green tea contains polyphenols, which include flavanols, flavandiols, flavonoids and phenolic acid; these compounds may account for up to 30% of the dry weight. Most of the polyphenols are flavanols, commonly known as catechins. Some major green tea catechins are (–)-epigallocatechin gallate (EGCG), (–)-epigallocathechin (EGC), epicatechin-3-gallate (ECG), (–)-epicatechin (EC), (+)-gallocatechin (GC), and (+)-catechin (C).

During fermentation, a series of complex chemical reactions take place; the most important one being the oxidation of polyphenols. The oxidized polyphenols are unstable, and other chemical reactions follow rapidly. As a result, catechins change to theaflavins (TF), thearubigins (TR) and other oxidized-polymerized compounds; and the various aroma compounds develop to form the special flavor of black tea. It has been demonstrated that the pharmacological activity of tea depends mainly on the catechins, and especially, on their esters, so BTE is less effective than GTE in most experiments due to the destruction of catechins.

The biochemical properties of catechins include the antioxidative activities, antiaggregation activity, free radical scavenging activity and others involved in cancer preventive effects. The biochemical mechanisms of tea in different diseases may vary due to a number of biochemical properties. The biochemical and cellular bases of tea and their relationship with various diseases will be discussed below.

1. ATHEROSCLEROSIS

Atherosclerosis is the most important of the cardiovascular diseases, and it may induce many serious complications. The etiology of atherosclerosis is very complicated. Although hypertension and hypercholesterolemia are thought to be involved in the formation of atherosclerosis, the exact mechanism is still not clear. Many studies have been carried out on the relationship between tea and atherosclerosis, and the possible mechanism of tea components against atherosclerosis hsa been discussed. Iked *et al.* (1991) found that a mixture of EC and EGC and their esters ECG and EGCG markedly lowered lymphatic cholesterol absorption in rats. Tea catechins, especially gallate esters, effectively reduced chloesterol absorption in the intestine by precipitating cholesterol in mixed bile salt micelles. Based on these results and previous studies, it is suggested that the hypocholesterolemic activity can be attributed to the inhibition of intestinal cholesterol absorption.

Sagesaka *et al.* (1991) reported the effect of hot water extract of green tea on the collagen-induced aggregation of washed rabbit platelet. The extract lowered submaximal aggregation and prolonged the lag time in a dose-dependent manner. After fractionation of the extract by Sephadex LH-20 column chromatography it was revealed that tea catechins are effective components for inhibition and that ester-type catechins are more effective than free-type catechins. One of the esters, EGCG, suppressed the collagen-induced platelet aggregation completely at a concentration of 0.2 mg/ml (= 0.45 mM). The IC50 value of EGCG is comparable to that of aspirin. Xu *et al.* (1991) carried out a series of *in vitro* and *in vivo* investigations on the biochemical effects of tea products and the relationship with the prevention of atherosclerosis.

The study on the antioxidative activity of tea products showed that the GTE containing 62.8% catechins and the BTE with 44.0% catechins had remarkable antioxidative activity on lard. The decreasing rate of peroxide value (POV) was 78.3% and 48.2% respectively after 14 days of storage. GTE is more effective than BTE. This may result from the fact that GTE contains more catechins. To study the anticoagulation effect and the inhibition of platelet aggregation (Pagt), four ethanol extracts of tea, GTE, BTE1, BTE2, and BTE3, were made from green tea and black tea fermented for different periods (30 min, 60 min and 90 min respectively). The amount of catechin in GTE, BTE1, BTE2 and BTE3 decreased gradually while that of oxides increased with longer fermentation. GTE showed the strongest anticoagulation effect. BTE1 and BTE2 showed decreased effects in accordance with the increase in fermentation time, and BTE3 failed to show any effect. These results demonstrated that catechins are major active components.

In *in vivo* experiments the ethanol extract of green tea (EEGT) and the extract of black tea made by 30 min fermentation (EEBT) were used to treat rabbits with atherosclerosis. The catechin amounts of EEGT and EEBT were 35% and 26%, and the catechin oxide amounts were 29.9% and 54.5% respectively. Being similar to the *in vitro* experiments, Pagt in atherosclerosis model rabbits treated with EEGT was remarkably inhibited, whereas the EEBT had no marked effect. EEGT and EEBT were able to reduce the lipid level and reduce the ratio of cholesterol to phosphatides (C/P), and they also played a role in decreasing the level of low density lipoprotein (LDL) and very low density lipoprotein (VLDL) in the plasma of rabbits. EEBT was more effective than EEGT, for which the catechin oxides TF and TR may also be responsible. In addition, both EEGT and EEBT were found to play an active role in protecting the visceral organs from damage due to the accumulation of lipid peroxidation (LPO). EEGT and EEBT are powerful depressors of the formation of aortic atherosclerosis, to some extent, and EEBT appears to be more effective than EEGT in protecting the aorta and coronary artery from atherosclerotic damage and obstruction.

Recently, more and more scientists have become interested in the antioxidative and free radical scavenging effects of tea and their relationship with cardiovascular diseases. It is well known that human low density lipoprotein (LDL) is the main carrier of cholesterol in the blood stream and can be oxidatively modified *in vivo* and *in vitro* as it has about 50% polyunsaturated fatty acid molecules. Many studies have provided strong evidence that oxidatively modified LDL (O-LDL) is the species involved in many pathophysiological processes, especially in cardiovascular diseases. Ding *et al.* (1995) reported that China green tea polyphenol (CGTP) was an effective antioxidant protecting LDL from oxidative modification. The antioxidant effect was shown by the mobility in agarose gel electrophoresis, the thiobarbituric acid reactive substances (TBARs), and the fluorescence emission spectra at 360 nm excitation as well as the degradation rate by mouse peritoneal macrophages. Salah & Catherine (1995) investigated quantitatively the antioxidant potential of the polyphenolic constituents of tea on the resistance of LDL to oxidation. The activity of the compounds was studied by measuring the inhibition of LDL peroxidation and the altered recognition properties of apoprotein B100. The effect of these compounds on conserving endogenous α-tocopherol within the LDL particles was measured by high performance liquid chromatography (HPLC). The antioxidant acitvity of the polyphenols in delaying oxidation and in their capability to conserve α-tocopherol was found to be in the order:

ECG = EGCG = EC = C > EGC > GA (gallic acid)

Taking into account their actual concentrations in green tea, the antioxidant effectiveness of the polyphenols would be in the order:

EGC > EGCG > EC > ECG > C

Tea extracts were able to inhibit LDL oxidation (0.275 ppm for 50% inhibition of oxidation.) In addition to the antioxidant effect, CGTP is also a strong lipid free radical scavenger. Xin *et al.* (1995) reported that CGTPs are effective scavengers of superoxide and hydroxyl radicals. The effect of CGTP on lipid free radicals was studied in different systems using electron spin resonance (ESR) techniques. The results indicated: (1) CGTP could effectively scavenge lipid free radicals produced from the gas phase of cigarette smoke in (GPCS)-treated rat liver microsomes. The scavenging capacities of different antioxidants were found to be in the order of EGCG (composing

50% in CGTP) > vitamin C > CGTP > vitamin E. (2) CGTP showed strong scavenging effects on lipid free radicals in the linoleic acid-lipoxygenase (LA-LPO) system; and the scavenging effects were dose-dependent. (3) In Fe^{2+} –$H_2 O_2$ induced erythrocyte membrane system, the increase of lipid peroxidation (LPO) was inhibited by adding 0.2 mg/ml CGTP. The results confirmed the scavenging effects of CGTP on lipid radicals induced in these systems and indicated that the initial reactive sites are the OH groups in the pyrogallol ring of catechin when CGTP reacts with lipid free radicals. The authors suggested that there may be several ways for CGTP to play their scavenging role. Firstly, CGTP may react in the initial stage of LPO and effectively scavenge O_2 in GPCS-induced LPO of the microsome system, CGTP can also scavenge Fe^{2+} or OH in Fe^{2+} –$H_2 O_2$-induced erythrocyte membrane system. Secondly, they may quench the lipid free radicals produced in the mid stage of LPO as chain-breaking agents. So it appears that CGTP may prevent both the initiation and propagation of LPO.

Based on the results described above, the possible role of tea polyphenols in preventing atherosclerosis can be summarized as follows:

1). Since tea polyphenols and their oxides can effectively inhibit LPO and scavenge the free radicals induced by LPO, they may play an important role in protecting arteries, especially the coronary artery and aorta, from the damage caused by LPO and free radicals.

2). Hypercholesterolemia and the accumulation of cholesterol on the damaged sites of arteries are the major cause of atherosclerosis. Tea polyphenols can reduce cholesterol absorption in the intestine by precipitating cholesterol solubilized in mixed bile salt micelles, thus decreasing the blood cholesterol level.

3). Tea polyphenols can effectively inhibit platelet aggregation, which may result in preventive and therapeutic effects on thrombus formation in arteries.

2. HYPERTENSION

Omori et al. (1991) reported the effect of Gabaron tea on the blood pressure of spontaneously hypertensive rats (SHR) loaded with common salt. Gabaron tea is made from anaerobically treated leaves, which is rich in gamma-aminobutyric acid (GABA). The amount of GABA in Gabaron tea is about 70 times as high as that in ordinary tea. The effect of Gabaron tea and ordinary tea on the blood pressure of spontaneously hypertensive rats fed with pellet chew containing 5% NaCl was examined. After 4 weeks treatment, the blood pressure of SHR was elevated, both in control rats due to aging, but much higher in NaCl treated rats. Gabaron tea was able to suppress the rise in blood pressure not only due to NaCl loading, but also due to aging of the rats, while the ordinary tea was able to suppress the rise in blood pressure due to NaCl loading only, it had no effect on that due to aging of the rats. The content of total and high density lipoprotein (HDL) cholesterol in blood increased and that of potassium and phosphorous decreased with NaCl loading. When Gabaron tea infusion was given, the content of these components returned to the normal level.

3. PREVENTION OF AGING-RELATED CENTRAL NERVOUS SYSTEM DAMAGE

Kuwabara *et al.* (1995) examined the protective effects of Rooibos tea (RT), *Aspalathus linealis,* against damage of the central nervous system associated with aging using the thiobarbituric-acid reaction (TBA) and magnetic-resonance imaging (MRI) technique in brains of chronically RT-treated rats. The TBA method was employed to measure the age-related accumulation of lipid peroxides, and the MRI technique was used to observe morphological changes. RT administration was begun in 3-month-old Wistar male and female rats and continued for 21 months. The content of lipid peroxides in the frontal cortex, occipital cortex, hippocampus and cerebellum in 24-month-old rats after administration with water was significantly higher than that in young rats (5 weeks old). However, no significant increase of lipid peroxides was observed in RT-treated aged rats. When MR images of the rat brains were taken, a decrease in signal intensity was observed in the cerebral cortex, thalamus and hippocampus of aged rats without RT treatment, whereas little change was observed in MR images of the same regions of aged rats treated with RT. The authors concluded that chronic administration of RT can prevent aging-related accumulation of lipid peroxides in several regions of the rat brain. Guo *et al.* (1995) presented evidence on the scavenging effect of green tea polyphenols on lipid free radicals generated from lipid peroxidation of synaptosomes. Brain tissue is particularly sensitive to free radical assault, partially due to its high content of polyunsaturated fatty acids and catalytically active metals (i.e., iron and copper). The scavenging effect of green tea polyphenols (GTP) on lipid free radicals generated from lipid peroxidation of synaptosomal membranes induced by Fe^{2+}/Fe^{3+} was studied and compared with those of Ginkgo-Biloba Extract (EGB) and Schisanhenol (Sal). Using spin trapping a-(4-pyridyl-l-oxide)-N-t-butylnitrone (4-POBN), carbon-centered radical adducts were detected. The electron paramagnetic resonance (EPR) spectra exhibited apparent hyperfine splittings characteristic of a POBN/alkyl radical ($a^N = 15.5G$ and $a^H = 2.7G$). GTP, EGB and Sal were shown to have scavenged lipid-derived free radicals. The concentrations of GTP, EGB and Sal needed to scavenge 50% free radicals (IC_{50}) were 0.0056 mg/ml, 0.24 mg/ml and 0.11 mg/ml, respectively.

4. ANTI-RADIATION EFFECTS

Du *et al.* (1995) reported the studies on anti-radiation effects of green tea polyphenols (GTP) in rats. GTP was administered orally to rats for one week. The rats were then irradiated with ^{60}Co and were given GTP for another three days. GSH-PX content in blood and the superoxide dismutase (SOD) content in liver were determined. GTP was found to enhance the activity of GSH-PX and SOD, reduce the amount of LPO in blood and liver, and reduce the amount of myocardial lipofuscin. Cao *et al.* (1995) reported the protective effect of GTP on radiation injury by ^{60}Co. After the mice were irradiated with ^{60}Co and treated with GTP, it was found that GTP had a significant effect on the loss of white blood cells and platelets by ^{60}Co irradiation and protected the normal immune function of the spleen and thymus. GTP also caused an increase in the immunocytes, colony forming unit of spleen (CFU-S) and mitosis index of granulocytes of bone marrow. In addition, GTP can remarkably reduce the micronucleus formation in polychromatophilic erythrocytes (PCE) induced by ^{60}Co irradiation in bone marrow.

5. CANCER PREVENTIVE EFFECTS

There are a large number of studies on the relationship between tea consumption and human cancer incidence. Whereas some studies may show a significant protective effect of tea consumption on certain types of cancer, the others do not. The effect may vary in different patients, different cancers and different causative factors of the cancer. Yang & Wahg (1993) reviewed this topic thoroughly, covering basic chemistry and biochemical activities of tea, epidemiological investigations and laboratory studies, as well as possible directions for further research. The inhibitory effects of tea preparations and tea polyphenols are believed to be mainly due to the antioxidative and possible antiproliferative effects of polyphenolic compounds in green tea and black tea. These compounds may also inhibit carcinogenesis by blocking endogenous formation of the N-nitroso compounds, suppressing the metabolic activation of precarcinogens, and trapping the genotoxic agents.

5.1. Antioxidative and Free Radical Scavenging Effects

Yang *et al.* (1993) pointed out that tea polyphenols possess strong antioxidant activity via three mechanisms:

1). Because of the presence of the "catechol" structure, most tea polyphenols are strong metal ion chelators. They can bind and thus decrease the level of free cellular ferric and ferrous ions, which are required for the generation of reactive oxygen radicals by the Fenton and Haber-Weiss reactions.
2). Tea polyphenols such as EGCG, EGC and ECG are strong scavengers against superoxide anion radicals and hydroxy radicals – two major reactive oxygen species that can damage DNA and other cellular molecules and can initiate lipid peroxidation reactions.
3). Tea polyphenols can react with peroxyl radicals and thus terminate lipid peroxidation chain reactions. Reactive oxygen species may play important roles in carcinogenesis through damaging DNA, altering gene expression, or affecting cell growth and differentiation.

Ruch *et al.* (1989) demonstrated that an antioxidant fraction of Chinese green tea (green tea antioxidant, GTA), containing several catechins, had antioxidant activity towards superoxide radical and hydrogen peroxide at a dose-dependent manner. GTA also prevented oxygen radical and H_2O_2 induced cytotoxicity and inhibition of intercellular communication in cultured B6C3F1 mouse hepatocytes and human keratocytes (NHEK cells). Zao *et al.* (1989) reported the free radical scavenging effects of the extracts of green tea and other natural foods by using the spin trapping technique. In the stimulated polymorphonuclear leukocyte system, the water extract fraction 6 (F6) from green tea and green tea polyphenols (GTP) showed a much stronger scavenging effect on the active oxygen radicals than vitamin C and vitamin E did. In a quantitative aqueous system, and irradiated riboflavin system, the scavenging effects of E6 and GTP were 74% and 72%, respectively, which were weaker than vitamin C, but much higher than vitamin E. Wang *et al.* (1992) reported that tobacco-specific nitrosamine 4-(methylnitrosamine)-1-(3-pyridyl)-1-butanone (NNK) and N-nitrosodiethylamine (NDEA)-induced cellular oxidative damage in A/J mice such as lipid peroxidation, DNA single strand breakage and the formation of 8-hydroxyl-deoxyguanosine (8-OH-dGuo) can be inhibited by green tea and black tea. Xu *et al.* (1992) reported that multiple doses of NNK cause a significant increase in 8-OH-dGuo level in the lungs of A/J mice, but not in the liver, and also cause lesions in lung cellular DNA. These changes can be suppressed by green tea and its polyphenols. In addition to the direct antioxidant effect, GTP may also play a role in inducing the activity of antioxidative enzymes. Khan *et al.* (1992) demonstrated that GTP can enhance the activity of several antioxidative enzymes, such as glutathione peroxides (GSH-Px), catalase and quinone reductase (QR), as well as phase II enzymes such as glutathion-S-transferase (GST). GTP fed (0.2%, W/V) to mice for 30 days significantly increased the activity of GSH-Px, catalase and QR in small bowel, liver and lung, and the activity of GST in small bowel and liver.

5.2. Antimutagenic Activity

Wang *et al.* (1989) reported that the antimutagenic activity of GTP and water extracts of green tea were found to inhibit the reverse mutation induced by benzo(a)pyrene (BP), aflatoxin B1 (AFB1), and methanol extract of coal tar pitch in *Salmonella typhimurium* TA100 and/or TA98 in the presence of a rat-liver microsomal activation system. GTP also inhibited gene forward mutation in V79 cells treated with AFB1 and BP, and decreased the frequency of sister-chromatid exchanges and chromosomal aberrations in V79 cells treated with AFB1. The addition of GTP during and after nitrosation of methylurea resulted in a dose-dependent inhibition of mutagenicity. The authors pointed out that multiple actions of GTP may contribute to inhibiting various mutagenic pathways:

1). GTP inhibits the p-450-dependent metabolic activation of precarcinogens. Precarcinogens like polycyclic aromatic hydrocarbons (PAHs) and AFB require metabolic activation by the cytochrome p-450-dependent enzymes to form highly reactive electrophilic metabolites, which can bind to macromolecules such as protein and nucleic acids in the target tissue to exert their mutagenic and carcinogenic effects. So the inhibition of metabolic activation of PAHs and AFB1 will prevent them from causing mutagenic effects.
2). Owing to their special chemical structure, GTP can bind directly to the active electrophilic sites of the carcinogens, thus prevent them from binding to macromolecules.

3). GTP inhibits the nitrosation reaction. Nitrosation of methylurea is known to result in the formation of a direct-acting mutagen. The addition of GTP during or after the nitrosation of methylurea resulted in dose-dependent inhibition of the mutagenicity.

4). GTP may exert its antimutagenic effects via the scavenging effect on PAH cation free radicals, which may attack the macromolecules to manifest mutagenesis.

5.3. Antipromotion Effects

The induction of ornithine decarboxylase (ODC) and protein kinase C (PKC) by 12-o-tetradecanoylphorbol-13-acetate (TPA) is believed to be closely related to the tumor promotion activity of this compound. Zhang *et al.* (1991) reported that in cultured BALB/3T3 cells treated with TPA, the ODC mRNA expression increased in 2 hours, and green tea extract (GTE) decreased ODC expression. GTE also inhibited PKC gene overexpression induced by croton oil in rat liver. In addition, GTE can block the TPA induced inhibition of intercellular communication (Ruch *et al.* 1989).

Recently, Nakamura *et al.* (1995) reported the antipromotion effect of a new preparation of tea, tea aqueous non-dialysates (TNDs). TNDs were prepared from the hot water infusion of green and black tea leaves, followed by extraction with chloroform, ethyl acetate, and n-butanol and then dialysis. The TNDs have a molecular weight of more than 12,000 Da and consist of mixed complex tannins, containing sugar(s), quinic acid and polyphenols such as gallates and catechins. Similar to tea catechins, TNDs inhibited tumor promotion in a model system of mouse duodenal carcinogenesis induced by N-methyl-N'-nitro-N-nitrosoguanidine (MNNG). In an *in vitro* system, TNDs reduced the neoplastic transformation and the cell-shape changes induced by TPA without particular cytotoxicity in mouse epidermal JB6 cell lines.

5.4. Inhibition of Tumor Cell Proliferation

Although many studies demonstrated that GTE or GTP inhibits tumor cell proliferation to a certain extent, the real mechanism is still not clear. The inhibition of tumor cell proliferation by GTE or GTP may related to their effects on intercellular communication, cellular transduction, cell cycle regulation and the expression of oncogenes, antionco-genes and cytokines. As it has been described above, GTE and GTP can block the TPA induced inhibition of intercellular communication thus leading to the restoration of contact inhibition, which may suppress cell proliferation. Suganuma *et al.* (1995) reported the inhibitory effects of EGCG and GTE on carcinogenesis. EGCG and GTE inhibited the growth of lung, mammary and stomach cancer cell lines. Analysis of DNA histogram by flow cytometry revealed that treatment of a gastric cancer cell line, KATO III cells, with EGCG for 24 hours increased the cells of G2/M phase from 8.2% to 21.6%, indicating that EGCG slightly induces G2/M block in the cells. EGCG inhibited the release of tumor necrosis factor-α (TNF-α), an endogenous tumor promoter in tissues. After administering [^{3}H]-EGCG into mouse stomach, radioactivity was found in various organs, such as the digestive tract, liver, lung, pancreas, skin, brain and blood, in

24 hours. These results suggest that oral administration of EGCG or green tea is effective against carcinogenesis of various organs. Lin (1995) reported a possible mechanism of GTP in the inhibition of carcinogenesis. It was found that the gene encoding EGF receptor was overexpressed in human epidermal carcinoma A-431 cells. The cell proliferation was significantly suppressed in a dose-dependent manner. Furthermore, GTP inhibited the autophosphorylation of EGF receptor and the phosphorylation of extracellular signal-related kinase (ERK-1 and ERK-2) remarkably. Therefore, it is suggested that the molecular mechanism of anticarcinogenesis by GTP might be mediated by proliferative signal blocking or differentiative signal modulating in the target cells. Zhang *et al.* (1991) reported the effects of GTE on the expression of certain oncogenes and antioncogenes in BALA/3T3 cell line. When the cells were treated with TPA, the expression of c-*fos* increased remarkably, and reached the maximum at 0.5 hour after exposure. The pretreatment of GTE reduced TPA-induced overexpression by 3-fold. TPA also enhanced the expression of c-*myc* about 2-fold in 2 hours, and GTE reduced the expression level almost to the control value. TPA down-regulated the expression of Rb, and preincubation with GTE slightly increased the Rb mRNA level.

REFERENCES

Cao, M.F. and Yang, X.Q. (1995) Studies on antiradiative effect of green tea polyphenols. *International Symposium on Natural Antioxidants*, Beijing, China. June 20. p. 332.

Chen, Z. (1991) Contribution of tea to human health. *International Symposium on Tea Science*, Shizuka, Japan. pp. 12–20.

Ding, Z.H., Chen, Y., Zhou, M., Fang, Y.Z. (1995) Inhibitory effect of China green tea polyphenols on LDL oxidative modification and fibroblast apoptosis. *International Symposium on Natural Antioxidants*, Beijing, China. June 20, p. 122.

Du, X.W. and Lin, Z.S. (1995) Studies on antiradiative effect of green tea polyphenols. *International Symposium on Natural Antioxidants*, Beijing, China. June 20. p. 208.

Guo, Q., Zhao, B., Hou, J., Xin, W. (1995) Scavenging effect of green tea polyphenols on lipid free radicals generated from lipid peroxidation of synaptosomes. *International Symposium on Natural Antioxidants*, Beijing, China. June 20. p. 315.

Iked, I., Imasato, Y., Sasaki, E., Nakayama, M., Nagao, H., Takeo, T., *et al.* (1991) Tea catechins decrease micellar solubility and intestinal absorption of cholesterol in rats. *Proceedings of the International Symposium of Tea Science*, Shizuoka, Japan. August. pp. 215–219.

Khan, S.G., Katiyar, S.K., Agarwal, R., Mukhtar, H. (1992) Enhancement of antioxidant and phase II enzyme activity by oral feeding of green tea polyphenols in drinking water to SKH-1 hairless mice: possible role in cancer chemoprevention. *Cancer Res.*, 52, 4050–4052.

Kuwabara, M., Inanami, O., Asanuma, T., Inukai, N., Jin, T., Shimokawa, S., *et al.* (1995) Rooibos tea (*Aspalathus linealis*) suppresses age-related accumulation of lipid peroxidation in rat brain. *International Symposium on Natural Antioxidants*, Beijing, China. June 20. p. 43.

Lin, J.K. (1995) Anticarcinogenesis of tea polyphenols. *International Conference of Food Factors: Chemistry and Cancer Prevention*, Hamamatsu, Japan. December 10, p. 99.

Nakamura, Y. and Tomita, I. (1995) Antitumor promoting effects of tea aqueous non-dialysates. *International Conference of Food Factors: Chemistry and Cancer Prevention*, Hamamatsu, Japan. December 10, p. 99.

Omori, M., Kato, M., Yano, T., Tushida, T., Murai, T., Fukatau, S., *et al.* (1991) Effect of anaerobically treated tea (Gabaron tea) on the blood pressure of spontaneously hypertensive rat loaded with common salt. *Proceedings of the International Symposium of Tea Science,* Shizuoka, Japan. pp. 230–234.

Ruch, R.J., Cheng, S.J., Klauning, J.E. (1989) Prevention of cytotoxicity and inhibition of intercellular communication by antioxidant catechins isolated from Chinese green tea. *Carcinogenesis,* **10,** 1003–1008.

Sagesaka, Y., Miwa, M., Okada, S. (1991) Platelet aggregation inhibitors in hot water extract of green tea. *Proceedings of the International Symposium of Tea Science*, Shizuoka, Japan. pp. 235–239.

Salah, N.R. and Catherine, A.R-E. (1995) The antioxidant effect of tea on low density lipoprotein oxidation. *International Symposium on Natural Antioxidants,* Beijing, China. June, p. 175.

Suganauma, M., Okabe, S., Oniyama, M., Sueoka, N., Kozu, T., Komori, A., *et al.* (1995) Mechanisms of EGCG and green tea in inhibition of carcinogenesis. *Internation Conference of Food Factors: Chemistry and Cancer Prevention,* Hamamatsu, Japan. December, p. 119.

Wang, Z.Y., Cheng, S.J., Zhou, Z.C., Athar, M., Khan, W.A., Bickers, D.R., *et al.* (1989) Antimutagenic activity of green tea polyphenols. *Mutation Res.* **223,** 273–285.

Wang, Z.Y., Hong, J.Y., Huang, M.T., Reuhl, K.R., Conney, A.H., Yang, C.S. (1992) Inhibition of N-nitrosodiethylamine and 4-(methylnitrosamino)-1-(3-pyridyl))-1-butanone induced tumorigenesis in A/J mice by green tea and black tea. *Cancer Res.,* **52,** 1943–1947.

Xin, W., Zhao, B., Shi, H., Yang, F., Hon, J. (1995) Studies on the effects of green tea polyphenols on lipid free radicals. *International Symposium on Natural Antioxidants*, Beijing, China. June. p. 126.

Xu, X. and Chen, R. (1991) Tea components effect on cardiovascular systemÄrelationship between the preventive effect of tea extracts on atherosclerosis and their compositions. *International Symposium on Tea Science*, Shizuka, Japan. pp. 225–229.

Xu, Y., Ho, C.T., Amin, S.G., Han, C., Chung, F.L. (1992) Inhibition of tobacco-specific nitrosamine-induced lung tumorigenesis in A/J mice by green tea and its major polyphenols as antioxidants. *Cancer Res,* **52,** 3875–3879.

Yang, C.S. and Wang, Z.Y. (1993) *Tea and cancer. J US Natl Cancer Inst.,* **85,** 1038–1049.

Zao, B.L., Li, X.J., He, R.G., Cheng, S.J., Xin, W.J. (1989) Scavenging effect of extracts of green tea and natural antioxidants on active oxygen radicals. *Cell Biophysics,* **14,** 175–185.

Zhang, C.Y., Zhao, Q.Z., Bai, J.F., Zhao, M., Guo, S.P., Wang, B., Hara, Y. Cheng, S.J. (1991) Studies on the antipromotion of green tea epicatechin compounds. *Proceedings of the International Symposium on Environmental Mutagrnesis and Carcinogenesis*, Shanghai, China. May, pp. 112–115.

7. PHARMACOLOGICAL EFFECT OF CAFFEINE AND RELATED PURINE ALKALOIDS

DE-CHANG ZHANG

Institute of Basic Medical Sciences, Chinese Academy of Medical Sciences and Peking Union Medical College, 5 Dong Dan San Tiao, Beijing 100005, China

Caffeine and its closely related purine alkaloids such as theophylline and theobromine are methylxanthines occurring in plants widely distributed geographically. The most popular caffeine-containing beverage is tea, prepared from the leaves of *Thea sinensis* (containing caffeine and small amounts of theophylline and thiobromine), a bush native to southern China and now extensively cultivated in other countries. More than half the population of the world consumes tea. The average citizen in China consumes some 2 to 3 cups of tea every day. Other examples are Cocoa and chocolate, from the seeds of *Theobroma cacao,* containing theobromine and caffeine. Coffee, the most important source of caffeine in the American diet, is extracted from the fruit of *Coffea arabica* and related species. Cola-flavored drinks usually contain considerable amounts of caffeine, in part because of their content of extracts of the nuts of *Cola acuminata*. Some statistic data shows that the consumption of caffeine in the Nordic countries and in Britain is close to 300 mg per day per inhabitant. The most important source of caffeine in China is tea. Caffeine consumption is of a similar magnitude in most countries. This means that caffeine is probably the most widely used of all psychoactive drugs. The only substance that can come anywhere close is ethanol.

Classical pharmacological studies, principally of caffeine, during the first half of this century confirmed that the methylxathines have stimulant and antisoporific actions that elevate mood, decrease fatigue, and increase capacity for work. Further research demonstrated that methylxanthines possess other important pharmacological properties as well. These properties were exploited for a number of years in a variety of therapeutic applications, many of which have now been replaced by more effective agents. However, in recent years there has been a resurgence of interest in the natural methylxanthines and their synthetic derivatives, principally as a result of increased knowledge of their cellular basis of action.

1. PHARMACOLOGICAL PROPERTIES OF CAFFEINE AND ITS RELATED METHYLXANTHINES

Caffeine, theophylline, and theobromine share in common several pharmacological actions. They relax smooth muscle, stimulate the central nervous system, stimulate cardiac muscle, and act on the kidney as a diuretic. Of the three agents, theophylline is most selective in its smooth muscle effects, while caffeine has the most marked central nervous system effects.

1.1. Effects on the Central Nervous System

Caffeine and theophylline are potent stimulants of the CNS. Theobromine is virtually inactive in this respect. Traditionally, caffeine is more clinically useful in this aspect. In low and moderate doses, caffeine causes mild cortical arousal with increased alertness and deferral of fatigue. People ingesting caffeine experience a more rapid and clearer flow of thought. Whereas, low doses of theophylline don't have the same effect. Increasing the dosages of caffeine or theophylline produces nervousness or anxiety, restlessness, insomnia, tremors, hyperesthesia and other signs of progressive CNS stimulation. At still higher doses, medullary simulation and convulsions are produced. Theophylline is clearly more potent than caffeine in this regard.

Methylxanthines are also used as stimulants of the medullary respiratory centers, particularly in pathophysiological states such as Cheyne-Stokes respiration, apnea of preterm infants, and when drugs such as opioids depress respiration. The methylxanthines appear to increase the sensitivity of medullary centers to the stimulatory actions of CO_2, and respiratory minute volume is increased at any given value of alveolar P_{CO2}. Nausea and vomiting are also induced by higher doses of theophylline because of their CNS actions.

Low doses of methylxanthines are effectively used to antagonize the CNS depression caused by certain agents. For example, aminophylline (2 mg/kg) can rapidly reverse the narcosis induced by as much as 100 mg of morphine given intravenously to produce anesthesia, and there is evidence that methylxanthines can specifically antagonize a number of the actions of opioids, including analgesia (DeLander and Hopkins, 1986). This effect apparently reflects the participation of adenosine in the actions of opioids.

1.2. Effects on the Cardiovascular System

The pharmacological effects of methylxanthines on the cardiovascular system include modest decreases in peripheral vascular resistance, sometimes-powerful cardiac stimulation, increased perfusion of most organs, and diuresis. Theophylline is more prominent in this aspect than caffeine. These effects are the results of both the action on the vagal and vasomotor centers in the brain stem, and the direct actions on vascular and cardiac tissues. Some indirect peripheral actions mediated by catecholamines and the renin-angiotensin system may contribute to the final effects, too. Furthermore, the resulting effects largely depend on the conditions prevailing at the time of their administration, the dose used, and the history of exposure to methylxanthines. For example, low doses of caffeine may induce a decrease in heart rate and modest increases in both systolic and diastolic blood pressure in methylxanthines-naive individuals, but the same low dose usually would have no effect on these parameters in those who consume caffeine regularly (Myers, 1988a).

At higher concentrations, both caffeine and theophylline produce definite tachycardia. Some sensitive individuals may experience other arrhythmias, such as premature ventricular contractions. However, it appears that the risk of inducing cardiac arrhythmias in normal subjects is quite low and that patients with ischemic heart disease or preexisting ventricular ectopy can usually tolerate moderate amounts of caffeine without provoking an appreciable increase in the frequency of arrhythmias (Myers, 1988b; Chou and Benowitz, 1994).

1.3. Effects on the Kidney

The methylxanthines, especially theophylline, are weak diuretics. They increase the production of urine and enhance excretion of water and electrolytes. This effect may involve increased glomerular filtration and reduced tubular sodium reabsorption. Clinically their effect on the kidney is too weak to be therapeutically useful.

1.4. Effects on Smooth Muscle

Caffeine and its related methylxanthines, especially theophylline, relax various smooth muscles. Their ability to relax the smooth muscles of the bronchi, especially if the bronchi have been constricted either experimentally by a spasmogen or clinically in asthma, is very useful in therapy. In addition to this direct effect of the airway smooth muscle, these agents inhibit antigen-induced release of histamine from lung tissue.

1.5. Effects on Skeletal Muscle

Caffeine increases the capacity for muscular work in human beings (Graham *et al.* 1994). It has been demonstrated that caffeine (6 mg/kg) improves the racing performance of cross-country skiers, particularly at high altitudes (Berglund and Hemmingsson, 1982). At therapeutic concentrations, both caffeine and theophylline have potent effects in improving contractility and in reversing fatigue of the diaphragm in both normal human subjects and in patients with chronic obstructive lung diseases. This effect on diaphragmatic performance, rather than an effect on the respiratory center, may account for theophylline's ability to improve the ventilator response to hypoxia and to diminish dyspnea even in patients with irreversible airflow obstruction.

2. CELLULAR AND MOLECULAR BASIS OF METHYLXANTHINES ACTION

Proposed mechanisms of xanthine-induced physiological and pharmacological effects have included (1) inhibition of phosphodiesterases, thereby increasing intracellular cyclic AMP, (2) direct effects on intracellular calcium concentration, (3) indirect effects on intracellular calcium concentrations via cell membrane hyperpolarization, (4) uncoupling of intracellular calcium increases with muscle contractile elements, and (5) antagonism of adenosine receptors. A large body of evidence suggests that adenosine receptor antagonism is the most important factor responsible for most pharmacological effects of methylxanthines in doses that are administered therapeutically or consumed in xanthine containing beverages.

Following a single effective dose of 100 mg, the peak levels of caffeine in body fluids are between 2 to 15 μM. On the other hand, in patients who have been admitted to hospital due to acute caffeine poisoning, the levels of caffeine are around a few hundred μM. The biochemical mechanism that underlies the actions of caffeine must hence be activated at concentrations between these extremes. From these facts it is not surprising that the direct release of intracellular calcium, which occurs only at millimolar concentrations, could be ruled out. Also the inhibition of cyclic nucleotide phosphodiesterases occurs at rather higher concentrations than those attained during human caffeine consumption. In fact, the only mechanism that is known to be significantly affected by the relevant doses of caffeine is binding to adenosine receptors and antagonism of the actions of agonists at these receptors (Fredholm, 1980).

2.1. Adenosine is a Normal Cellular Constituent

As all the other autocoids, adenosine is a normal cellular constituent. The balance of several enzymes regulates its intracellular level. The extracellular level of adenosine is controlled by means of different transporters.

Adenosine is formed by the action of an AMP selective 5'-nucleotidase and the rate of adenosine formation via this pathway is mainly controlled by the amount of AMP. Adenosine kinase and adenosine deaminase are the most important enzymes for the removal of adenosine. Adenosine deaminase is present mostly intracellularly, and its preferred substrate is not adenosine but 2-deoxyadenosine (Fredholm & Lerner, 1982). Its K_m for adenosine is much higher than 5 μmolar and it is believed that this enzyme is more important when adenosine levels are high (Arch & Newsholme, 1978). Adenosine kinase, by contrast, has a K_m in the range of physiological intracellular adenosine concentrations. Blockade of adenosine kinase has an even larger effect on the rate of adenosine release than does blockade of adenosine deaminase (Lloyd & Fredholm, 1995). S-Adenosyl homocysteine hydrolase is another enzyme of importance in the regulation of adenosine. This enzyme sets the equilibrium between adenosyl homocysteine and adenosine+L-homocysteine. When the level of the amino acid is low this enzyme serves to generate adenosine. On the other hand, when the level of L-homocysteine is raised it can trap adenosine formed via AMP breakdown as S-adenosyl homocysteine inside the cell.

Intracellular and extracellular adenosine concentrations are controlled by means of equilibrate transporters, and many other transporters. Several agents such as nitrobenzylthioinosine, propentofylline, dipyridamole and dilazep block the equilibrate transporters. The level of adenosine rises in the CNS when the inhibitors are given. On the other hand, some sodium-dependent, concentrating transporters can move extracellular adenosine into cells. The above agents do not block them. Their precise role in the CNS is still not very clear.

From these considerations it can be expected that adenosine levels in the extracellular fluid should be raised whenever there is a discrepancy between the rate of ATP consumption and ATP synthesis. In addition, it is expected that drugs that interfere with the key enzymes and with the transporters should affect adenosine levels. Therefore, it is not very easy to determine the basal level of adenosine (both intracellular and extracellular) in the brain because it is controlled by many very active metabolic pathways. It has been reported that the adenosine level of awake, unrestrained rats is estimated to be between 30 and 300 nM. The level of adenosine can increase dramatically to 10 μM or more following ischemia (Andine *et al.*, 1990). However, even the basal levels are sufficient to cause a tonic activation of at least some types of adenosine receptors.

2.2.   Adenosine Acts on Several Types of Adenosine Receptors

At present 4 distinct adenosine receptors, A_1, A_{2A}, A_{2B} and A_3 have been cloned and characterized. Of these receptors A_{2A} and A_{2B} are coupled with G_s protein, whereas, A_1 and A_3 are coupled with G_i. The overall and transmembrane domain amino acid identity among the four human adenosine receptors is approximately 30 and 45%, respectively. They all fit the structural motif typical of G protein-coupled receptors, of which the adrenergic and muscarinic receptor G protein families have been the most extensively studied and from which most generalities of receptor structure-function relationships are based. Briefly, this architecture features seven stretches of hydrophobic α-helical regions. The regions connecting the membrane-spanning domains 2–3, 4–5, and 6–7 (typically referred to as extracellular loops 1, 2 and 3, respectively) and the amino terminus of the receptor are oriented into the extracellular space. Regions connecting transmembrane domains 1–2, 3–4 and 5–6 (intracellular loops 1, 2 and 3, respectively) and the carboxy-terminal of the receptor are directed cytoplasmically. Portions of these segments are believed to interact directly with the α-subunit of G proteins in order to transmit the signal of receptor activation, and the segments also contain sites involved in receptor regulatory processes such as phosphorylation. Compared to many G protein coupled receptors, the A_{1A}, A_{2B} and A_3 adenosine receptors are small in size, and all adenosine receptors possess a rather short amino terminus. The A_1 and A_{2A} receptors in native tissues have been shown to be glycoproteins, and consensus sites for N-linked glycosylation exist on all adenosine receptors. Other structural features common to many G protein coupled receptors that are present in adenosine receptors include an aspartate residue in transmembrane domain 2 that may be involved in receptor regulation by sodium ions, an Asp-Arg-Tyr sequence in the second intracellular loop, and cysteines in extracellular regions that may be involved in intrareceptor disulfide bond formation. A conserved cysteine residue that may be a site for receptor palmitylation exists in the carboxy-terminal tail of all adenosine receptor subtypes, with the exception of the A_{2A}.

A_1 and A_{2A} receptors are activated at the low basal concentrations of adenosine measured in resting rat brain. Thus, these receptors are likely to be the major targets of caffeine and its relative methylxanthines. The A_1 receptor is coupled to the pertussis toxin sensitive G-proteins G_{i-1}, G_{i-2}, G_{i-3}, G_{o-1} and G_{o-2}. Therefore, activation of A_1 receptors can cause inhibition of adenylyl cyclase and some types of voltage sensitive Ca^{2+}-channels such as the N- and Q-channels, activation of several types of K^+-channels, phospholipase C and phospholipase D. Consequently a host of different cellular effects can ensue. A_{2A} receptors associate with G_s-proteins, activation of these receptors cause the activation of adenylyl cyclase and perhaps also activation of some types of voltage sensitive Ca^{2+}-channels, especially the L-channel. Thus, it is apparent that A_1 and A_2 receptors have partly opposing actions at the cellular level. This is interesting because the two types of receptor are co-expressed in the same cell.

It was thought that the A_{2B} receptor is unlikely to provide an explanation for the actions of caffeine and other methylxanthines because the concentration of adenosine for activation of this receptor is higher than those found in resting animal tissues. In general, A_{2B} receptors are recognized to have a lower affinity for agonists compared with other receptor subtypes, this is not true for antagonists. The structure-activity relationship of A_{2B} receptors for adenosine antagonists has not been completely characterized but at least some xanthines are as potent antagonists at A_{2B} receptors as they are at other adenosine receptors (Feoktistov and Biaggioni, 1993; Brackett and Daly, 1994). A_{2B} receptors have been implicated in the regulation of vascular smooth muscle tone, cell growth, intestinal function and neurosecretion. A_3 receptors are poorly affected by many methylxanthines including caffeine. This receptor is not a target of caffeine actions in man.

Results of receptor autoradiography and *in situ* hybridization showed that A_1 receptors are widely distributed in the brain and the highest levels are in cortex, hippocampus and cerebellum, whereas the A_{2A} receptors are found to be concentrated in the dopamine rich regions of the brain.

2.3. Adenosine A_1 Receptors Regulate Transmitter Release and Neuronal Firing Rates

The inhibitory effect of adenosine on transmitter release in both the peripheral and central nervous system has been demonstrated (Fredholm and Hedqvist, 1980). There is some evidence that the release of excitatory transmitters is more strongly inhibited by adenosine than that of inhibitory neurotransmitters (Fredholm and Dullwiddie, 1988). This is in agreement with a proposed role of adenosine as a homeostatic regulatory factor that serves to match the rate of energy consumption to the rate of metabolite supply. The receptors involved are adenosine A_1 receptors. Several mechanisms of synaptic vesicle docking, such as decreased the calcium entry via different kinds of calcium channels, interacting with small G proteins rea3A, known to be involved in transmitter release and to be pertussis toxin insensitive, are involved in A_1 receptor mediated inhibition of transmitter release (Fredholm *et al.* 1989, Thompson *et al.* 1992).

There is also some evidence that increases in cyclic AMP in nerve endings are associated with an increase in transmitter release (e.g. Chavez-Noriega and Stevens, 1994). Since activation of adenosine A_l receptors is known to cause a decrease in cyclic AMP formation it is conceivable that this may also be a mechanism for decreased transmitter release at least under some circumstances.

Adenosine also acts to decrease the rate of firing of central neurons (Phillis & Edstrom, 1976) through activation of potassium channels via adenosine A_1 receptors (Dunwiddie, 1985). When the effect of endogenous adenosine at these receptors on glutamatergic neurons is blocked by caffeine it leads to epileptiform activity *in vitro* (Dunwiddie, 1980; Dunwiddie *et al.* 1981) and this is likely to provide a basis for the well-known seizure inducing effect of caffeine *in vivo*. It is also known that caffeine increases the turnover of several monoamine neurotransmitters including dopamine, noradrenaline and 5-hydroxytryptamine (Bickford *et al.* 1985; Fredholm and Jonzon, 1988; Hadfield and Milio, 1989). There is evidence that methylxanthines increase the rate of firing of noradrenergic neurones in the locus coeruleus (Grant and Redmond Jr., 1982) and of dopaminergic neurons in the ventral tegmental area (Stoner *et al.* 1988). Quite recently it was shown that the mesocortical cholinergic neurons are tonically inhibited by adenosine and that caffeine consequently increases their firing rate (Rainnie *et al.* 1994). It was postulated that this effect is of importance in the EEG arousal following caffeine ingestion. Since dopamine and noradrenaline neurons also are involved in arousal there is ample neuropharmacological basis for the central stimulatory effect of caffeine being related to inhibition of adenosine A_1 receptors.

2.4. Adenosine A_{2A} Receptors Inhibit Postsynaptic Dopamine D_2 Receptors

Adenosine A_{2A} receptors are located preferentially in the dopamine-rich regions of the brain, i.e. nucleus caudatus/putamen, nucleus accumbens and tuberculum olfactorium. This association was in fact noted a long time ago when it was shown that in cell-free homogenates from these regions, and only from these regions, adenosine stimulated adenylate cyclase (Fredholm, 1977). Now it has been convincingly shown that adenosine A_{2A} receptor mRNA is co-localised with dopamine D_2 receptor mRNA in a subpopulation of the medium sized spiny GABAergic neurons, namely those that also express enkephalin mRNA. These neurones in the striatum project to globus pallidus. Interestingly, these co-localized receptors have been shown to interact functionally. Thus, activation of A_{2A} receptors has been shown to decrease the affinity of dopamine binding to dopamine D_2 receptors (Ferre *et al.* 1992). Furthermore, adenosine A_2, receptor stimulation has been shown to block post- but not presynaptic D_2 receptor actions (Ferre *et al.* 1992).

This interaction is very interesting since it could provide a mechanism for several actions of adenosine receptor antagonists on dopamine activity. Thus, an inhibition of A_{2A} receptors would be expected to increase the activity of dopamine at D_2 receptors (Ferre *et al.* 1992). There is indeed ample evidence that caffeine (and other adenosine receptor antagonists) can increase behaviors related to dopamine. The first demonstration of an adenosine-dopamine interaction on behavior was the finding that several adenosine receptor antagonists could increase dopamine receptor activated rotation behavior. This type of finding has since been repeatedly confirmed and elaborated (Ferre *et al.* 1992).

2.5. Caffeine Induces Immediate Early Genes in the Striatum

Expression of so-called immediate early genes such as *c-fos, c-jun, junB, junD, zif*268 (NGFI-A) and NGFI-B is often accompanied with an increase of neuronal activity. Caffeine causes a concentration-dependent increase in *c-fos* expression, which is confined to the striatum (Johansson *et al.* 1994). Concentrations of caffeine higher than about 50 mg/kg are required in order to see any increase at all. These concentrations are clearly higher than those required for behavioral stimulation. This could mean that the caffeine-induced increase in immediate early genes is related to the second phase of caffeine action, which involves a behavioral depression. Alternatively the dose-response relationship could indicate that many times higher concentrations are required to observe a generalized *c-fos* increase than are needed to produce activation of a sufficient number of neurons to produce a behavior stimulation.

The other members of the same family of immediate early genes such as *c-jun* and *junB* are also increased by caffeine. Furthermore, *c-fos, c-jun* and *junB* can form an increased expression of the AP-1 transcription factor. Moreover, there are later changes in the expression of neuropeptides that are known to have AP-1 sensitive regulatory elements, notably preproenkephalin. These results suggest that caffeine can induce changes in gene expression that could lead to adaptive changes in the brain.

2.6. There are Marked Adaptive Changes Following Long-term Treatment with Caffeine

The chronic effects of caffeine have been studied extensively as well as the acute effects, and the phenomena of tolerance and withdrawal are well-documented in animals and humans. The only known biochemical targets for caffeine, at which it has significant activity at concentrations achieved during normal human use, are A_1 and A_2 adenosine receptors. Xanthines were envisaged to have therapeutic potential as central stimulants and cognitive enhancers, as cardiac stimulants, as anti-asthmatics, as anti-Parkinson's disease agents, as antiobesity agents, as analgesic adjuvants, and as diuretics. Many of the xanthines are selective for adenosine receptor subtypes, and many are more potent than caffeine. Conversely, a variety of adenosine analogues have been developed with envisaged therapeutic potential as analgesics, antipsychotics, anticonvulsants, or for treatment of stroke. However, over the past few years it has become apparent that the effect of acute or chronic treatment with caffeine, or other adenosine receptor ligands, can be qualitatively different. Thus, long-term treatment with adenosine receptor antagonists can have effects that resemble the acute effects of adenosine receptor agonists, and vice versa. Such diametrically opposed actions of chronic versus acute treatments have important implications in the development of adenosine receptor ligands as therapeutic agents.

2.7. Changes after Chronic Administration of Caffeine

After chronic administration of caffeine the A_1 receptors were found to be up-regulated in most studies, without any changes in adenosine A_1 receptor mRNA (Johansson *et. al.* 1993). In contrast, A_{2A} receptor binding and mRNA expression were unaltered.

In addition to A_1 adenosine receptors, 5-HT receptors, acetylcholine receptors, GABA receptors and δ-opioid receptors were up-regulated following chronic caffeine ingestion by mice (Shi *et al.* 1995). Only β-adrenoceptors were downregulated.

However, it should be noted that an apparent up-regulation of nicotinic acetylcholine receptors probably represents conversion to a desensitized, high-affinity state, and thus, actually represents a downregulation, as has been reported after chronic nicotine treatment (Marks *et al.* 1993). Adenosine A_{2A} receptors, α-adrenoceptors, dopamine receptors, and excitatory amino acid receptors were unaltered. Among other biochemical alterations, the apparent increase in levels of L-type Ca^{2+} channels after chronic caffeine administration is noteworthy.

There is now good evidence that treatment with adenosine receptor antagonists can alter immediate early gene expression and the expression of secondary gene products, such as neuropeptides. High doses of caffeine (> 50 mg kg^{-1}) are known to induce *c-fos*, *junB*, *c-jun*, nerve growth factor-induced clone A (NGFI-A), and NGFI-B mRNA (Johansson *et al.* 1994, Svenningsson *et al.* 1995). The first three can form the AP-1 transcription factor, which also increases after caffeine administration, and the latter two are transcriptional factors in their own right. Lower doses of caffeine can cause a decrease in the expression of certain gene products, particularly in the striatopallidal neurones in the cortex. Thus, there are marked phenotypic changes in selected neurones following acute and long-term caffeine administration. It is reasonable to assume that these changes are also manifested in adaptations of their functional characteristics. Indeed, the various biochemical changes after chronic caffeine administration are paralleled by altered behavioral responses to caffeine and to other agents.

2.8. Changes in Behavioral Responses after Chronic Administration of Caffeine

Chronic ingestion of caffeine by humans can lead to tolerance, and withdrawal syndromes can occur, including apathy, drowsiness, headaches, nausea and anxiety. True dependence, that is addiction, probably does not occur in humans and has been difficult to demonstrate in animals. In rats, chronic caffeine administration appears to result in almost complete tolerance to the motor stimulant effects of caffeine (Holtzman *et al.* 1991). One explanation of complete tolerance to the behavioral stimulant effects of caffeine may be based on the biphasic concentration-response curve for caffeine. At low doses of caffeine, there is an increase in locomotor activity, presumed to be the result of antagonism at adenosine receptors, while higher doses cause depression. An increase in adenosine receptor function, after chronic administration of caffeine, might be expected to increase the threshold for the stimulatory action of caffeine to the point where the depressant action, which occurs at a different site (possibly inhibition of phosphodiesterases), overrides the stimulatory effects of adenosine receptor antagonism. It is also possible that alteration in other receptors, in pathways modulated by adenosine receptors, accounts for the tolerance to caffeine.

The behavioural responses to cholinomimetic agents are also altered after chronic caffeine administration in mice. This can be explained in terms of desensitization of nicotinic acetylcholine receptors, that is, a tolerance to locomotor effects of nicotine, and in terms of up-regulation of muscarinic receptors – an increase in the threshold for locomotor stimulant effects of the muscarinic acetylcholine receptor antagonist, scopolamine (Nikodijevic *et al.* 1993a, b). The lack of alteration in striatal dopamine receptors

(Shi *et al.* 1995) and the responses to drugs such as amphetamine or cocaine, which affect dopamine systems (Nikodijevic *et al.* 1993), after chronic caffeine administration is remarkable, as there is considerable evidence implicating dopamine systems as a major target for the pharmacological effects of caffeine and theophylline (Ferr, *et al.* 1992). A possible clue to both the tolerance to caffeine and the apparent lack of alteration in dopamine-mediated responses has now been forthcoming. Thus, rats rendered tolerant to caffeine are also tolerant to agonists selective for either D_1 and D_2 receptors, but remain responsive to agonists that activate both D_1, and D_2 receptors (Marks *et al.* 1993).

Further correlative studies on behavioral and biochemical alterations after chronic caffeine administration may lead to a more integrated view of the central sites of action of this agent. Although complete, or nearly complete, tolerance to stimulant locomotor effects of caffeine occurs in rats (Holtzman *et al.* 1991, Lau & Falk, 1994), it does not occur in NIH Swiss strain mice (Nikodijevic *et al.* 1993) or CD-1 mice (Kaplan *et al.* 1993). Instead, chronic caffeine ingestion results in a behavioral depression of activity in the mice, in marked contrast to the behavioral stimulation caused by acute administration of caffeine.

It is known that tolerance develops to some, but not to all effects of caffeine in man and experimental animals (Holtzman and Finn, 1988; Robertson *et al.* 1981). Over the past few years it has become apparent that the effect of acute or chronic treatment with caffeine, or other adenosine receptor ligands, can be qualitatively different. Thus, long-term treatment with adenosine receptor antagonists can have effects that resemble the acute effects of adenosine receptor agonists, and vice versa. Such diametrically opposed actions of chronic versus acute treatments have important implications in the development of adenosine receptor ligands as therapeutic agents.

The ingestion of 85 to 250 mg of caffeine produces an increased capacity for sustained intellectual effort and decreases reaction time; however, tasks involving delicate muscular coordination and accurate timing or arithmetic skills may be adversely affected (Curatolo and Robertson, 1983; Arnaud, 1987). Similarly, the ability of asthmatic children to perform repetitive tasks requiring concentration declines during periods of medication with theophylline (Furukawa *et al.* 1988). Patients with panic disorders may be particularly sensitive to the effects of the methylxanthines. In one study, many such individuals given doses of caffeine that resulted in plasma concentrations of about 8 μg/ml experienced anxiety, fear, and other symptoms characteristic of their panic attacks (Charney *et al.* 1985). Since the long-term ingestion of caffeine (and presumably theophylline) can produce tolerance and evidence of physical dependence, history of exposure to methylxanthines will influence the effects of a given dose. Hence, enhanced alertness, energy, and ability to concentrate could reflect alleviation of withdrawal symptoms in some instances.

3. CAFFEINE AND HEALTH

Caffeine is one of many constituents in foods that can exert pharmacological and physiological effects. It was placed on the U.S. FDA's Generally Recognized As Safe (GRAS) list in 1958. In 1978, the agency recommended additional research be conducted to resolve any uncertainties about its safety (Institute of Food Technologists 1987, Lecos, 1988). Since then, a great deal of research has been conducted on caffeine and its association with the development of various diseases and health concerns. Although caffeine is one of the most comprehensively studied ingredients in the food supply, some questions and misperceptions about the potential health effects associated with this ingredient still persist.

People differ greatly in their sensitivity to caffeine. When analyzing caffeine's effects on an individual, many factors must be weighed: the amount ingested, frequency of consumption, individual metabolism and sensitivity (Dews, 1986).

3.1. Caffeine and Children

Most people believe that children are more sensitive to caffeine, but the fact is that children, including those diagnosed as hyperactive, are no more sensitive to the effects of caffeine than adults (Dews, 1986). After reviewing 82 papers, examining the behavioral effects of caffeine in children, Leviton found the result was reassuring. Except for infants, children metabolize caffeine more rapidly than adults; and children in general consume much less caffeine than adults, even in proportion to their smaller size (Leviton, 1992). Also a study by Rapoport demonstrated that caffeine was not a cause of attention deficit disorder with hyperactivity (Rapoport *et al.* 1984).

3.2. Caffeine and Cancer

Over the years, both caffeine and coffee have been linked to certain cancers, but these associations are no longer supported by medical research.

A case-control study on the relationship of coffee consumption to digestive tract cancers (LaVecchia *et al.* 1989) found no correlation between coffee consumption and the incidence of digestive tract cancer. The study included patients with confirmed cases of oral, rectal, stomach, liver and colon cancers, as well as patients who did not suffer any digestive tract disorders. After reviewing 13 epidemiological and clinical studies that examined the link between bladder, rectal, colon and pancreatic cancers and coffee and tea consumption, Rosenberg found no relationship between coffee or tea consumption and the incidence of bladder, rectal, colon or pancreatic cancers (Rosenberg, 1990). A 1991 study by Jain *et al.* examined the association between coffee and alcohol with pancreatic cancer. The population-based study included 750 subjects and adjusted for smoking as well as caloric and fiber intake. After calculating lifetime tea and coffee consumption and the variety of coffee consumed, Jain confirmed the results of an earlier review, involving several thousand subjects, which stated that current epidemiological evidence does not suggest any significant increase in risk of pancreatic cancer with coffee consumption (Jain *et al.* 1991; Gordis, 1990).

A 1993 meta-analysis by Yale University School of Medicine researchers critically reviewed 35 case control studies of coffee and lower urinary tract cancer that had been published since 1971. They showed no evidence of an increase in the risk of lower urinary tract cancer with coffee consumption after adjustment for the effects of cigarette smoking (Viscoli *et al.* 1993).

A scientific review by Lubin and Ron examined all the data of 11 case-control studies linking caffeine consumption and malignant breast tumors. None established a significant link between caffeine intake and breast cancer incidence (Lubin and Ron, 1990). Specifically, three separate well-controlled studies performed in Israel, the United States and France established no association between coffee consumption and breast cancer (National Research Council, 1989).

The American Cancer Society's Guidelines on Diet, Nutrition and Cancer state there is no indication that caffeine is a risk factor in human cancer; and the National Academy of Sciences' National Research Council reports there is no convincing evidence relating caffeine to any type of cancer (National Research Council, 1989; American Cancer Society's Medical and Scientific Committee, 1996).

3.3. Caffeine and Cardiovascular Diseases

The relationship between caffeine and cardiovascular disease is an area that has been extensively examined.

A 1986 study cited a link between excessive coffee consumption and heart disease, but the investigators failed to control for other significant risk factors such as diet and smoking (LaCroix *et al.* 1986).

On the other hand, many studies could not find the link between cardiovascular disease and consumption of coffee and caffeine. A prospective study conducted by Harvard University researchers concluded that caffeine consumption causes "no substantial increase in the risk of coronary heart disease or stroke." The study included 45,589 men between the ages of 40 to 75 years old and adjusted for major cardiovascular-risk indicators including dietary intake of fats, cholesterol and smoking (Grobbee *et al.*, 1990). Additionally, a case-control study on the effect of filtered-coffee consumption on plasma lipid levels indicated that coffee consumption led to a small increase in the level of high-density lipoprotein cholesterol, that is believed to protect against and lower the risk for coronary heart disease (Fried *et al.* 1992). Results from the Scottish Heart Health Study, published in 1993, support the finding that filtered-coffee consumption was not linked to an increase in cholesterol concentrations or coronary heart disease (CHD). This study of 9740 men and women in the United Kingdom, concluded that neither tea nor coffee consumption was linked to CHD. The majority of coffee consumed in the United Kingdom is instant, and the researchers noted that previous studies indicating a positive relationship between coffee and CHD were focusing on unfiltered and boiled coffee which are consumed in Scandinavia, but rarely in the US (Brown *et al.* 1993). The result of a study published in the Journal of the American Medical Association, which is the largest study on caffeine and CHD ever conducted on women, found no evidence of a positive relationship between coffee consumption (regular or decaffeinated, current or past consumption) and risk of CHD. The study also pointed out that there was no observable difference in effects between genders (Willett *et al.* 1996).

The effects of caffeine on blood pressure have been the subjects of various hypotheses, many of which have been disproved. A number of studies have shown that any temporary rise in blood pressure due to caffeine consumption is less than the elevation produced by normal, daily activities. The result of a double-blind randomized trial including 69 healthy participants indicated that caffeine has no adverse effect on cardiovascular risk by inducing unfavorable changes in blood pressure or serum lipids (Bak and Grobbee, 1991). Recent analysis from a multiple year intervention trial indicates an inverse correlation between caffeine intake and both systolic and diastolic blood pressure (Stamler *et al.* 1997).

3.4. Caffeine and Women's Health

The effect of caffeine on health of women, especially during childbearing ages, is a very important question. Many women wonder if it is safe to consume caffeine-containing foods and beverages. Women's health issues, reproductive effects and osteoporosis, for example, continue to be actively investigated. Recent research continues to support moderate consumption of caffeine during pregnancy and in post-menopausal women.

The effect of coffee consumption on fertility has been studied extensively. A study conducted by the Centers for Disease Control and Prevention and Harvard University examined 2800 women who had recently given birth and 1800 with the medical diagnosis of primary infertility. The researchers found that caffeine consumption had little or no effect on the reported time to conceive in those women who had given birth, and that caffeine consumption was not a risk factor for continued infertility in women being treated for infertility (Joesoef *et al.* 1990). These findings were confirmed in an epidemiological study of more than 11,000 Danish women published in 1991 and further strengthened in 1995 by University of California-Berkeley researchers who evaluated 1300 pregnant women. The studies examined the relationship between the number of months to conception, cigarette smoking and coffee and tea consumption. The research found no association between delayed conception and the consumption of caffeinated beverages among nonsmokers (Olsen, 1991, Alderete *et al.* 1995). In a study of 210 women, published in the American Journal of Public Health in 1998, the researchers did not find a significant association between total caffeine consumption and reduced fertility. In fact, they found that women who drank more than one-half cup of tea per day had a significant increase in fertility. This was particularly true with caffeine consumption in the early stages of a woman's attempt at conception (Caan *et al.* 1998).

After reviewing the results of 26 human studies on the effect of caffeine consumption on reproduction, Leviton concluded, "caffeine as currently consumed by pregnant women has no discernible adverse effects on fetuses" (Leviton, 1995). A seven-year prospective study comprising more than 1500 women examined caffeine use during pregnancy and infant outcome. Caffeine consumption, as determined by self-reporting, averaged 193 mg of caffeine daily during early pregnancy and decreased to 152 mg per day by mid-term. The study showed no relationship between caffeine intake and birth weight, birth length or head circumference. Follow-up examinations of the children at ages eight months, four and seven years also revealed no relationship between caffeine intake during pregnancy and early childhood development measures of motor skills or intelligence (Barr & Streissguth, 1991).

In a study, conducted in 1990–91, 5,342 pregnant women were interviewed, and the researchers concluded that there was no increased risk for spontaneous abortion associated with caffeine consumption. A research team headed by the US National Institute of Child Health and Human Development published the results of their prospective study on 431 women during pregnancy. The researchers carefully monitored the women and the amount of caffeine they consumed from conception to birth. After accounting for nausea, smoking, alcohol use and maternal age, the researchers found no correlation between caffeine consumption of up to 300 mg per day and adverse pregnancy outcomes, including spontaneous abortion (miscarriage) (Mills *et al.*, 1993).

Stein and Susser hypothesized that the nausea commonly seen in pregnancy may create an erroneous association between caffeine consumption and miscarriage. Nausea is associated with increasing hormone levels during a normal pregnancy and is significantly less common in pregnancies that end in miscarriage. A National Institutes of Health study pointed out that women who experienced nausea consumed significantly less caffeine than women who did not encounter nausea. The reduced caffeine consumption in women with nausea compared with women already destined to abort, who have less nausea and thus less reduction in caffeine consumption, can be misconstrued as an adverse effect of caffeine (Stein & Susser, 1991). The Stein-Susser nausea hypothesis may explain the findings of a 1993 study, a retrospective case-control study of 331 cases with fetal loss and 993 controls with a normal pregnancy. Caffeine intake before and during pregnancy was shown to be associated with increased fetal loss, but the authors failed to measure the effects of nausea or to assess the impact of nausea on fetal loss. In addition, caffeine consumption was not measured and adjustments for smoking and alcohol consumption were not made (Infante-Rivard *et al.* 1993). A study involving almost 900 cases provided further evidence against an effect of caffeine consumption on pre-term delivery (Pastore & Savitz, 1995). There's no evidence that moderate caffeine intake has adverse effects on pregnancy or pregnancy outcome (McDonald *et al.* 1992; Armstrong *et al.* 1992).

The relationship between caffeine intake and osteoporosis is a relatively new area of investigation. Because caffeine has been shown to impact on calcium excretion slightly, it has been suggested as a risk factor for osteoporosis. An array of studies has been conducted in recent years. In 1994, an NIH convened federal advisory panel at the Consensus Development Conference on Optimal Calcium Intake concluded that caffeine has not been found to affect calcium absorption or excretion significantly (National Institutes of Health, 1994). Another study conducted from 1984 to 1990, followed 145 healthy college-aged women, concluding there was no association between caffeine consumption and bone density (Packard and Recker, 1996).

A double-blind, placebo controlled crossover trial conducted by the same Creighton University researchers who first demonstrated the association, found that caffeine (400 mg/day) had no appreciable effect on external calcium balance in premenopausal women consuming at least 600 mg of calcium per day, less than half of the RDA. The researchers concluded that moderate caffeine intake was not a threat to bone health (Barger-Lux *et al.* 1990).

A 1992 study examined the lifetime intake of caffeinated coffee in 980 postmenopausal women, and showed there was an association between lifetime caffeinated coffee intake (equivalent to two cups/day) and reduced bone mineral density. However, this observation was only seen among women who had low intake of milk suggesting that coffee replaced milk consumption in these women. Supplementation of calcium intake by consuming at least one glass of milk per day eliminated the relationship between coffee intake and decreased bone density (Cooper *et al.* 1992).

Lloyd *et al.* examined the effects of long-term habitual caffeine intake on bone status of healthy postmenopausal women. Caffeine content was measured from diet records then analytically tested, and bone density measurements were taken from both hips. To reduce confounding variables, women aged 55–70 who had minimal or no exposure to hormone replacement therapy were studied. Caffeine intake, from none to 1400 mg per day in this study population, was not associated with any changes in bone density (Lloyd & Rollings, 1997).

3.5. Caffeine and Withdrawal

Depending on the amount ingested, caffeine can be a mild stimulant to the central nervous system. Although sometimes colloquially referred to as "addictive," moderate caffeine consumption is safe and should not be classified with addictive drugs of abuse. When regular caffeine consumption is abruptly discontinued, some individuals may experience withdrawal symptoms, such as headaches, fatigue or drowsiness. These effects are usually temporary, lasting up to a day or so, and can often be avoided if caffeine cessation is gradual (Hughes *et al.* 1992; Silverman *et al.* 1992; Strain *et al.* 1994). Moreover, most caffeine consumers do not demonstrate dependent, compulsive behavior, characteristic of dependency to drugs of abuse (Hughes *et al.* 1988). Although pharmacologically active, the behavioral effects of caffeine are typically minor. As further elaborated by the American Psychiatric Association, drugs of dependence cause occupational or recreational activities to be neglected in favor of drug-seeking activity (Hughes *et al.* 1988). Clearly, this is not the case with caffeine.

REFERENCES

Alderete, E., Eskenazi, B., Sholtz, R. (1995) Effect of cigarette smoking and coffee drinking on time to conception. *Epidemiology*, 6, 403–408.

American Cancer Society's Medical and Scientific Committee (1996) Guidelines on diet, nutrition, and cancer. *CA-A Cancer Journal for Clinicians*, 41, 334–338.

Andine, P., Rudolphi, A., Fredholm, B.B., Hagberg, H. (1990) Effects of propentofylline (HWA 285) on extracellular purines and excitatory amino acids in CAI of rat hippocampus during rransient ischemia. *Brit. J. Pharmacol.*, 100, 814–818.

Arch, J.R.S. and Newsholme, E.A. (1978) The control of the metabolism and the hormonal role of adenosine. *Essays Biochem*, 14, 88–123.

Armstrong, B.G., McDonald, A.D., Sloan, M. (1992) Cigarette, Alcohol, and coffee consumption and spontaneous abortion. *American Journal of Public Health*, 82, 85–90.

Arnaud, M.J. (1987) The pharmacology of caffeine. *Prog. Drug Res.*, 31, 273–313.

Bak, A.A.A. and Grobee, D.E. (1991) Caffeine, blood pressure, and serum lipids. *American Journal of Clinical Nutrition*, **53**, 971–975.

Barger-Lux, M.J., Heaney, R.H., Stegman, M.R. (1990) Effects of moderate caffeine intake on the calcium economy of premenopausal women. *American Journal of Clinical Nutrition*, **52**, 722–725.

Barr, H.M. and Streissguth, A.P. (1991) Caffeine use during pregnancy and child outcome: a 7-year prospective study. *Neurotoxicology and Teratology*, **13**, 441–448.

Berglund, B. and Hemmingsson, P. (1982) Effects of caffeine ingestion on exercise performance at low and high altitudes in cross-country skiers. *Int. J. Sports Med.*, **3**, 234–236.

Bickford, P.C., Fredholm, B.B., Dunwiddie, T.V., Freecman, R. (1985) Inhibition of Purkinje cell riring by systemic administration of phenyl-isopropyl adenosine: effect of central noradrenalin depletion by DSP4. *Life Sci.*, **37**, 289–297.

Brackett, L.E. and Daly, J.W. (1994) Functional characterization of the A2b fibroblasts adenosine receptor in NIH 3T3 *Biochem. Pharmacol.*, **47**, 801–814.

Brown, C.A., Bolton-Smith, C., Woodward, M., Tunstall-Pedoe, H. (1993) Coffee and tea consumption and the prevalence of CHD in men and women: results from the Scottish Heart Health Study. *Journal of Epidemiological Community Health*, **47**, 171–175.

Caan, B., Quesenberry, C.P., Coates, A.O. (1998) Differences in Fertility Associated with Caffeinated Beverage Consumption. *American Journal of Public Health*, **88**, 270–274.

Charney, D.S., Heminger, G.R., Jatlow, P.I. (1985) Increased anxiogenic effects of caffeine in panic disorders. *Arch. Gen. Psychiatry*, **42**, 233–243.

Chavez-Noriega, L.E. and Stevens, C.F. (1994) Increased transmitter release at excitatory synapses produced by direct activation of adenylate cyclase in rat hipposampal slices. *J. Neurosci.*, **14**, 310–317.

Chou, J.M. and Benowitz, N.L. (1994) Caffeine and coffee: effects on health and cardiovascular disease. *Compl. Biochem. Physiol.*, **109c**, 173–189.

Cooper, C., Atkinson, E.J., Wahner, H.W., O'Fallon, W.M., Riggs, B.L., Judd, H.L., *et al.* 3d. (1992) Is caffeine consumption a risk factor for osteoporosis? *Journal of Bone and Mineral Research*, 7, 465–471.

Curatolo, P.W. and Robertson, D. (1983) The health consequences of caffeine. *Ann. Intern. Med.*, **98**, 641–653.

DeLander, G.E. and Hopkins, C.J. (1986) Spinal adenosine modulates descending antinociceptive pathways stimulated by morphine. *J. Pharmacol. Exp. Ther.*, **239**, 88–93.

Dews, P.B. (1986) Caffeine Research: An International Overview. Paper presented at a meeting of the International Life Sciences Institute, Sidney.

Dunwiddie, T.V. (1980) Endogenously released adenosine regulates excitability in the *in vitro* hippocampus. *Epilepsia*, **21**, 541–548.

Dunwiddie, T.V., Hoffer, B.J., Fredholm, B.B. (1981) Alkylxanthines elevate hippocampal excitability. Evidence of a role of endogenous adenosine. *Naunym-Schmiedeberg's Arch. Pharmacol.*, **316**, 326–330.

Dunwiddie, T.V. (1985) The physiological role of adenosine in the central nervous system. *E. Int. Rev. Neurobiol.*, **27**, 63–139.

Fenster, L., Hubbard, A.E., Swan, S.H., Windham, G.C., Waller, K., Hiatt, R.A., Benowitz, N. (1997) Caffeinated Beverages, Decaffeinated Coffee, and Spontaneous Abortion. *Epidemiology*, 8, 515–522.

Feoktistov, I. and Biaggioni, I. (1993) Characterization of adenosine receptors in human erythroleukemia cells and platelets: further evidence for heterogeneity of adenosine A2 receptor subtypes. *Mol. Pharmacol.*, **43**, 909–914.

Ferr, S., Fuxe, K., von Euler, G., Johansson, B., Fredholm, B.B. (1992) Adenosine-dopanine interactions in the brain. *Neuroscience*, **51**, 501–512.

Fredholm, B.B. (1977) Acrivation of adenylate cyclase from rat striatum and tuberculum olfactorium by adenosine. *Med. Biol.*, **55**, 262–267.

Fredholm, B.B. (1980) Are methylxanthine effects due to antagonism of endogenous adenosine? *Trends Pharmacol. Sci.*, **1**, 129–132.

Fredholm, B.B. and Hedqvist, P. (1980) Modulation of neurotransmission by purine nucleotides and nucleosides. *Biochem. Pharmacol.*, **29**, 1635–1643.

Fredholm, B.B. and Lerner, U. (1982) Metabolism of adenosine and $2'$-deoxy-adenosine by fetal mouse calvaria in culture. *Med. Biol.*, **60**, 267–271.

Fredholm, B.B. and Dullwiddie, T.V. (1988) How does adenosine inhibit transmitter release? *Trends Pharmacol. Sci.*, **9**, 130–134.

Fredholm, B.B. and Jonzon, B. (1988) Adenosine receptors: Agonists and antaginists. In: *Role of adenosine in cerebral metabolism and circulation*. Eds.: V. stefanovich & I. Okyayuz-Bakluti. *VSP BV*, Utrecht, pp. 17–46.

Fredholm, B.B., Proctor, W., Van der Ploeg, I., Dunwiddie, T.V. (1989) *In vivo* pertusis toxin treatment attenuates some, but not all, adenosine A_1 effects in slices of the rat hippocampus. *Eur. J. Pharmacol. Mol. Pharmacol. Sect.*, **172**, 249–262.

Fried, R.E., Levine, D.M., Kwiterovich, P.O., Diamond, E.L., Wilder, L.B., Moy, T.F., earson, T.A. (1992) The effect of filtered-coffee consumption on plasma lipid levels. *Journal of the American Medical Association*, **267**, 811–815.

Furukawa, C.T. (1988) Comparative trials including a β_2 adrenergic agonist, a methylxanthine, and a mast cell stabilizer. *Ann. Allergy*, **60**, 472–476.

Gordis, L. (1990) Consumption of methylxanthine-containing beverages and risk of pancreatic cancer. *Cancer Letters*, **52**, 1–12.

Graham, T.E., Rush, J.W. and van Soeren, M.H. (1994) Caffeine and exercise: metabolism and performance. *Can. J. Appl. Physiol.*, **19**, 111–138.

Grant, S. and Redmond, D.E., Jr. (1982) Methylxanthine activation of noradrenergic unit activity and reversal by clonidine. *Eur. J. Pharmacol.*, **85**, 105–109.

Grobbee, D.E., Rimm, E.B., Giovannucci, E., Colditz, G., Stampfer, M., Willett, W. (1990) Coffee, caffeine, and cardiovascular disease in men. *The New England Journal of Medicine*, **323**, 1026–1032.

Hadfield, M.G. and Milio, C. (1989) Caffeine and regional brain monoanine utilization in mice. *Life Sci.*, **45**, 2637–2644.

Holtzman, S.G. and Finn, I.B. (1988) Tolerance to behavioral effects of caffeine in rats. *Pharmacol. Biochem. Behav.*, **29**, 411–418.

Holtzman, S.G., Mante, S. and Minneman, K.P. (1991) Role of adenosine receptors in caffeine tolerance. *J. Pharmacol. Exp. Ther.*, **256**, 62–68.

Hughes, J.R., Higgins, S.T., Bickel, W.K. (1988) Caffeine self-administration, withdrawal, and adverse effects among coffee drinkers. *Archives of General Psychiatry*, **48**, 611–617.

Hughes, J.R., Oliveto, A.H., Helzer, J.E., Higgins, S.T., Biclel, W.K. (1992) Should caffeine abuse, dependence, or withdrawal be added to DSM-IV and ICD-10? *American Journal of Psychiatry*, **149**, 33–40.

Infante-Rivard, C., Fernandez, A., Gauthier, R., David, M., Rivard, G-E. (1993) Fetal loss associated with caffeine intake before and during pregnancy. *Journal of the American Medical Association*, **270**, 2940–2943.

Institute of Food Technologists (IFT) Expert Panel on Food Safety & Nutrition. Caffeine (1987) A Scientific Status Summary.

Jain, M., Howe, G.R., St. Louis, P., Miller, A.B. (1991) Coffee and alcohol as determinants of risk of pancreas cancer: a case-control study from Toronto. *International Journal of Cancer*, 47, 384–389.

Joesoef, M.R., Beral, V., Rolfs, R.T. (1990) Are caffeinated beverages risk factors for delayed conception? *The Lancet*, 335, 136–137.

Johansson, B., Ahlberg, S., Van der Ploeg, I., Brene, S., Lindefors, N., Persson, H., Fredholm, B. B. (1993) Effect of long terj caffeine treatment on A_1 and A_2 adenosine receptor binding and on mRNA levels in rat brain. *Naunyn-Schmied. Arch. Pharmacol.*, 347, 407–414.

Johansson, B., Lindstràm, K. and Fredholm, B.B. (1994) Differences in the regional and cellular localization of *c-for* mRNA induced by amphetamine, cocaine and caffeine in the rat. *Neuroscience*, 59, 837–849.

Kaplan, G.B., Greenblat, D.J., Kent, M.A., Cotreau-Bibbo, M.M. (1993) Caffeine treatment and withdrawal in mice: relationships between dosage, concentrations, locomotor activity and A_1 adenosine receptor binding. *J. Pharmacol. Exp. Ther.*, 266, 1563–1572.

La Vecchia, C., Monica, F., Negri, E., D'Avanzo, B., Decarli, A., Levi, F., et al. (1989) Coffee consumption and digestive tract cancers. *Cancer Research*, 49, 1049–1051.

La Croix, A.Z., Mead, L.A., Lian, G.K.Y. (1986) Coffee consumption and the incidence of coronary heart disease. *The New England Journal of Medicine*, 315, 977–982.

Lau, C.E. and Falk, J.L. (1994) Tolerance to oral and IP caffeine: locomotor activity and pharmacokinetics. *Pharmnacol. Biochem. Behav.*, 48, 337–344.

Lecos, C. (1988) Caffeine jitters: some safety questions remain. *FDA Consumer*, 21, 22–27.

Leviton A. (1992) Behavioral correlates of caffeine consumption by children. *Clinical Pediatrics*, 31, 742–750.

Leviton, A. (1995) Does coffee consumption increase the risk of reproductive adversities? *JAMWA*, 50, 20–2.

Lloyd, H.G.E., Fredholm, B.B. (1995) Involvement of adenosine deaminase and adenosine kinase in regulation extracellular adenosine concentration in rat hippocampal slices. *Neurochem. Int.*, 26, 387–395.

Lloyd, T., Rollings, N. (1997) Dietary caffeine intake and bone status of postmenopausal women. *American Journal of Clinical Nutrition*, 65, 1826–1830.

Lubin, F. and Ron, E. (1990) Consumption of methylxanthine-containing beverages and the risk of breast cancer. *Cancer Letters*, 53, 81–90.

Marks, M.J., Grady, S.R. and Collins, A.C. (1993) Down regulation of nicotinic receptor function after chronic nicotine infusion. *J. Pharmacol. Exp. Ter.*, 266, 1268–1276.

McDonald, A.D., Armstrong, B.G., and Sloan, M., (1992) Cigarette, alcohol, and coffee consumption and congenital defects. *American Journal of Public Health*, 82, 91–93.

Mills, J.L., Holmes, L.B., Aarons, J.H., Simpson, J.L., Brown, Z.A., Jovanovic-Poterson, L.G., et al. (1993) Moderate caffeine use and the risk of spontaneous abortion and intrauterine growth retardation. *Journal of the American Medical Association*, 269, 593–597.

Myers, M.G. (1988b) Caffeine and cardiac arrhythmias. *Chest*, 94, 4–5.

Myers, M.G. (1988a) Effects of caffeine on blood pressure. *Arch. Intern. Med.*, 148, 1189–1193.

National Institutes of Health (1994) Optimal calcium intake. NIH Consensus Development Conference, June 6–8. Washington, DC.

National Research Council (1989) Diet and Health: Implications for Reducing Chronic Disease Risk. Washington, D.C.: National Academy Press.

Nikodijevic, O., Jacobson, K.A. and Daly, J.W. (1993) Locomotor activity in mice during chronic treatment with caffeine and withdrawal. *Pharmacol Biochem Behav.*, 44, 199–216.

Normile, H.J. and Barraco, R.A. (1991) *Brain Res. Bull.*, 27, 101–104.

Olsen Jorn (1991) Cigarette smoking, tea and coffee drinking and subfecundity. *American Journal of Epidemiology*, **133**, 734–739.

Packard, P.T. and Recker, R.R. (1996) Caffeine Does Not Affect the Rate of Gain in Spine Bone in Young Women. *Osteoporosis International*, **6**, 149–152.

Pastore, L.M. and Savitz, D.A. (1995) Case-control study of caffeinated beverages and pre-term delivery. *American Journal of Epidemiology*, **141**, 61–69.

Phillis, J.W. and Edstrom, J.P. (1976) Effects of adenosine analogs on rat cerebral cortical neurons. *Life Sci.*, **19**, 1041–1053.

Rainnie, D.G., Grunze, H.C.R., McCarley, R.W., Greene, R.W. (1994) Adenosine inhibition of mesopontine cholinergic neurons: implications for EEG arousal. *Science*, **263**, 689–692.

Rapoport, J.L., Berg, C.J., Ismond, D.R. (1984) Behavioral effects of caffeine in children. *Archives of General Psychology*, **41**, 1073–1079.

Robertson, D., Wade, D., Workman, R., Woosley, R.L., Oates, J.A. (1981) Tolerance to the humoral and hemodynamic effects of caffeine in man. *J. Clin. Invest.*, **67**, 1111–1117.

Rosenberg, L. (1990) Coffee and tea consumption in relation to the risk of large bowel cancer: A Review of Epidemiologic Studies. *Cancer Letters*, **52**, 163–171.

Shi, D., Nikodijivi, O., Jacobson, K.A., and Daly, J.W. (1995) Effects of chronic caffeine on adenosine, dopamine and acetylcholine systems in mice. *Arch. Int. Pharmacodyn. Ther.*, **328**, 261–287.

Silverman, K., Evans, S.M., Strain, E.C. (1992) Withdrawal syndrome after the double-blind cessation of caffeine consumption. *The New England Journal of Medicine*, **327**, 1109–14.

Stamler, J., Caggiula A.W., Grandits G.A. (1997) Chapter 12. Relation of body mass and alcohol, nutrient, fiber, and caffeine intakes to blood pressure in the special intervention and usual care groups in the Multiple Risk Factor Intervention Trial. *American Journal of Clinical Nutrition*, **65**, Supplement: pp. 338–365.

Stein, Z., and Susser, M. (1991) Miscarriage, caffeine, and the epiphenomena of pregnancy: the causal model. *Epidemiology*, **2**, 163–7.

Stoner, G.R., Skirboll, R., Werkman, S. and hommer, D.W. (1988) Preferential effects of caffeine on limbic an cortical dopamine systems. *Biol. Psychiatry*, **23**, 761–768.

Strain, E.C., Mumford, G.K., Silverman, K., Griffiths, R.R. (1994) Caffeine dependence syndrome: evidence from case histories and experimental evaluations. *JAMA*, **272**, 1043–8.

Svenningsson, P., Nomikos, G.G. and Fredholm, B.B. (1995) Increased expression of c-jun, junB, AP-1, and preproenkephalin mRNA in rat striatum following a single injection of caffeine. *J. Neurosci.*, **15**, 7612–7624.

Thompson, S.M., Haas, H.L. and GÆhwiler, B.H. (1992) Comparison of the actions of adenosine at pre- and postsynaptic receptors in the rat hippocampus *in vitrol*. *J. Physiol. (London)*, **451**, 347–363.

Viscoli, C.M., Lachs, M.S., Horwitz, R.I. (1993) Bladder cancer and coffee drinking: a summary of case-control research. *Lancet*, **341**, 1432–1437.

Willett, W.C., Stampler, M.J., Manson, J.E. (1996) Coffee consumption and coronary heart disease in women: a ten-year follow-up. *JAMA*, **275**, 458–462.

Zarrindast, M.R. and Shafaghi, B. (1994) *Eur. J. Pharmaawl*, **256**, 232–239.

8. THE EFFECTS OF TEA ON THE CARDIOVASCULAR SYSTEM

ZONG-MAO CHEN

Tea Research Institute, Chinese Academy of Agricultural Sciences, 1 Yunqi Road, Hangzhou, Zhejiang 310008, China

1. INTRODUCTION

Cardiovascular diseases, together with cancers, are the main killing diseases of humans in the world. Of the cardiovascular diseases, atherosclerosis is one of the most prevalent. Atherosclerosis is primarily caused by hypercholesterolemia in which excess cholesterol accumulates in the blood vessels and oxidation of low-density cholesterol (LDL) leads to foci of endothelial abnormalities associated with the process of atherosclerosis (Weisburger, 1994). It deteriorates further with the oxidation of lipids in the blood. Therefore, in order to maintain the cardiovascular system in good condition, it is very important to prevent not only an excessive increase of cholesterols in the blood, but also the oxidation of lipids in the blood. Hypertension is another major factor that can affect the health of the cardiovascular system. In this article, the antioxidative, hypolipidemic, hypotensive and the obesity-depressing activity of tea will be discussed.

2. ANTIOXIDATIVE ACTIVITY OF TEA

The role of free radical and active oxygen in the pathogenesis of certain human diseases, including aging, cardiovascular disease and cancer is becoming increasingly recognized. Lipid peroxidation has been regarded as one of the mechanisms of senescence of humans and the cause of atherosclerosis. Because of their very high chemical reactivity, free radicals show very short lifetimes in biological systems. However, the excessive amounts of free radicals are able to produce metabolic disturbances and to damage membrane structures in a variety of ways. Therefore, much attention has been focused on the use of antioxidants, especially natural antioxidants, to inhibit lipid peroxidation or to protect the damage of free radicals.

Many investigations indicated that intake of certain amounts of fruits and vegetables that contain a large quantity of vitamin C and vitamin E showed antioxidative activity. Tea is not only rich in vitamin C and E, but also contains an important group of polyphenols, i.e., catechins, which display obvious antioxidative activity. The polyphenols are able to act as antioxidants by virtue of the hydrogen-donating capacity of their phenol groups, as well as their metal-chelating potential

(Rice-Evans *et al.* 1995). As early as in 1963, Kajimoto first reported the antioxidative activity of tea. After that, many researchers demonstrated that catechins possess more potent antioxidative activity than that of vitamin C and E as well as other synthetic antioxidants (such as BHA, BHT etc). By adding the catechins, DL-α-tocopherol (vitamin E) or BHA (butylhydroxy anisole) to lard or vegetable oils, Hara (1994) reported that the catechins reduced the formation of peroxides more effectively than vitamin E or BHA. In an experiment using linoleic acid as the material of antioxidative activity, it was shown that among various kinds of tea, the antioxidative activity decreased in the order of semi-fermented tea > unfermented tea > fermented tea (Yen and Chen, 1995). According to the results of rancimet method of pure lard and the lipoxygenase assay method, among the various tea components, the theaflavin (TF) compounds isolated from black tea showed the strongest antioxidative activity. The IC$_{50}$ of TF monogallate B and TF digallate toward soybean 15-lipoxygenase enzyme was 0.4 and 0.2 μg respectively. EGC, ECG and EGCG showed moderately IC$_{50}$ values ranging from 4.6–7.7 μg. By using the red cell membrane system *in vitro*, the peroxidation of the rabbit red blood cell membrane was induced and different compounds were introduced to this system to test their antioxidative activity.

Among the tested tea compounds, all TFs exhibited much stronger antioxidative effects than vitamin E or propyl gallate. TF digallate showed the most potent antioxidative activity, inhibiting about 80% of the peroxidation, followed by TF monogallate B, TF monogallate A and TF (Hara, 1994). Li *et al.* (1993) showed that the inhibition rate of 200 μg/ml green tea polyphenol on the lipid peroxidation of red cell membrane ranged from 32.9% to 55.3%. 1mg of green tea polyphenol showed an ability to scavenge superoxide anion radicals the same as 9 mg bovine cupro-zinc superoxide dismutase (SOD) (Fang, 1995). Serafini *et al.* (1996) evaluated the *in vitro* antioxidant activity of green and black tea as well as their *in vivo* effect on plasma oxidation potential in man. Results showed that both teas inhibited the *in vitro* peroxidation in a dose-dependent manner. Green tea was sixfold more potent than black tea. The addition of milk to tea did not appreciably modify their *in vitro* antioxidant potential. *In vivo*, the ingestion of tea produced a significant increase of TRAP (Human Plasma Antioxidant Capacity) (P < 0.05).

The antioxidant activity of aqueous extracts of green tea, black tea and Oolong tea was compared by von Gadow *et al.* (1987) with the β-carotene bleaching method and α,α-diphenyl-β-picrylhydrazyl (DPPH) radical scavenging method. Results by these two methods showed the same tendency: green tea > black tea > Oolong tea. In a comparative experiment of the antioxidative activity of various kinds of tea, Xie *et al.* (1994) reported that the antioxidative activity on autooxidation of lard was positively related to the contents of catechins, especially EGCG. The antioxidative activity of various teas was decreased in the order of Oolong tea > green tea and black tea (Xie *et al.* 1994). Salah *et al.* (1995) investigated the relative antioxidative activities *in vitro* of tea components. Results showed that those compounds with more hydroxy groups appear to exert the greatest antioxidative activity. The antioxidative potentials (Trolox Equivalent Antioxidant Activity, TEAC) of catechin and epicatechin were 2.4 mM and 2.5 mM, respectively. When the structure of catechin was modified to the catechin-gallate esters – ECG and EGCG, the TEAC was enhanced to 4.93 mM and 4.75 mM, respectively. The quercetin has an identical numbers of hydroxyl groups in the same position as catechin, but also contains the 2,3-double bond in the C ring and the 4-oxo group. This structure confers an enhancement of the TEAC value to 4.72 mM. With regards to the relationship of structure and antioxidative activity, it was noted that the conjugation between A and B rings via a planar C ring is the important structure for antioxidative activity. In addition, the contribution of the 3',4'-dihydroxyl structure substitution in the B ring is highly significant for the antioxidative activity. And the presence of a 3-hydroxyl group in the C ring and a 5-hydroxyl group in the A ring is highly important for maximal radical scavenging potential (Rice-Evans *et al.* 1995).

Salah *et al.* (1995) used the oxidation of low-density lipoproteins (LDL) as a model for investigating the efficacy of the polyphenols as lipid chain-breaking antioxidants. The relative effectiveness of the catechins and catechin-gallate esters in inhibiting LDL oxidation was determined. As shown, the gallic acid is the least effective, requiring about 1.2 μM for 50% inhibition of maximal oxidation. The IC_{50} value of EGC for inhibiting the oxidation was 0.75 μM. The IC_{50} values for catechin, EC, ECG and EGCG were in the ranged of 0.25–0.38 μM. Regarding the antioxidant activity of the polyphenolic components, the contribution of the components to the antioxidative effectiveness in green tea was determined as follows: EGC = EGCG >> ECG = EC. C. Zhang *et al.* (1997) reported that tea polyphenols from jasmine tea and catechins showed antioxidative activity on the Cu mediated LDL peroxidation and protected the oxidative degradation of unsaturated fatty acids in LDL of humans. Fiala *et al.* (1996) in an experiment adding peroxynitrite DNA and L-tyrosine showed that EGCG was a significantly better inhibitor of peroxynitrite-mediated oxidation of deoxyguanosine and tyrosine nitration. The 50% inhibition activity of the oxidation of the former was 4 times higher than that of vitamin C. He *et al.* (1997) conducted an investigation on antioxidative effects of various tea polyphenols and catechins in a fish meat model system. Results showed that the antioxidative activity of green tea, tea polyphenols and catechin on fish lipid was higher than that of vitamin E, BHT, BHA and TBHQ.

Among the catechins, EGCG and ECG possessed the strongest activity. Yen *et al.* (1995, 1997) used calf thymus DNA as the experimental material and the antioxidative effects of various tea extracts including the green tea, oolong tea and black tea were investigated. Results showed that the oxidation of deoxyribose was markedly decreased by various tea extracts in higher dosage, especially oolong tea, which inhibited 73.6% peroxidation of linoleic acid. The antioxidative activity decreased in the order of semifermented tea > nonfermented tea > fermented tea. In a similiar experiment *in vitro*, it was demonstrated that tea polyphenols suppressed the oxidative modification of porcine serum LDL which is assumed to be an important step in the pathogenesis of atherosclerosis lesion. The activity was in the order of (–)-EGCG > (–)-ECG > (–)-EC > (–)-EGC. It was found that the Cu mediated cholesterol ester degradation in LDL was almost completely inhibited by 5.0 μM EGCG (Miura *et al.* 1994). Tomita (1997) investigated the IC_{50} of various antioxidants and tea catechins on the BHP (t-Butylheteroperoxide) induced peroxidative reaction in rat liver homogenerate. The IC_{50} of ECG, EGC, EGCG and

TF, TF monogallate A was 10–80 times lower than that of vitamin C and vitamin E and 10 times lower than that of BHT and BHA. The IC_{50} on TBARS (Thioburbituric Acid Reactive Substances) formation of LDL was 0.95, 1.03, 1.13, 1.36, 2.74 and 3.09 μM for ECG, EGCG, EC, C, EGC, and BHT, respectively.

An experiment conducted in China showed that the green tea polyphenols had an inhibitory effect on iron-induced lipid peroxidation in synaptosomes. Among the various components of green tea polyphenols, the inhibitory effect decreased in the order of EGCG > ECG > EGC > EC, similar to the results from Miura *et al.* (1994). However, the free radical-scavenging activity was decreased in the order of ECG > EGCG > EC > EGC. It was regarded that the preventive activity of these catechins on lipid peroxidative damage induced by Fe^{++}/Fe^{+++} is not only depended on the complexing ability with iron and free radical-scavenging ability, but also the stability of formed semiquinone free radicals (Guo *et al.* 1996). Shen *et al.* (1997) reported the threshold values for catechins to protect red cell membrane from injury caused by free radicals in the presence of iron ions. The values were 0.02 mM/L for EGCG, 0.025 mM/L for ECG, 0.028 mM/L for EGC and 0.05 mM/L for EC, respectively, with the same order reported by Guo *et al.* (1996).

By using mice as the experimental animal, Chang *et al.* (1993) indicated that tea inhibited the formation of peroxidative lipid in the mice liver-brain tissue homogenate *in vitro*. By oral administration of tea infusion, the peroxidative lipid contents in heart, liver and brain tissue of young and adult mice decreased. The decreasing rates in old and adult BALB/c mice ranged from 23.7–41.95% and 12.7–23.2% in the low dosage group (3 g tea/kg/d) and 24.2–44.0% and 12.5–31.3% in the high dosage group (9 g tea/kg/d), respectively. Hara (1995) administered excessive lipid to rats and observed the antioxidative activity of tea catechins on peroxidated lipid in the plasma of the rat. Results showed that the TBARS (Thiobarbituric Acid Reactive Substance) value of plasma was decreased (P < 0.05) in the perilla oil + 1% catechin diet. At the same time, the plasma α-tocopherol content was increased significantly (P < 0.001). Yoshino *et al.* (1994) examined the effects of tea polyphenols on the contents of lipids and lipid peroxidation in rat plasma, kidney and liver *in vivo*. The supplementation of tea polyphenols (0.5% and 1%) in the diet was performed from weanling (3 weeks of age) to 19 months old. The TBARS (Thiobarbituric Acid Reactive Substances) in the plasma of the 1.0% tea polyphenol group was significantly lower than that of the *in vivo* control group (P < 0.01). The content of plasma lipids in the 1.0% tea polyphenols group of 19 months old rats were significantly lower than the control group (P < 0.05), indicating the hypolipidemic activity and antioxidative effect of tea polyphenols. Sano *et al.* (1991, 1995) reported that feeding 3% tea leaf powder to rats for 50 days protected the BHP-induced and tert-butyl hydroperoxide-induced lipid peroxidation in rat liver and kidney slices. Feeding with black tea resulted in an excellent antioxidant effect against lipid peroxidation, which was similar to that observed after feeding with green tea.

A recent experiment on the antioxidant effect of green tea and black tea in man conducted in Italy showed that both teas possessed a potent antioxidative activity *in vitro* and *in vivo* (Sefafini *et al.* 1996). The ingestion of tea causes a significant increase of human plasma antioxidant capacity that represents the μmoles of peroxyl radicals trapped by one liter of plasma. Both green and black teas show the peak increase at 30–50 min; however, the antioxidative ability of green tea was about five times more potent than that of black tea. Moreover, the pro-oxidant property for some antioxidants including the tea polyphenols has been increasingly studied in recent years. The pro-oxidant activity is a result of the ability to reduce metals, such as Fe^{3+}, to forms that react with O_2 or H_2O_2 to form initiators of oxidation. The tea extracts also showed pro-oxidant effects at lower dosages (Yen *et al.* 1997). So, in evaluating antioxidant activity of tea polyphenols, it must be assessed for pro-oxidant properties *in vivo*.

The above experiments clearly demonstrate that tea drinking possesses an antioxidative ability on lipid, which is believed to be exerted by the EGCG, ECG and related catechins in green tea as well as the TFs and thearubigins (TRs) in black tea. Even the ingestion of one large cup of tea could produce an appreciable increase in plasma TRAP values in man (Serafini *et al.* 1996). It can be deduced that the inhibition of lipid peroxidation might be one of the mechanisms in preventing cardiovascular disease in humans.

3. BLOOD-PRESSURE LOWERING ACTIVITY OF TEA

Hypertension is a common disorder in humans. Tea drinking can lower blood pressure. There are many Chinese traditional prescriptions, with tea as a major constituent, used in the treatment of hypertension and coronary disease in Chinese traditional medicine. A survey on the relationship between hypertension and tea drinking in 964 adults was carried out by Zhejiang Medical University of China during the 1970s. Results showed that the average rate of hypertension was 6.2% in the group who drank tea as habit, and 10.5% in the group who did not. Clinical experiments showed that hot water extract of green tea possessed a degree of blood pressure lowering effect. An experiment *in vivo* carried out on rats fed with diet supplemented with 0.5% crude catechins showed that the blood pressure in treated rats was 10–20 mm Hg lower than that in the control group (Hara, 1990). A clinical experiment using green tea on high blood pressure patients was conducted at the Anhui Medical Research Institute of China. Results showed that a 10 g tea intake daily treatment over half a year, decreased the blood pressure by 20–30% (Chen, 1994).

A study was conducted to determine whether the effect *in vitro* is reflected in the lowering of blood pressure of animals after oral feeding of tea. Crude catechins (extracted from green tea and composed mainly of EGCG) were fed in the diet to spontaneously hypertensive rats (SHR) or stroke-prone SHR (SHRSP) and their blood pressure was measured by the tail cuff method. The mean systolic blood pressure of SHR fed a diet containing 0.5% crude catechins from the age of 5 weeks increased with time, but remained significantly lower than that in the control group (P < 0.05). At the age of 16 weeks, the diets of the two groups were exchanged. As a result, the blood pressure curves of these two groups crossed in 2 weeks and this situation remained. In SHRSP given 1% saline as drinking water, addition of 0.5%

crude catechins to the diet not only tended to suppress the blood pressure to a non-significant extent relative to the control group, but also clearly prolonged the survival period before death due to stroke. Ishigaki and Hara (1991) carried out the blood pressure lowering experiment by using tea polyphenols (dosage: 400 mg/kg, × 2 in succession of three months) on 21 volunteer-adults (10 male and 11 female). The blood was sampled. Results showed that the blood pressure of those patients whose pressure was around 160 mmHg was decreased significantly. Yokozawa et al. (1994) showed that rats orally administered with 2 mg green tea tannin tended to have a lower systolic blood pressure. A further increase to 4 mg produced a significant decrease in the systolic blood pressure, 7% lower than that in the control rats. Similar changes produced by green tea tannin were observed in the diastolic pressure value. Kobayashi et al. (1996) reported that high doses of theanine (1500–2000 mg/kg) significantly decreased the blood pressure of spontaneously hypertensive rats (SHR). The dose-dependent changes in the blood pressure values (systolic, diastolic and mean) resulted from the administration of theanine to experimental animals. Theanine entered the brain via the blood-brain barrier after the intragastric administration. Results showed that brain serotonin concentration was significantly decreased by theanine.

The effects of caffeine on blood pressure depend on dose and route of administration. Intravenous administration frequently produces an initial fall in blood pressure (seen only after larger doses) followed by a secondary rise. In contrast, following oral caffeine intake in animals and human, the initial blood pressure fall is only rarely seen and the maximal plasma concentration are much smaller than those seen following intravenous administration (Robertson et al. 1984; Pincomb et al. 1988). Caffeine may influence blood pressure to a greater extent during stress; however, studies on the blood pressure effects of caffeine in humans indicate that it is not deleterious in essential hypertension. It was suggested that the plasma levels of caffeine needed to exacerbate renovascular hypertension in their study could be reached by moderate to heavy caffeine users. Omori (1987) first developed a new type of tea (Gabaron tea) by means of anaerobic treatment of fresh leaves and found it contained large amounts of Υ-aminobutyric acid (170–270 mg %), 8–10 times higher than that in common green tea (Omori, 1995; Li and Chang, 1993). Clinical experiments conducted in Japan and China Taiwan showed that Gabaron tea possesses a significant blood pressure lowering effect (Saito et al. 1995), the blood pressure in the treated animals was 14–17% (25–30 mmHg) lower 20 days after treatment. According to a clinical experiment on 13 hypertension patients, the blood pressure in 7 patients was reduced significantly, and 6 were insignificantly reduced (Omori, 1995). However, the blood pressure rose rapidly to that of the control rats when the administration of Gabaron tea ceased. Thus, it was recommended that the administration of Gabaron tea should be conducted continuously for the purpose of lowering the blood pressure of hypertensive patients.

Taniguchi et al. (1987) reported that hot water extracts of green tea showed a prolonged hypotensive effect in anesthetized rabbits. The major active principle was (–)- gallocatechin gallate [(–)-GCG]. (–)-GCG at a dose of 0.1 mg/kg (i.v.) effectively reduced the blood pressure of anesthetized rabbits and at 0.5 mg/kg lowered it by 20–40 mmHg for an extended period of time.

Regarding the mechanism of hypertension-depressing effect, hypertension is regulated by angiotensin, the inactive angiotensin I is converted into angiotensin II with potent blood pressure raising action by angiotensin I transferase (ACE). So, those compounds inhibiting the activity of ACE also showed the blood pressure lowering effect. Investigation proved that tea components, especially TF digallate and EGCG, showed obvious inhibiting effects (Hara *et al.,* 1987. Horie *et al.,* 1996). In an experiment comparing the inhibiting ability of various kinds of tea on the angiotensin I transferase, it was shown that green tea extracts possessed the most potent inhibitory ability. The ACE activity in the green tea treatment (1 g in 200 ml hot water) was 1% in comparison with 100% in the control (Horie *et al.* 1996). Of course, there are many mechanisms of action of antihypertensive drugs, such as β-blockers, central α-stimulants, angiotensin converting enzyme inhibitors, Ca antagonists (Williams, 1992). Based on the experimental data reported by Tollins and Raiz (1990), angiotensin enzyme inhibitors may be superior to Ca antagonists in halting the progression of renal dysfunction. Besides, tea polyphenols and caffeine may ameliorate the development of hypertension by improving the renal circulatory state. So, it was regarded that the blood pressure depressing effect of tea is resulted from its direct action in the kidney, inducing the activation of the kinin-kallikrein-postaglandin system in the kidney (Yokozawa *et al.* 1994).

4. BLOOD LIPID AND CHOLESTEROL LOWERING EFFECT

Excessive lipids in blood is a common disorder of middle aged or old aged men and women. High serum-lipid includes high cholesterol and triglyceride content in blood. The cholesterol includes low-density cholesterol (LDL), ultralow-density cholesterol (VLDL) and high-density cholesterol (HDL). Among those, LDL and VLDL have harmful roles in promoting the formation of atherosclerosis. On the other hand, HDL is a kind of beneficial cholesterol and plays a role in preventing the occurrence of atherosclerosis. Tea drinking showed the effect of decreasing the serum lipid level, the contents of LDL and VDL cholesterol. Green & Harari (1992) in their study of 650 men in six factories in Israel in 1986, found a statistically significant inverse relationship between tea drinking and the level of both total and LDL cholesterol. A study on 4700 men and 4500 women in Finland revealed that increasing tea consumption was associated with slight reductions in plasma cholesterol level (Tuonilento *et al.* 1987). An epidemiological survey including 7710 men and 8222 women was carried out in Norway and showed a negative relationship between tea drinking and the serum lipid level (Stensvold *et al.* 1992).

Over the past ten years, several investigations have been conducted on rats, mice and rabbits to test the protective effect of tea on plasma lipids, plasma cholesterol level and/or atherosclerosis. As reported (Muramatsu *et al.* 1986; Matsuda *et al.* 1986; Fukuyo *et al.* 1986; Matsumoto *et al.* 1995; Deng *et al.* 1998), in animals fed with high fat/high cholesterol atherogenic diets, the administration of tea (green tea or black tea) which was added in the drink or the diet resulted in reducing the plasma level of cholesterol.

A clinical experiment on hyperlipidemia patients was carried out by a research group in Quanming Medical Institute of China, after daily drinking of 15 g Yunnan Tucha (a kind of compressed tea) for one month, the lipid-lowering effect was comparable to that of Atromid-S (a popular lipid-lowering medicine) (Chen, Z.M., 1994). Iwata et al. (1991) reported that the plasma level of triglyceride and phospholipid on healthy human female volunteers was decreased significantly after drinking Oolong tea. Recently, Nakachi et al. (1995) reported the results of a cohort study conducted in Japan. The survey covered 8553 people aged over 40 years old living in a town in Saitama, Japan, and asked about 90 lifestyle factors, such as present and past eating habits, history of consumption of tea, tobacco and alcohol, history of disease, present state of health. Out of the 8553 persons surveyed by the questionnaire, 3625 people gave blood samples that were subjected to biochemical and immunological assays during 1986–1990. Results showed that the consumption of green tea was significantly associated with lower serum lipid concentration and low density lipoprotein. An increase in green tea consumption substantially decreased serum total cholesterol and triglyceride concentration. This strong association remained almost unaltered even after age, cigarette smoking, and alcohol consumption were adjusted for using a statistical method. Increased consumption of green tea was associated with an increased serum concentration of HDL, which is often referred to as good cholesterol. As a result, a significant decrease in the atherogenic index, which is an important index for atherosclerosis, was observed in the group that drank the most green tea. A similar cross sectional study of effects of drinking green tea on cardiovascular disease was conducted on 1371 men aged over 40 years by Imai et al. (1995) and similar results were reported. The administration of 0.5–1.0% EGCG, crude catechin or 1.5–2.0% Oolong tea supplemented diet can lead to the decrease of total cholesterol, free cholesterol, LDL cholesterol and triglyceride in plasma and body fluid significantly; moreover, it can increase HDL cholesterol simultaneously (Ohtsuru et al. 1991; Hara, 1991; Ishigaki and Hara, 1991). ECG and EGCG were more effective in increasing the serum cholesterol excretion via feces (Ishigaki and Hara, 1991).

The atherogenic index can be decreased significantly after the feeding of 1–2% catechins (Muramatsu et al. 1986) as well as ECG and EGCG (Matsuda, 1986) in experimental animals with a high cholesterol diet. A similar relationship was established between jasmine tea drinking and the inhibition of LDL-oxidation (Zhang et al. 1997). Further research showed that tea catechins not only decrease the serum lipid and cholesterol as well as increase the excretion of body lipid, but also effectively reduce the cholesterol absorption from the intestine by reducing the solubility of cholesterol in mixed micelles (Muramatsu et al., 1986. Ohtsuru et al. 1991). Among the various catechins, the gallate ester type catechins (EGCG, ECG) was most effective in precipitating micellar cholesterol than those free catechins (EC, EGC) (Ohtsuru et al. 1991). Thus, the hypocholesterolemic activity observed with catechin feeding can be also attributed to the inhibition of intestinal cholesterol absorption (Nanjo et al. 1992).

According to the investigation of Ikeda *et al.* (1992), it is likely that EGCG forms insoluble coprecipitates with cholesterol in mixed micellar solution (Ikeda *et al.* 1992). Thus, the insoluble coprecipitation of cholesterol and catechins cannot be absorbed from the intestine. However, the mode of binding between cholesterol and EGCG needs to be clarified by further investigation. The effects of green tea polyphenols on body weight and the contents of lipids and peroxidized lipid in rat plasma, kidney and liver were investigated. The diet of young rats from 3 weeks of age until they were 19 months old was supplemented with green tea polyphenols (0.5% and 1.0%). In the older animals, peroxidized lipids were significantly lower in the 1-% group (at 13–19 months old). At 19 months old, triglycerides, total cholesterol and phospholipids were significantly decreased in the serum of the 1-% feed group. This suggests a hypocholesterolemic effect from long-term feeding of green tea constituents (Yoshino *et al.* 1994). In a similar investigation carried out in China, the results showed that the serum total cholesterol and triglyceride levels were significantly reduced and serum HDL level was significantly increased compared to the controls (Shen *et al.* 1993). Based on these results, some antihyperlipid agents have been developed in Japan, in which tea catechins were used as the active principles (Isota *et al.* 1990; Mori, 1990).

There are also several negative relationships between the tea drinking and lipoprotein levels. In 1985, Klatsky *et al.* conducted a large epidemiological study of serum lipids and the tea intake in 22,000 males and 25,000 females and reported no statistically significant relationship. In the same experiment, coffee was found to be positively and significantly associated with serum cholesterol levels. Similar results were reported by Kark *et al.* in 1985 from their study of 1000 males and 500 females in Jerusalem. There was no relationship between tea drinking and cholesterol levels before adjustment for covariables, although there was a slight rise (6 mg/dl) in plasma cholesterol at the highest levels of tea drinking in males after covariable adjustment. On the other hand, female tea drinkers had a slightly lower plasma cholesterol level after similar analysis. In a study of 900 males and 1200 females, Haffner and his colleagues (1985) confirmed the lack of relationship between tea drinking and LDL level. Coffee once again was found to be positively correlated. Data from the heart study of 5858 Japanese men in Honolulu were essentially identical to those reported by Haffner *et al.* that tea drinking was not related to the cholesterol levels.

5. OBESITY-DEPRESSION AND PREVENTION OF CARDIOVASCULAR DISORDERS

Excessive lipid induces obesity. This is a physiologically abnormal phenomenon in modern society. Obesity is closely related to excessive serum lipid. Experiments show that tea drinking plays an obesity-depressing role via an increase of fundamental metabolic rate and the degradation of fat. Investigations carried out by French, Japanese and Chinese scientists have also shown that Pu-Er tea and Oolong tea possess a significant obesity depressing effect (Ishigaki *et al.* 1991; Chen, 1994). Researches using different kinds of tea revealed that the serum lipid depressing and obesity depressing effects of compressed tea was greater than that of green tea and black tea (Matsumoto and Hara, 1992).

High levels of blood cholesterol induce the deposit of lipid on the vessel wall and cause the constriction of coronary arteries, atherosclerosis and thrombus formation. It is related to the fact that tea drinking decreases the serum lipid and cholesterol level. In the past, atherosclerosis was thought to result from a level of serum cholesterol above 200dl and a relatively low level of HDL and high level of LDL. Current views are that it is induced by the oxidation of LDL cholesterol that leads to foci of endothelial abnormalities associated with the process of atherosclerosis (Witztum and Steinberg, 1991). Tea and tea polyphenols possess the ability to prevent the oxidation of LDL cholesterol (Fang, 1995) and anticoagulation of blood platelets and anti-athero-sclerosis effects (Segesaka-Mitane et al. 1990; Watanabe, 1990; Namiki et al. 1991). From the viewpoint of hemorheology, a high hemagglutination situation favors thrombus formation. Research on 120 patients with hyperlipidemia also found that tea pigment (tea polyphenols and their oxidative products are the major components), catechins, thearubigin (TR) and TF possess anticoagulation and antihemagglutination activity, and the ability to promote fibrinolysis, indicating a therapeutic effect on the atherosclerosis (Lou et al. 1989). Scientists from China also showed that Oolong tea drinking decreased the viscosity of blood. Newly emerging data suggests that those blood lipids that have previously undergone oxidation may be more likely to promote the development of atherosclerosis. Hence, inhibition of fat oxidation, either in vitro or in vivo, may reduce the risk of developing atherosclerosis and ultimately cardiovascular diseases (Watanabe, 1990; Mitscher et al. 1997). Epidemiological studies showed that individuals consuming four or more cups of tea per day have a lower risk of atherosclerosis and coronary heart disease (Green and Harari, 1992; Kono et al. 1992; Stensvold et al. 1992; Hensrud and Heimburger, 1994). Tea also showed a certain degree of efficiency in the control of coronary heart diseases. According to a survey of coronary heart disease in the population of a tea growing area by Zhejiang Medical University in China, the coronary heart disease percentage in the population who drank no tea or occasionally drank tea was on average 5.7%, and that in the population who frequently drank tea averaged 1.07%, that was one-fifth of the above group. There were some prescriptions used in the control of coronary heart disease in Chinese traditional medicine. For example, the extracts of old tea plant root with glutinous rice wine was effective in curing the coronary heart disease. As mentioned above, tea polyphenols showed antioxidative activity (Kajimoto, 1963; Matsuzaki and Hara, 1985; Spernius et al. 1989; Pascual et al. 1992; Yen and Chen, 1995), thus suggesting a role in the prevention of atherosclerosis.

Namiki et al. (1991) carried out an investigation on the inhibitory effect of hot water extracts prepared from various kinds of tea and catechins on platelet aggregation induced by collagen by using turbidimetry. Results showed that Japanese steamed tea and Kenya black tea possessed strong inhibitory activity (more than 50%) on the aggregation of blood platelets. Among the 6 catechins and 3 TFs as well as caffeine, (–)-EC, (+)-C, (+)-GC and TF monogallate showed a marked inhibitory activity (Namiki et al. 1991). A compound of 4-hydroxy angular furocoumarin was isolated from the tender leaves of tea plant and showed an anticoagulating property of blood (Banerjee and Gnguly, 1993). An inhibiting effect of a hot water extract of green tea on collagen-induced aggregation of washed rabbit platelet was shown and suggested the catechin and EGCG were the most active components (0.45 mM). This potency is in a comparable range to aspirin. Thrombin- and PAF-induced aggregation was also shown to be inhibited by EGCG (Sagesaka-Mitane *et al.* 1990). A further experiment showed that ECG and EGCG were found to be more potent inhibitors for thrombin by mode of noncompetitive inhibition. The 50% inhibition of thrombin amidolysis by ECG and EGCG were 1.2×10^{-6} M and 1.1×10^{-6} M and of fibrin formation were 2.5×10^{-5} M and 5.8×10^{-6} M, respectively (Kinoshita and Horie, 1993). It is regarded that the aggregation of blood platelets was related to the activity of proteinases in fibrinolysis and kallikrein-kinin systems (Kinoshita and Horie, 1994). Research showed that the ester type catechins (ECG and EGCG) are more potent inhibitors to those enzymes in the fibrinolytic and the kallikrein-kinin systems, thus showing the ability to prevent thrombosis and thrombolysis (Kinoshita and Horie, 1994). Ali *et al.* (1990) isolated a potent thromboxane formation inhibitor from fresh tea leaves. A further investigation identified the active compound as 2-amino-5-(N-ethylcarboxyamido)-pentanoid acid. It has the same chemical structure as theanine, just a different nomenclature system. It is apparently unique to tea when fed orally to mice or rats for eight weeks, a significant *in vivo* diminution in thromboxane level was measured in the serum. The black tea did not produce this effect. It was a potent inhibitor of thrombin-stimulated thromboxane formation in rabbit whole blood. The inhibitory activity was 100 times higher than that of caffeine (Ali *et al.* 1990). A concentration as low as 50 μM suppresses thromboxane formation by 84% (Ali and Afzal, 1987).

Imai *et al.* (1995) have conducted epidemiological investigations on the relationship between tea drinking and cardiovascular disorders in Saitama region since 1986. In total, 8553 inhabitants aged over 40 years old were surveyed, the blood samples from 3625 inhabitants were collected and 36 items including total serum cholesterol, HDL, LDL, neutral lipid were analysed. The investigated people were divided into three groups according to the number of cups of green tea (< 3 cups/day, 4–9 cups/day and >10 cups/day). Results showed that the total serum cholesterol, neutral lipids, and the atherosclerosis index were decreased with the increasing of amount of tea drinking. It is also interesting that the average value of peroxidized lipid in smokers who drank more green tea was lower than that in those who drank less. The peroxidized lipid value was 9.1, 9.0 and 8.5 nmol/ml in < 3 cups, 4–9 cups and > 10 cups/day. Results showed that for the green tea, the more drunk, the less cardiovascular disorders occur (Imai *et al.* 1995). An epidemiological investigation among 5910 non-tea drinking and non-smoking women (> 40 years of age) was conducted in Sendai city, Taijiri and Wakuya villages in Miyagi region in Japan. The medical history of strokes was less frequently observed among those who consumed more green tea. These populations were followed for four additional years and the incidence of stroke and cerebral hemorrhage was two to three times lower in those who drank more green tea (5 cups was the dividing line) (Sato *et al.* 1989). A cohort of 552 men aged 50 to 69 years old was selected in 1970 and followed up for 15 years for the epidemiological survey on dietary flavonoids and the incidence of stroke. During the 15 years of follow-up, 42 first fatal or nonfatal strokes occurred in 552 men. Results showed that dietary flavonoids including tea consumption were inversely associated with stroke incidence. The relative risk of the highest vs the lowest amount of flavonoids intake (> 28.6 mg/d vs < 18.3 mg/d) was 0.27 (0.11 to 0.70). In the Nertherlands, it was regarded that black tea contributed about 70% of the flavonoid intake. The relative risk for a daily consumption of 4.7 cups of tea *vs* less than 2.6 cups of tea was 69% reduced risk of stroke (Keli *et al.* 1996).

Thus, it can be concluded that the possible mechanisms for the antiatherosclerosis effect of tea drinking can be explained as follows: 1. Reducing plasma lipid and cholesterol formation (Muramatsu *et al.* 1986b; Hara, 1991; Iwata *et al.* 1991); 2. Elevating the high-density cholesterol (HDL) level in plasma (Muramatsu *et al.* 1986b; Hara, 1991; Ohtsuru *et al.* 1991; Ishigaki *et al.* 1992); 3. Inhibiting the cell's absorption of lipid as well as accelerating the elimination and decomposition of the cholesterol that had deposited on the artery wall (Muramatsu *et al.* 1986b; Ohtsuru *et al.* 1991; Ikeda *et al.* 1992); 4. Improving the hemorheological state and microcirculation of blood as well as reducing thrombosis (Lou *et al.* 1989); 5. Decreasing the viscosity of blood and prolonging the coagulation time of blood platelets as well as increasing the blood fluidity (Lou *et al.* 1989; Namiki *et al.* 1991); 6. Antioxidative activity on blood lipids (Kajimoto, 1963; Matsuzaki and Hara, 1985; Zhao *et al.* 1989; Spernius *et al.* 1989; Pauscal *et al.* 1992; Yen and Chen, 1995); 7. Inhibiting the enzymes in the fibrinolytic and the kallikrein-kinin system (Kinoshita and Horie, 1993, 1994).

6. PROSPECTS

As mentioned above, tea is not only a nutritional and flavored beverage, but also a physiologically function-modulating one. The obesity-depressing, anti-atherosclerosis, cardiotonic, blood-lipid depressing, blood-pressure depresssing activities from tea drinking can be traced to medical books in ancient China (Lin, 1996). A large amount of investigation has shown that tea and tea components protect against cardiovascular diseases as a result of their cholesterol-lowering and antioxidative actions (Graham, 1992; Stensvold *et al.* 1992; Weisburger, 1996; Hertog *et al.* 1993; Keli *et al.* 1996; Serafini *et al.* 1996). Also, some epidemiological studies have demonstrated a protective effect of flavonoid intake of tea consumption on mortality from cardiovascular diseases (Stensvold *et al.* 1992; Hertog *et al.* 1993; Imai *et al.* 1995) and a reduction of lipid risk factors of cardiovascular disease (Green and Jucha, 1986; Sato *et al.* 1989; Green and Harari, 1992; Stensvold *et al.* 1993; Keli *et al.* 1996). However, some studies cannot establish these associations (Kark *et al.* 1985; Klatsky *et al.* 1985; Brown *et al.* 1993; Haffner *et al.* 1985). There may be various reasons for these conflicting results. Atherosclerosis, cardiovascular disease and associated disorders are polygenic and multifactorial in their origins. In addition to abnormal lipid metabolism as a risk factor, there are other risk factors present. Another possibility may be the total flavonoids and other dietary components, and not tea *per se*, that is responsible for the beneficial effects, thus diminishing the effects of tea. Further investigations and epidemiological studies under carefully controlled conditions are needed. Until now, although the clear-cut conclusion on the positive effect of tea and tea components on cardiovascular disease cannot be drawn, a lot of encouraging results have been obtained. The huge worldwide consumption of three billion cups per day, apparent lack of toxicity and cheap price may attract to the development of tea drinking as an adjunctive therapy for prevention of some chronic diseases.

REFERENCES

Ali, M. and Afzal, M. (1987) A potent inhibitor of thrombin stimulated platelet thromboxane formation from unprocessed tea. *Prostag. Leuko. and Medicine*, **27**, 9–14.

Ali, M., Afzal, M., Gubler, C.J. and Burka, J.F. (1990) A potent thromboxane formation inhibitor in green tea leaves. *Prost. Leuko. Essent. Fatty Acids*, **40**, 281–285.

Banerjee, J. and Ganguly, S.N. (1993) A new anticoagulating compound from *Camellia sinensis* (L.) O. Kuntze. *Proc. of the Internat. Symp. on Tea Science and human Health*, Jan. 1993, Calcautta, India, pp. 145–147.

Bingham, S.A., Vorster, H., Jerling, J.C., Mugee, E., Mulligan, A., Runswick, S.A. *et al.* (1997) Effect of black tea drinking on blood lipids, blood pressure and aspects of bowel habits. *Brit. J. of Nutrition*, **78**, 41–55.

Brown, C.A., Balton-Smith, C., Woodcoard, M. and Tunstall-Pedoe, H. (1993) Coffee and tea consumption and the prevalence of coronary heart diseases in men and women. Results from the Scottish heart health study. *J. Epidemiol. Commun. Health*, **47**, 171–175.

Chang, S.Y., Shi, H.G., Nie, J.R. and Chang, R.M. (1993) Effect of 88,104 health care tea on lipid *in vivo* and *in vitro* peroxidation in mouse.. *Tea J.*, **19** (2), 33–35.

Chen, Z.M. (1994) The physiologically modulating function of tea to human (Chinese). *Tea abstracts*, 8(1), 1–8; 8(2), 1–8.

Deng, Z.Y., Tao, B.Y., Li, X.L., He, J.M., Chen, Y.F. and Chen, F. (1998) Effects of tea on blood glucose, blood lipids and antioxidative activity in old rats (Chinese). *J. Tea Sci.*, **18** (2), 74–77.

Fang, Y.Z. (1995) Antiatherosclerotic and anticarcinogenic effects of green tea polyphenol. *Proc. Of '95 Internat. Tea-Quality-Human Health Symp.*, Nov, 1995, Shanghai, China, pp. 45–48.

Fiala, E.S., Sodum, R.S., Bhattacharya, M. and Li, H. (1996) (–)-Epigallocatechin gallate, a polyphenolic tea antioxidant, inhibits peroxynitrite-mediated formation of 8-oxodeoxyguanosine and 3-nitrotyrosine. *Experientia*, **52**, 922–926.

Fredholm, B.B. (1984) Cadiovascular and renal actions of methylxanthines. In *The methylxanthine beverage and foods: Chemistry, consumption and health effects*. Alan, R. Liss Inc., New York, pp. 303–330.

Fukuyo, M., Hara, Y. and Muramustu, K. (1986) Effect of tea catechin, (–)-epigallocatechin gallate, on plasma cholesterol level in rats. *J. Jp. Soc. Nat. Fd. Sci.*, **39**, 495–500.

Graham, H.N. (1992) Green tea: composition, consumption and polyphenol chemistry. *Prev. Medicine*, **21**, 334–350.

Green, M.S. and Jucna, E. (1986) Association of serum lipids with coffee, tea and egg consumption in free-living subjects. *J. Epidemiol. Commun. Health*, **40**, 324–329.

Green, M.S. and Harari, G. (1992) Association of serum lipoproteins and health-related habits with coffee and tea consumption in free-living subjects examined in the Israeli CORDIS study. *Prev. Med.*, **21**, 526–531.

Guo, Q., Zhao, B.L., Li, M.F., Shen, S.R. and Xin, W.J. (1996) Studies on protective mechanisms of four components of green tea against lipid peroxidation in synaptosomes. *Biochem. et Biophy. Acta*, **1304**, 210–222.

Haffner, S.M., Knapp, J.A., Stern, M.P., Hazuda, H.P., Rosenthal and M., Franco, L.F. (1985) Coffee consumption, diet and lipids. *Amer. J. Epidemiology*, **122**, 1–12.

Hara, Y. (1991) Blood-pressure depressing and cholesterol decreasing effect of green tea. *Fd. Chem.*, **9**, 122–127.

Hara, Y. (1993) The antioxidative function of tea polyphenol components. *J. Act. Oxyg. Free Rad.*, **4**(3), 307–313.

Hara, Y. (1994) Antioxidative action of tea polyphenols. *Internat. Biotech. Lab.*, **12**, 14–15.

Hara, Y. (1995) Action of tea on cardiovascular disease. *Proc. of '95 Internat. Tea-Quality-Human health Symp.*, Nov. 1995, Shanghai, China, pp. 16–31.

Hara, Y., Matsuzaki, T. and Suzuki, T. (1987) Angiotensin I converting enzyme inhibiting activity of tea components. *Japan J. Agricul. Chem.*, **61**(7), 803–808.

Hara, Y. and Tono-Oka, T. (1990) Hypotensive effect of tea catechins of black tea on rats. *J. Nutrit. Sci. Vitaminol.*, **43**(5), 345–348.

He, Y.H. and Shahidi, F. (1997) Antioxidant avtivity of green tea and its catechins in a fish meat model system. *J. Agricult. Fd. Chem.*, **45**, 4262–4267.

Hensrud, D.D. and Heimburger, D.C. (1994) Antioxidant status of fatty acids and cardiovascular disease. *Nutrition*, **10**, 170–175.

Hertog, M.G.L., Feskens, E.J.M., Hollman, P.C.H., Katan, M.B. and Kromhout, D. (1993) Dietary antioxidant flavonoids and risk of coronary heart disease: the Zatphen elderly study. *Lancet*, **342**, 1007–1011.

Horie, H., Goto, T. and Kohata, K. (1996) Comparison of inhibitory activity of angiotensin I-converting enzyme among various kinds of teas. *Res. Bullt. of Veget. & Tea Experim. Stat. B (Tea)*, **9**, 37–40.

Ikeda, I., Imasato, Y., Sasaki, E., Nakayama, M., Nageo, H., Takeo, T., Yayabe, F. and Sugano, M. (1992) *Tea* catechins decrease micellar solubility and intestinal absorption of cholesterol in rats. *Biochim. Biophys. Acta*, **1127**, 141–146.

Imai, K and Nakachi, K. (1995) Cross sectional study of effects of drinking green tea on cardiovascular and liver diseases. *Brit. Med. J.*, **310**, 693–696.

Imai, K., Suga, K. and Nakachi, K. (1995) Epidemiological survey on the Tea drinking and prevention of cancer, heart disease (Japanese). *Tea*, **48**(7), 6–10.

Ishigaki, F. and Hara, Y. (1991) Blood pressure depressing effect of tea. *Fd. Ind.*, **38**, 20–25.

Ishigaki, K., Takakuwa, T. and Takeo, T. (1991). Suppression of the accumulation of body and liver fat by tea catechins. *Proc. of the Internat. Symp. on Tea Science*, Aug. 1991, Shizuoka, Japan, pp. 240–242.

Isoda, Y., Nishizawa, S., Kashima, N., Akusawa, K. and Kokai, M. (1990) Constituents of blood cholesterol-depressing oil. *JP Bin-2-243622*, Sept. 1990.

Iwata, K., Inayama, T., Miwa, S., Kawaguchi, K. and Koike, G. (1991) Effect of Oolong tea on plasma lipids and lipoprotein lipase activity in young women. *Jap. J. Nutrit. Sci. Vitamal.*, **44**(4), 251–259.

Kajimoto, G. (1963) On the antioxidative components and antiseptic components in tea. Part I. Antioxidant action and antiseptic action of materials extracted with alcohol and water from tea (Japanese), *Jpn. J. Fd. Ind.*, **10**, 1–6.

Kark, J.D., Friedlander, Y., Kaufmann, N.A. and Stein, Y. (1985) Coffee, tea and plasma cholesterol: The Jerusalem lipid research clinic prevalence study. *Brit. Med. J.*, **21**, 659–704.

Keli, O., Hertog, M.G.L., Feskens, J.M. and Kromhout, D. (1996) Dietary flavonoids, antioxidant vitamins and incidence of stroke. *Arch. Internat. Med.*, **156**, 637–641.

Kinoshita, A. and Horie, N. (1993) Inhibitory activity of green tea catechins on thrombin. *Jpn. J. Throm. & Hemostas.*, **4**(6), 417–422.

Kinoshita, A. and Horie, N. (1994) Inhibitory activity of green tea catechins on enzymes in the fibrinolytic and the kallinkrein-kinin systems, *Jpn. J. Throm. & Hemostas.*, **5**(2), 98–104.

Klastky, A.L., Petitti, D.B., Armstrong, N.I.A. and Friedman, G.D. (1985) Coffee, tea and cholesterol. *Amer. J. Cardicol.*, **55**, 577–578.

Kobayashi, M., Mochizuki, T., Terashima, T. and Yokogoshi, H. (1996) Hypotensive effect and activation of dopaminergic neurons by theanine in rats. *Proc. of the Internat. Symp. on Tea Culture and Health Sci.*, Oct. 1996, Kakegawa, Japan pp. 164–169.

Kono, S., Shincli, K., Ikeda, N., Yanai, F., and Imanishi, K. (1992) Green tea consumption and serum lipid profiles. A cross sectional study in Northern Kyseshu, Japan. *Prev. Med.*, **21**, 526–531.

Li, M.X. and Chang, R.H. (1993) Gabaron tea and its hypotensive effects (Chinese)., *Food Ind. (Taiwan)*, **25**(10), 38–44.

Lin, K.L. (1996) Ancient tea thearpy in China. *Proc. of the Internat. Symp. on Tea Culture and Health Sci.*, Oct. 1996, pp. 42–49.

Lou, F.Q., Zhang, M.F., Zhang, X.G., Liu, J.M. and Yuan, W.L. (1989) A study on tea pigment in prevention of atherosclerosis. *Chin. Med. J.*, **102** (5), 579–583.

Matsuda, H., Chisaka, T., Kubomura, Y., Yamahara, J., Sawada, T., Fujimura, H. and Kimura, H. (1986) Effects of crude drugs on experimental hypercholesterolemia. I. Tea and its active principles. *J. Ethnopharmac.*, **17**, 213–224.

Matsumoto, N. and Hara, Y. (1995) Blood-cholesterol depressing activity of tea catechin. *Food Chem.*, **13**, 81–84.

Matsuzaki, T. And Hara, Y. (1985) Antioxidative activity of tea leaf catechins. *J. Agricul. Chem. Soc. Jpn.*, **59**, 129–134.

Mitscher, L.A., Jung, M., Shankel, D., Dou, J.M., Steele, L. and Pillai, S.P. (1997) Chemoprotection: A review of the potential therapeutic antioxidant properties of green tea (*Camellia sinensis*) and certain of its Constituents., *Medic. Res. Rev.*, **17** (4), 327–365.

Miura, S., Watanabe, J., Tomita, T., Sano, M. and Tomita, I. (1994) The inhibitory effects of tea polyphenols (flavan-3-ol derivatives) on Cu^{2+} mediated oxidative modification of low density lipoprotein. *Biol. Pharm. Bull.*, **17** (12), 1567–1572.

Mori, M. (1990) Hypolipidemia agent. *JP Bin-2-286620*, Nov. 1990.

Muramatsu, K., Fukuyo, M. and Hara, Y. (1986) Effect of green tea catechins on plasma cholesterol level in cholestero-fed rats. *J. Nutr. Sci. Vitaminol.*, **32**, 613–622.

Nakachi, K. Suga, K. and Imai, K. (1995) Preventive effects of drinking green tea on cardiovascular disease and cancer. *Proc. of the 3rd Internat. Symp. on Green Tea, Sept. 1995*, Seoul, Korea, pp. 13–20.

Namiki, K., Yamanaka, M., Tateyama, C., Igarashi, M. and Namiki, M. (*1991*) Platelet aggregation inhibitory activity of tea extracts. Proc. of Internat. Symp. on Tea Sci., Aug. 1991, Shizuoka, Japan. pp. 327–331.

Nanjo, F., Hara, Y. and Kikuchi, Y. (1992) Effects of tea polyphenols on blood rheology in rats fed high-fat diet. In Ho, C.T. *et al.*, (eds.), Food *phytochemicals for cancer prevention. II. "Teas, spices and herbs"*, *ACS Symp. Series* 547, pp. 76–82.

Ohtsura, M., Nishimura, K., Makita, T., Yayabe, F. and Kakuda, T. (1991) Biochemical examination of the effect of chronic Oolong tea consumption in the rabbit. *Jpn. J. Fd. Ind. Soc.*, **38** (7), 52–54.

Omori, M. (1995) Effect of anaerobic treated tea (Gabaron tea) on the blood pressure of spontaneously hypertensive rats. *Proc. of the 3rd Internat. Symp. on Green Tea*, Sept. *1995*, Seoul, Korea, pp. 53–58.

Omori, M., Yano, T., Okamoto, J., Tsushida, T., Murai, T. and Higuchi, M. (1987) Hypertensive-*inhibitory* activity of anaerobically treated green tea (Gabaron tea) on the spontaneously hypertensive rats. *Jpn. J. Agricul. Chem.*, **61**(11), 1449–1451.

Pinocomb, G.A., Lovallo, W.R., Passey, R.B. and Wilson, M.F. (1988) Effect of behaviour state on caffeine's ability to alter blood pressure. *Amer. J. Med.*, 77, 54–60.

Rice-Evans, C.A., Miller, N.J., Bolwell, P.G., Bramley, P.M. and Pridham, J.B. (1995) The relative antioxidant activities of plant-derived polyphenolic flavonoids. *Free Rad. Res.*, **22** (4), 375–383.

Robertson, D., Hollister, A.S., Kincaid, D., Workman, R., Goldberg, M.R., Tung, C.S. *et al.* (1984) Caffeine and hypertension. *Amer. J. Med.*, 77, 54–60.

Sagesaka-Mitane, Y., Miwa, Y. and Okada, M. (1990) Platelet aggregation inhibitors in hot water extract of green tea. *Chem. Pharm. Bull.* 38, 790–795.

Saito, H., Omori, M., Saijo, R., Fukatsu, O., Hakamadu, K. and Hashizume, K. (1995) Effect of Gabaron tea on the hypertension correlation with kidney function. *Proc. of '95 Internat. Tea-Quality-Human Health Symp.*, Nov. 1995, Shanghai, China, pp. 49–52.

Salah, S., Miller, N.J., Paganga, G., Tizburg, L., Bolwell, P. and Rice-Evans, C. (1995) Polyphenolic flavanols as scavengers of aqueous phase radicals and as chain-breaking antioxidants. *Achiev. Biochem. & Biophy.*, **322** (2), 339–346.

Sano, M., Takabashi, Y., Komutsu, C., Nakamura, Y., Shimoi, K., Tomita, F., *et al.* Antioxidative activities of green tea and black tea in rat organ and blood plasma. *Proc. of the Internat. Symp. on Tea Sci. (ISTS)*, Aug. 1991, Shizuoka, Japan, pp. 304–308.

Sano, M., Takabashi, Y., Yoshino, K., Shimoi, K., Nakanura, Y., Tomita, I., *et al.* (1995) Effect of tea (*Camellia sinensis*) on lipid peroxidation in rat liver and kidney: a comparison of green tea and black tea feeding. *Biol. Pharm. Bull.*, 18 (7), 1006–1008.

Sato,Y., Nakatsuka, H., Watanabe, T., Hisamichi, S., Shimiza, H., Fujisaku, S., *et al.* (1989) Possible contribution of green tea drinking habits to the prevention of stroke. *Tohoku J. Exp. Med.*, 157, 337–343.

Serafini, M., Ghiselli, A. and Ferro-Luzzi, A. (1996) *In vivo* antioxidant effect of green and black tea in man. *Eup. J. of Clin. Nutrition*, 50, 28–32.

Shen, S.R., Zhao, Y.F. and Zhao, B.L. (1997) Inhibitory effect of catechins to free radicals at the present of iron ion (Chinese). *J. Tea Sci.*, 17 (Suppl.), 119–123.

Spernius, V.C., Venegas, P.L. and Wattenberg, C.W. (1982) Glutathione S-transferase activity: enhancement by compounds inhibiting chemical carcinogenesis by tea and by dietary constituents. *J. Nat. Cancer Inst.*, **68**, 493–496.

Stensvold, I., Tverdal, A., Solvoll, K. and Foss, O.P. (1992) tea consumption: relationship to cholesterol, blood pressure and coronary and total mortality. *Prev. Med.*, 21, 546–553.

Tanigushi, S., Miyashita, Y., Ueyama, T., Haze, K., Hirase, J., Takemoto, T., *et al.* (1987) A hypotensive constituent in hot water extracts of green tea. Yakugaku Zasshi, **108** (1), 77–81.

Tomita, I. (1997) Peroxidation and its prevention in respect of food hygienic science (Japanese). *J. Fd. Hyg.*, **38** (2), 39–47.

Tuomilento, J., Tanskanen, A., Pietinen, P., Aro, A. Salonen, J.T., Happonen, P., Nissinen, A. and Puska, P. (1987) Coffee consumption is correlated with serum cholesterol in middle-age Finnish men and women. *J. Epidemiol. & Common Health*, **41**, 237–242.

von Gadow, A., Joubert, E. and Hansmann, C.F. (1987) *Comparison* of the antioxidant activity of rooibos tea (*Aspalathus linearis*) with green, Oolong and black tea. *Fd. Chem.*, **60** (1), 73–77.

Vorster, H., Jerling, J., Oosthuizen, W., Cummings, J., Bingam, S., Magee, L., *et al.* (1996) Tea drinking and haemostasis: a randomized, placebo-controlled, crossover study in free-living subjects. *Haemostasis* **26**, 58–64.

Weisburger, J.H. (1996) Tea antioxidants and health. In E. Cadenas and L. Packer (eds.), *Handbook of Antioxidants*, Marcel Dekker Inc., NY publishers, pp. 469–486.

Williams, G.H. (1992) In *Harrison's Principles of Internal Medicine*, J.D. Wilson, E. Braunwald, K.J. Isselbacher (eds), N.Y. pp. 1001–1015.

Witztum, J.L. and Steinberg, D. (1991) Role of oxidized low density lipoprotein in atherosclerosis. *J. Clin. Invest.*, **88**, 1785–1792.

Xie, B.J., Shi, H. and Hu, W.H. (1994) Antioxidant properties and the inhibitory properties on lipoxygenase activity of major fractions and polyphenol constituents from green, Oolong and black teas. *Natural Prod. Res. & Develop.*, **6** (4), 19–25.

Yen, G.C. and Chen, H.Y. (1995) Antioxidant activity of various tea extracts in relation to their antimutagenicity. *J. Agricul. Fd. Chem.*, **43**, 27–32.

Yen, G.C., Chen, H.Y. and Peng, H.H. (1997) Antioxidant and pro-oxidant effects of various tea extracts. *J. Agricult. Fd. Chem.*, **45**, 30–34.

Yokozawa, T., Oura, H., Sakanaka, S., Ishigukis, Y. and Kim, M. (1994) Depressor effect of tannin of green tea on rats with renal hypertension. *Biosci. Biotech. Biochem.*, **58** (5), 855–858.

Yoshino, K., Tomita, I, Sano, M., Oguni, I., Hara, Y. and Nakano, M. (1994) Effects of long-term dietary supplement of tea polyphenols on lipid peroxide levels in rats. Age, **170**, 79–85.

Zhang, A.Q., Chan, P.T. and Luk, Y.S. (1997) Inhibitory effect of jasmine green tea epicatechin isomerss on LDL-oxidation. *J. Nutr. Biochem.*, **8** (6), 334–340.

Zhang, A.Q., Zhu, Q.Y., Luk, Y.S., Ho, K.Y., Fung, K.P. and Chen, Z.Y. (1997) Inhibitory effects of jasmine green tea epicatechin isomers on free radical-induced lysis of red blood cells. *Life Sci.*, **61** (4), 383–394.

Zhao, B.L., Li, X.J., He, R.G., Cheng, S.J. and Xin, W.J. (1989) Scavenging effect of extracts of green tea and natural antioxidants on active oxygen radicals. *Cell Biophys.*, **14**, 175–185.

9. ANTIMICROBIAL ACTIVITY OF TEA PRODUCTS

PEI-ZHEN TAO

Institute of Medicinal Biotechnology, Chinese Academy of Medical Sciences and Peking Union Medical College, 1 Tiantan Xili, Beijing 100050, China

In old China tea was first used as an antidote and in India as a kind of medicine for treatment of diarrheal disease. Tea is classified into four types, green tea, oolong tea, black tea, and pu-erh tea, on the basis of the manufacturing process. All of these teas are prepared from leaves of *Camellia sinensis* and its varieties. Tea has multifunctions as indicated in different chapters of this book. The antimicrobial activities, including antibacteria, antifungi, anti-caries and antivirus, of tea are the focus of this chapter.

1. ANTIBACTERIAL ACTIVITY OF TEA

1.1. Bacteriostatic and Bactericidal Activity of Tea

One of the earliest reports (Anonymous, 1923) recommended use of tea as a prophylactic against typhoid. In recent years there are some reports, mostly from Japan, dealing with the antibacterial activity of tea. It has been reported (Ryu, 1980; Eugster, 1981; Ryu *et al.* 1982) that the incorporation of 0.5–1% tea powder of oolong tea into nutrient agar could inhibit the growth of pathogen bacteria including *Staphylococcus aureus*, *Salmonella typhimurium*, *Salmonella paratyphi* A, *Salmonella paratyphi* B, *Vibrio cholerae*, *Vibrio parahaemorrhagiae*, *Shigella dysenterie*, *Klebsiella pneumonia*, *Proteus mirabilis*, *Pasteurella multocida*, *Pseudomonas aeruginosa*, and *Streptococcus zooepidemicus*. 0.5% tea powder of green tea and black tea showed similar inhibition effects against the first seven pathogenic bacteria. All three tea powders could not inhibit the growth of *Escherichia coli*, the *E. coli* was just further diluted. They also found that 3% suspension of oolong tea, green tea and black tea, respectively could kill *V. cholerae* and *V. parahaemorrhagiae* in 30 minutes. 3% suspension of oolong tea and green tea, could each kill *Sal. paratyphi* B in 2 hr.

Das (1962) reported that the infusions of green and black tea could inhibit the growth of *Staphylococcus aureus*, *Salmonella typhosa*, *Shigella dysenteriae* and *El Tor vibrio* (an allied strain of *Vibrio cholerae*) by the agar diffusion method. Toda *et al.* (1989a, 1989b) using the same method tested the inhibition effects of extracts of black tea, five kinds of Japanese green tea, Pu-erh tea (Chinese tea) and coffee against various bacteria known to cause diarrheal diseases. They found that Gram-positive bacteria were more sensitive than Gram-negative bacteria and all extracts of tea showed more inhibitory effects than extract of coffee. *Staphylococcus aureus*, *Staph. epidermidis*, *Vibrio cholerae* non 01 and *Plesiomonas shigelloides* were sensitive to all the tea and coffee extracts tested. *Escherichia coli* and *Pseudomonas aeruginosa* were resistant. They also found that these extracts were bactericidal against one strain of each of *Staph. aureus*, *V. parahaemolyticus* and *V. cholerae* 01. All tea and coffee extracts killed *V. parahaemolyticus* within 3 hrs. In the case of *Staph. aureus*, black and pu-erh tea showed stronger bactericidal activity than green tea or coffee. *V. cholerae* 01 was killed within 1 hr by all four extracts. The bactericidal activity was shown even at the drinking concentration in daily life. Recently Shetty *et al.* (1994) confirmed the above results by the agar diffusion method and reported the same conclusion with extracts of black tea, Japanese green tea, Chinese tea and coffee. 44 strains of Gram-positive and Gram-negative bacteria including 25 strains of *Salmonella typhimurium,* a pathogen causing infantile diarrhea, and one strain of fungus *Candida albicans* were tested. All the 25 strains of *S. typhimurium* were sensitive to black tea, Japanese green tea and coffee but not Chinese tea. The viable cell count method has also been used to study whether the growth inhibition by these extracts was bactericidal or bacteriostatic. In the case of one strain of *V. cholerae* all the tea and coffee extracts were bactericidal within 2 hrs. In the case of one strain of *Staph. Aureus,* black tea and Japanese green tea showed stronger bactericidal activity than Chinese tea and coffee. In the case of one strain of *S. typhimurium* black tea, Chinese tea showed stronger bactericidal activity than the Japanese green tea and coffee.

Okubo (1991) reported that 2.5% infusion of black tea completely inhibited the growth of both *Trichophyton mentagrophytes* and *T. rubrum.* The infusion also showed a fungicidal activity against *Trichophyton* in a dose- and contact time-dependent manner. The infusion neither inhibited the growth of *Candida albicans* nor killed both the *C. albicans* and *neoformans*.

1.2. Antibacterial Activity of Tea Components

It is well known that tea polyphenols are responsible for the antibacterial activities of various tea products. Polyphenolic compounds make up some 30% of the dry weight of flush and black tea leaves. The simplest compounds are catechins which are well-characterized isoflavanoids, mainly consisting of four compounds, (–)-epicatechin (EC), (–)-epigallocatechin (EGC), (–)-epicatechin gallate ECG, and (–)-epigallocatechin gallate (EGCG). About 5% dry weight of black tea and 10% dry weight of green tea as well as its aqueous extracts is made up of catechins. Catechins may present in a cup of tea at a concentration of 1 mg/ml. The larger molecules in the class include theaflavins and thearubigins, which are oxidation and polymerization products of simple isoflavanoids and main polyphenolic compounds in black tea.

One of the early reports (Das, 1962) indicated that EC, ECG, EGC and EGCG inhibited the growth of *Staphylococcus aureus*, Shigella dysenteriae, Salmonella typhosa, EIT or vibrio. Hara *et al.* (1989, 1989) estimated the inhibitory effects of catechins (crud catechin, EC, ECG, EGC, EGCG, crud theaflavin, TF1, TF2A, TF2B and TF3) against food-borne pathogenic bacteria. The results showed that these compounds had antibacterial activities against *Vibrio parahaemolyticus*, *V. fluvialis*, *V. metschnikovii*, *Staphylococcus aureus*, *Clostridium perfringens*, *Bacillus cereus*, *Bacillus stearothermophilus*, *Plesiomonas shigelloides* and *Aeromonas sobris* with the MIC values of 100–800 ppm.

EGCG and theaflavin digallate (TF3) had been tested for their antifungal and fungicidal activities against *Trichophyton mentagrophytes*, *T. rubrum*, *Candida albicans* and *Cryptococcus neoformans* (Okubo *et al.* 1991). EGCG at 2.5 mg/ml could not inhibit the growth of *Trichophyton*, whereas TF3 at 0.5 mg/ml inhibited the growth. TF3 at 1 mg/ml killed *Trichophyton* within 72–96 hr contact.

Tea also contains flavor components in which green tea flavor contains more than 100 volatile compounds and black tea contains more than 300 volatile compounds (Flament, 1991). Kubo *et al.* (1992) first tested the antibacterial activities of 10 main nonpolar substances in green tea against 13 selected microorganisms including *Bacillus subtilis*, *Brevibacterium ammoniagenes*, *Staphylococcus aureus*, *Streptococcus mutans*, *Propionibacterium acnes*, *Escherichia coli*, *Pseudomonas aeruginosa*, *Enterobacter aerogenes*, *Saccharomyces cerevisiae*, *Candida utilis*, *Pityrosporum ovale*, *Trichophyton mentagrophytes*, and *Penicillium chrysogenum*. These compounds are linalool, δ-cadinene, geraniol, nerolidol, α-terpineol, *cis*-jasmone, indole, β-ionone, 1-octanol, and caryophyllene. Most of the teas seem to consist of almost the same components, but the compositions differed as a result of the manufacturing process. The results showed that δ-cadinene inhibited *P. acnes* with MIC of 3.13 µg/ml, caryophyllene inhibited *P. acnes* with MIC of 6.25 µg/ml, nerolidol inhibited *S. mutans* with MIC of 25 µg/ml. For other compounds the MICs against various microorganisms were 50–800 µg/ml, so the activities were moderate or weak. It is said that a cup of tea prepared with 2 g (the usual amount for a commercial tea bag) of the tea leaves in 100 ml of hot water contains a total of 7 µg/ml of volatiles. This concentration does not seem to be strong enough to control these microorganisms including *S. mutans* which is responsible for causing dental caries.

In addition to antibacterial activities, most of the green tea flavor compounds tested exhibited antifungal activities against *P. ovale*, *S. cerevisiae*, *C.utilis*, *T. mentagrophytes*, and *P. chrysogenum*. Nerolidol inhibited the growth of *T. mentagrophytes* at 12.5 µg/ml, other volatiles except δ-cadinene and caryophyllene also inhibited the growth of *T. mentagrophytes* with MICs between 50 and 200 µg/ml. *T. mentagrophytes* and *P. ovale* occur primarily on human hair and cause human dermatomycosis. So the authors suggest that these flavor compounds may be considered potential anti-microbial agents for cosmetic and food products because of the safe use of tea for thousands of years.

1.3. Anti-*Vibrio Cholerae* 01 Activity of Tea

In the first paragraph of this chapter it is described that tea has antibacterial and bactericidal activities against various bacteria that cause diarrheal diseases including various Vibrios. It was also indicated that tea could agglutinate *Vibrio cholerae* and render them non-mobile (Toda *et al.* 1991). Black tea extract (500 µl/ml) could kill *Vibrio cholerae* 01 V86 within 1 hr and strain 569B of *Vibrio cholerae* was usually killed immediately after contact with the extract. As we are aware that cholera still occurs in developing countries, it would be interesting to know more about the inhibition effects of tea against *Vibrio cholerae* 01.

Okubo *et al.* (1989) first demonstrated that tea has direct anti-toxin activities. The extracts of black tea, green tea and pu-erh tea (China tea) were examined for the inhibition of hemolytic activity of *Staphylococcus aureus* α-toxin and *V. parahaemo-lyticus* thermostable direct haemolysin (Vp-TDH). All the extracts inhibited the hemolytic activity of Vp-TDH by 100% in the experimental conditions. The inhibition of the hemolytic activity of α-toxin was in the order of black tea > green tea > pu-erh tea. Toda *et al.* (1990) reported the MIC values of catechins and analogs against the growth of *Vibrio cholerae* 01 and *Staphylococcus aureus*. The strongest inhibitor in the catechins was ECG with MICs of 125 and 16 μg/0.1 ml respectively. In the analogs pyrogallol was the strongest with MICs of 25 and 12.5 μg/0.1 ml respectively. Rutin and caffeine did not show any inhibitory activity against the growth of these two bacteria at concentrations of 1000 and 2000 μg/0.1 ml. The gallate moiety of the molecule is important for the growth inhibitory activity of catechins. EGCG, ECG and EGC at the concentration of 33 μg/ml could inhibit the hemolysis of Vp-TDH by 100%, 65% and 69% respectively and inhibit the hemolysis of *V. cholerae* 01 V86 by 100%, 100% and 0% respectively. For the analogs only tannic acid could inhibit the hemolysis of *V. cholerae* 01 V86 by 100% at the concentration of 33 μg/ml. Beside the anti-cholera haemolysin activity of tea components, the black tea extract could inhibit the toxic activity of purified cholera toxin. In Chinese hamster ovary cells (CHO) the black tea extract at non-cytotoxic concentration of 0.3 μl/ml significantly inhibited morphological changes of CHO cells induced by cholera toxin by 83.5% (Toda *et al.* 1991).

Ikigai *et al.* (1990) investigated the structure-activity relationship of anti-hemolysin activity of catechins and theaflavins. The inhibitory effects of catechins and theaflavins against the hemolysis of *Vibrio cholerae* 01 hemolysin and *Staphylococcus aureus* α-toxin were examined *in vitro*. The results indicated that both catechins and theaflavins showed anti-hemolysin activities in a dose-dependent manner. Among the catechins examined, (−)catechin gallate, (−)epicatechin gallate and (−)epigallocatechin gallate with galloyl groups in their molecules showed more potent anti-hemolysin activities against both toxins by inhibition rates of 40–90% at a concentration of 100 μM. For cholera hemolysin the most potent inhibitor was (−)gallocatechin gallate and for α-toxin was (−)epigallocatechin gallate. Among dextrocatechins, (+)epicatechin and (+)epigallocatechin proved to be more effective than (+)catechin and (+)gallocate-chin. The anti-hemolysin activities of theaflavins against both toxins were dependent on the number of the galloyl group in their structure. Theaflavin digallate (10 μM) inhibited the hemolysis of cholera hemolysin by 75% and inhibited the hemolysis of α-toxin by 92% at a concentration of 2 μM. These results suggest that the tertiary structure of the catechin or theaflavin and the active site of hemolysin play an important role in the anti-hemolysin activity.

The black tea extract and tea catechins had been tested in animals for their anti-cholera toxin and anti-cholera infection effects (Toda *et al.* 1991; Toda *et al.* 1992). The purity of tea catechins was 91.2% containing (+)gallocatechin (1.6 w/w%),

(–) epicatechin (6.4%), (–) epigallocatechin (19.2%), (–) epicatechin gallate (13.7%) and (–) epigallocatechin gallate (59.1%). Both products displayed similar results. Inhibitory activity of tea catechins and black tea extract against cholera toxin was determined using sealed adult mice. Cholera toxin was administered to mice 5 min before administration of tea catechins (5 mg) or tea extract. Tea catechins and tea extract significantly inhibited fluid accumulation induced by cholera toxin with statistical significance ($P < 0.001$). Inhibitory activity of tea catechins and tea extract against cholera infection was determined in a rabbit model. cholera vibrios were injected into the rabbit intestinal loops 5 min before injection of tea catechins (10 mg) or tea extract. The catechins and tea extract completely inhibited fluid accumulation induced by strain 569B in the rabbit intestinal loops.

These results indicate that black tea extract and tea catechins can inhibit the growth of V. cholerae 01 and kill the bacteria. Both products can directly inhibit the cholera hemolysin in vitro and cholera toxin in vitro and in vivo. A cup of tea beverage (about 100–150 ml) contains about 100–150 mg of catechins, a level that could be expected to be effective in the human small intestine. The potential usage of catechins in prevention and treatment of cholera and other food-borne diseases need to be examined in humans in the future.

1.4. Effect of Tea on Intestinal Microflora

Intestinal microorganisms not only participate in normal physiological function, but are also involved in various diseases. Previous investigations indicate that there are age- and disease-associated differences in intestinal bacteria (Finegold et al. 1974; Finegold et al. 1975; Gorbach et al. 1967). Diet, stress and other factors may also influence the composition of the microflora. The normal microflora is found to be predominantly composed of lactic acid bacteria. On the other hand, the microflora of cancer patients is mainly composed of clostridia and eubacteria with few lactic acid bacteria. Elderly persons also harbor fewer bifidobacteria but larger numbers of clostridia than younger persons do. Disturbance of the microflora may cause various diseases or abnormal physiological states. Tea has been reported to inhibit the growth of clostridia. Ahn et al. (1990) examined the effects of methanol extract of green tea on 45 strains of bacteria. The tea product showed moderate growth-promoting activity for Bifidobacterium adolescentis (strain E194a), B. longum, B. breve, B. infantis, Lactobacillus casei and L. salivarius at a concentration of 0.1% (w/v) but not at the higher concentration (1% w/v). Growth of other test bacteria was not affected by addition of the extract to PYF broth. At a concentration of 10 mg extract/disk it inhibited the growth of all of the clostridial species (17 strains) tested with the exception of C. butyricum, C. coccoides and one of the two strains of C. ramosum (strain C-oo). No activity was apparent against any of the other test bacteria. The green tea extract not only enhanced growth of some lactic acid bacteria but also specifically inhibited various clostridia strains including C. difficile and C. perfringens, causative agents of a wide variety of human diseases, and C. paraputrificum. The same authors (Ahn et al. 1991) examined the inhibition of tea catechins on C. difficile and C. perfringens. Only (–)-epicatechin gallate and (–)-epigallocatechin gallate at a concentration of 5 mg/disk showed strong inhibition effects. The results showed that the gallate moiety of polyphenols might be required, but their stereochemistry did not appear critical for the inhibitory activity. Hara *et al.* (1989) also proved that the MICs of ECG, EGCG and theaflavins were between 150 to 200 ppm against *C. botulinum*.

Okubo *et al.* (1992) reported the human trial of polyphenol. Effects of tea polyphenol intake (0.4 g/volunteer, 3 times per day, for four weeks) on fecal microflora, bacterial metabolites and pH were investigated in eight healthy volunteers. The intestinal microflora examined included *Bacteroidaceae*, Eubacteriun spp., *Peptococcaceae*, Bifidobacterium spp., Veillonella spp., Megasphaera spp., Curved rods Clostridium spp., *C. perfringens*, Lactobacillus spp., *Enterobacteriaceae*, *Streptococcaceae*, *Micrococcaceae*, Bacillus spp., *P. aeruginosa*, Corynebacterium spp., Yeasts and Molds. The tea polyphenols used were Sunphenon containing about 70% polyphenols including (+)-catechin, (–)-epicatechin, (+)-gallocatechin, (–)-epigallocatechin, (–)-epicatechin gallate, (–)-gallocatechin gallate and (–)-epigallocatechin gallate. The results showed that only *Clostridium spp.* were seriously affected by the Sunphenon intake. Their counts (in $\log_{10}$) were reduced from 5.81 for *C. perfringens* and 8.89 for other Clostridium spp. to 4.36 and 7.74 respectively. The frequency of occurrence of *C. perfringens* also decreased significantly from 12/16 to 5/16. After two weeks of cessation of intake of Sunphenon the counts and frequency of occurrence of *C. perfringens* returned to original states. Other bacteria affected were Bifidobacterium spp. (increased) and *Peptococcaceae* (decreased), but the differences were not significant. The percentage composition of Clostridium spp. in total counts decreased from 2.0% to 0.5%, then returned to 1.8%. *Peptococcaceae* decreased from 9.9% to 3.9%. In contrast, Bifidobacterium spp. increased from 10.2% to 16.0%.

Clostridia are considered to cause a variety of diseases such as sudden death, toxicity, mutagenesis, carcinogenesis, Alzheimer's disease and aging (Hentges, 1983). They participate in biotransformation of a variety of ingested or endogenously formed compounds to yield harmful products such as N-nitroso compounds or aromatic steroids which may be carcinogenic. The above findings suggest that tea polyphenols could selectively inhibit clostridial growth while promote the growth of Bifidobacterium spp, very important in reversing the harmful effects of *Clostridial* bacteria. These results are consistent with the *in vitro* effects of tea polyphenols on these bacteria, i.e., Clostridium spp. (inhibited) and Bifidobacterium spp. (promoted).

Bacterial metabolites such as volatile fatty acids, including acetic and propionic acids, were significantly increased by the Sunphenon intake, but it had no marked effects on putrefactive products, ammonia or enzyme activities. It also decreased the fecal pH from about 6.2 to 5.8 which was related to the increase of Bifidobacterium spp. (the acid forming bacteria). Low intestinal pH will help to improve intestinal condition and to reduce the formation of harmful compounds by bacteria. The administered quantity of tea polyphenols is equivalent to the content of 10 cups of concentrated green tea. The large quantities administered during the four weeks did not show any undesirable effects.

Ishigami *et al.* (1996) reported similar results in a clinical study. Fifteen elderly inpatients who were suffering from gastroenteral liquid alimentation received the same diet. Tea polyphenols used in this study were Polyphenon 60^r containing (–)-epigallocatechin (EGC) 18.5%, (–)-epicatechin (EC) 6.5%, (–)-epigallocatechin gallate (EGCG) 30.5% and (–)-epicatechin gallate (ECG) 7.0%. A total of 300 mg of tea polyphenols (467 mg of Polyphenon 60^r), divided into three doses, was administered daily. The study was conducted for one month. The fecal specimens were collected and analyzed periodically. The results showed that the levels of lactobacilli, bifidobacteria and coagulase-negative staphylococci increased significantly during tea polyphenol administration. The levels of *Enterobacteriaceae*, lecithinase-negative clostridia, total bacteria, *Bacteroidaceae* and eubacteria decreased significantly during tea polyphenol administration. The detection rate of bifidobacteria increased, whereas the detection rate of lecithinase-positive and negative clostridia decreased during tea polyphenol administration. Percentage of bifidobacteria and lactobacilli in total bacteria showed an increase. The pH values and the concentrations of ammonia decreased significantly. Fecal concentrations of sulfide increased on day 7 of administration, but decreased significantly on day 21 of administration. Fecal phenol, cresol, ethylphenol, indole and skatol decreased significantly while the amount of total fecal organic acids increased significantly. From this study it is inferred that tea polyphenols may work favorably to improve elderly inpatients' flora. The amount of polyphenols given each patient was equivalent to that contained in about 3 cups a day of green tea. One hundred mg of polyphenol would be obtained from 1.5 gram of tea leaves extracted in 150 ml of hot water. From the above studies, tea polyphenols have many favorable effects on the intestinal flora such as growth-inhibitory activity against a wide variety of putrefactive bacteria such as *Clostridium difficile* or *C. perfringens* and growth-promoting activity on some bifidobacteria and lactobacilli. Thus, tea polyphenols are considered to improve the condition of the human intestinal flora and have a deodorizing effect on feces.

1.5. Anti-MRSA Activity of Tea

Toda *et al.* (1991) investigated the inhibitory effects of black tea and green tea extracts and tea catechins on MRSA (methicillin-resistant *Staphylococcus aureus*) and food poisoning strains of *S. aureus*. 52 strains of clinically isolated MRSA and 20 strains of food poisoning strains of *S. aureus* were tested. They found that tea infusion (50 μl each), (–)-epigallocatechin gallate (EGCG, 63 μg) and theaflavin digallate (TF3, 125 μg) added to one ml of culture medium each inhibited the growth of all strains of MRSA, food poisoning *S. aureus* and standard *S. aureus* significantly. Black tea at 2.5% and 5.0% showed a bactericidal activity against MRSA even at the same concentration as in ordinarily brewed tea. EGCG at a concentration of 250 μg/ml also showed a bactericidal activity against MRSA but not against food poisoning *S. aureus*. EGCG at 500 μg/ml reduced markedly the viable number of *S. aureus* within 48 hrs.

Recently Yamazaki (1996) investigated the combined effects of Japanese green tea extract and 17 kinds of antibiotics against clinically isolated MRSA by a disk method. The growth-inhibition of MRSA was clearly observed in distinct areas where sub-MIC concentrations of the tea extract and β-lactam antibiotic had been applied. This indicated that there was an enhancing or co-operative effect between the tea extract and the β-lactams. The MICs (μg/ml) against MRSA of 7 β-lactams were significantly decreased by combining the tea extract with cefazolin from 200 to 1.56, cefotiam from 400 to 3.13 ~ 6.25, cefmetazole from 100 to 6.25, ceftizoxime from > 3200 to 1.56 ~ 12.5, flomoxef from 100 to 1.56, imipenem from 50 to < 0.1, ampicillin from 50 to 3.13. Lower levels of growth-inhibition or no inhibition were seen with oxacillin, carumonam and non-β-lactam antibiotics including clindamycin, vancomycin, netilmicin, levofloxacin, chloramphenicol, minocycline, erythromycin and fosfomycin. It would be worthwhile further examining the combined effects of tea catechins and these β-lactam antibiotics against MRSA and the mechanisms involved.

1.6.   Anti-Phytopathogenic Bacteria Activity of Tea

Fukai *et al.* (1991) reported the inhibitory effects of tea catechins against phytopathogenic bacteria which tended to infect commonly cultivated vegetables. The tea catechins included crude catechins and its four components, *i.e.*, (–)-epicatechin (EC), (–)-epigallocatechin (EGC), (–)-epicatechin gallate (ECG) and (–)-epigallocatechin gallate (EGCG) from green tea, as well as the crude theaflavins from black tea. The phytopathogenic bacteria included 8 strains of Erwinia, 10 strains of Pseudomonas and one strain each of Clavibacter, Xanthomonas and Agrobacterium. The MIC values were estimated. All tea catechins showed marked inhibitory effects on the phytopathogenic bacteria tested. Pyrogallol catechins (EGC and EGCG) were more effective than catechol catechins (EC and ECG) with MIC values being mostly below 100 ppm. In the case of theaflavins, crude theaflavins showed a slightly weaker inhibitory potency than pyrogallol catechins with MICs from 50 to 200 ppm and there were not many noticeable differences of MIC values observed among the theaflavins. These results indicated that tea polyphenols could be safe potential agricultural chemicals against vegetable diseases.

In the same year Kodama *et al.* (1991) also reported their research dealing with the anti-phytopathogenic bacterium and fungus activity of tea catechins. Tea catechins included EC, EGC, ECG, EGCG and M-EGCG (methylated-EGCG, purity 91%, obtained by methylation of eight hydroxyl radicals of EGCG). Eight strains of plant pathogenic bacteria and six strains of plant pathogenic fungi were used in this study. Pyrogallol type of catechins (EGC and EGCG) showed antibacterial activities at a concentration of 100 ~ 200 μg/ml against *Agrobacterium tumefaciens*, *Clavibacter michiganensis* subsp. *michiganensis*, *Erwinia carotovora* subsp. *carotovora* and *Pseudomonas syringae pv. pisi*. The activities of catechol type catechins (EC and ECG) were lower than those of EGC and EGCG. On the other hand, ECG and EGCG, which have a galloyl radical, showed significant activity against *P. syringae pv. lachrymans*, *P. solanacearum*, *X. campestris pv. vesicatoria* and *X. campestris pv. citri* with MICs of 25~50 μg/ml. Among these catechins, EGCG demonstrated the highest inhibitory activities against all plant pathogenic bacteria tested. However, M-EGCG had no inhibitory activity against these bacteria showing the hydroxy radicals were very important for the inhibitory activities of tea catechins. Both EGCG and M-EGCG had been tested for their inhibitory effects against plant pathogenic fungi. M-EGCG showed strong activity against *P. oryzae* at a concentration of 50 μg/ml, which inhibited the growth by 69.2%. For other fungi, M-EGCG showed weak activities. EGCG showed moderate activity against *C. miyabeanus* at the concentration of

400 μg/ml, which inhibited the growth by 56.8%. For other fungi EGCG showed little or no activity.

Ester type catechins mixture (ETC, containing ECG 21.1% and EGCG 72.1%) had been evaluated in the field test at the concentration of 2000 μg/ml against bacterial leaf spot of tomato (BLST) and citrus canker of Natsudaidai (CCN). As a positive control a copper compound (CC) was also tested. In the first test all leaves of tomato plants about 30 cm in height were sprayed with solution of the test samples with a hand sprayer and left outdoor for 24 hrs. Liquid culture of *Xanthomonas campestris pv. vesicatoria* was diluted to $1 \sim 2 \times 10^8$ cells/ml and sprayed over leaves of each tomato plant. The plants were maintained outdoors for two weeks and the disease severity was measured by the disease indices. The results indicated that the disease severity values of the plants pretreated with ETC and CC were 50 and 45 respectively, whereas that of the control plant was 70. In the second test one year-old seedlings of Natsudaidai were pretreated with ETC or CC in the same manner as in the case of BLST. The suspension of *X. campestris pv. citri* was sprayed over young leaves. The disease severity was measured by disease indices after a month. The disease severity values of the plants pretreated with ETC and CC were 36 and 56 respectively, whereas that of the control plant was 76. These results suggest that tea catechins are useful and safe for prevention of these bacterial diseases of plant.

1.7. The Bactericidal Mechanism of Tea Catechins

There is not so much data about the bactericidal mechanism of tea catechins. Tea catechins show bactericidal and anti-toxin activities indicating that the compounds deactivate proteins. Catechins also exert their effect on mammalian cells inducing lymphocyte proliferation, immunoglobulin synthesis, mitogenicity of B-lymphocytes and stimulating the interleukin production of human leukocytes at a low concentration as indicated in different chapters of this book. The Gram-positive bacteria are more sensitive to the growth inhibition and bactericidal activities of tea catechins than Gram-negative bacteria. Ikigai *et al.* (1993) investigated the damaging effect of (–)-epigallocatechin gallate (EGCG, the strongest inhibitor, bactericidal) and (–)-epicatechin (EC, the weakest inhibitor, non-bactericidal) on the lipid bilayer. The results showed that EGCG at 1.25 mM (606 μg/ml) caused rapid leakage of 5,6-Carboxyfluorescein (CF) entrapped in the intraliposomal space made of egg-yolk phosphatidylcholine (PC) and the extent of CF-release reached 50% within 5 min and slowly increased thereafter in a concentration-dependent manner. Whereas EC at 1.25 mM (363 μg/ml) caused CF-release equivalent to or less than CF-release by 0.16 mM EGCG at 5 min, the extent of CF-release was nearly parallel with bactericidal activity of both compounds. This result indicated that EGCG first damaged the bacterial membrane resulting in leakage of intracellular materials. This membrane damage may enhance the penetration of catechin itself into the interior of the cell. Gram-negative bacteria are generally more resistant to catechins than Gram-positives, which is probably due to the tight penetration barrier of the outer membrane. They also found that phosphatidylserine (PS) and dicetyl phosphate (DCP), both negatively charged compounds entrapped in the liposomes, significantly reduced the catechin-mediated CF release, but stearylamine (SA, positive charged) had no effect on catechin-mediated CF release. These results indicated that the surface charge of the target membrane is important in catechin susceptibility. The low catechin susceptibility of Gram-negative bacteria may be due partially to the presence of a strong negative charge of lipopolysaccharide at the exterior of the outer membrane. It is unknown how uncharged catechins interact with negatively charged groups of membrane lipids. It was found that 1.25 mM EGCG induced aggregation of liposomes containing negatively charged DCP or PS, but not positively charged SA. Tests of the intact bacteria for EGCG-mediated aggregation showed that *Staphylococcus aureus* cells aggregated to a significant extent in the presence of 1.25 mM EGCG but the aggregation was undetectably low in *E. coli*. It was also found that *S. aureus* absorbed about 2.5 times more EGCG than *E. coli* per unit weight of bacterial cells.

Earlier reports showed that the catechins were incorporated into the plasma membrane of rat hepatocyte and caused reduced flux of thiourea and cycloleucine in mouse ascites tumor cells. It is likely that the catechins interact with the membrane and damage the lipid bilayers possibly by the catechin directly penetrating the lipid bilayer and disrupting the barrier function. However, the question of how bactericidal catechins damage the bacterial cell membrane still remains unanswered.

2. ANTICARIES ACTIVITY OF TEA

2.1. Anti-Mutans Streptococci and Glucosyltransferase Activity of Tea

Mutans streptococci have been implicated as primary causative agents of dental caries in humans and experimental animals (Hamada *et al.* 1984). These bacteria are known to synthesize glucan of two types: water-soluble and water-insoluble glucans. The two glucans are synthesized by two different groups of glucosyltransferases (GTase; EC 2.4.1.5.) Cariogenicity is considered to be strongly associated with the ability of these organisms to synthesize extracellular water-insoluble glucans on the surface of the bacteria. The water-insoluble, highly branched glucan is synthesized from sucrose by cooperative actions of GTases and it is highly adherent to various solid surfaces, including the tooth surface. This biochemical process results in firm, irreversible adherence of mutans streptococci and other microorganisms to the tooth surface, which eventually leads to formation of dental plaque. The bacteria grown there, metabolize various saccharides and produce organic acids, especially lactic acid which is retained in dental plaque eventually decalcifying the tooth enamel to develop dental caries (Koga *et al.* 1986). As early as 1981 a clinical test of the effect of drinking green tea extract on dental caries was conducted at primary schools over a year. The incidence of dental caries among children who took a cup of tea immediately after lunch was found to be significantly lower (Onisi *et al.* 1981). This result suggested that green tea contained certain unknown substances effective in the prevention of dental caries. Initially, it was suggested that the efficacy of green tea infusion for the prevention of dental caries was due to a considerably high concentration of fluorides. However, one could not attribute this result to the action of fluoride alone, because it was shown that tea extract was much more effective in the preventive effect than fluoride of the same concentration (Onisi *et al.* 1981).

Various tea extracts including green tea, Japanese green tea, black tea, Po-lei tea and Oolong tea have been tested for their inhibitory effects on oral bacteria and mutans streptococci (Hamada *et al.* 1996; Nakahara *et al.* 1993; Sakanaka *et al.* 1989). The growth of oral bacteria isolated from human whole saliva collected from two young girls was inhibited by the extracts of Po-lei tea, green tea and black tea using the agar diffuse method. Inhibition of *Streptococcus sobrinus* (both *S. mutans* and *S. sobrinus* are the primary oral pathogens responsible for producing caries) adhesion by the Po-lei tea extract was examined by checking the weight of S. sobrinus accumulated on the inside of the test tube. The results showed that the extract of Po-lei tea decreased the weight significantly. It was suggested that the active substances responsible for the anti-bacteria and anti-adhesion activity were thermostable, low molecular weight substances, which could be dialyzed. The extract of Japanese green tea showed bactericidal activity against *S. mutans*. By incubation of *S. mutans* suspension for 30 min with the extract, the colony forming units (CFU) of *S. mutans* were reduced to about one tenth of the initial amount. After eight hrs of incubation with the extract the CFU decreased to 10^2 CFU/ml (the control was $10^{6.2}$ CFU/ml).

Seven catechins purified from Japanese green tea were tested for their growth inhibitory effects against three strains of cariogenic bacteria (*S. mutans* MT8148, *S. mutans* IFON13955 and *S. sobrinus* 6715DP). The minimum inhibitory concentrations (MIC) were estimated. (+)-gallocatechin (GC) and (–)-epigallocatechin (EGC) completely inhibited the growth of three strains of cariogenic bacteria at 250 and 250 or 500 μg/ml respectively. The MIC of (–)-epigallocatechin gallate (EGCG) whose amount in the green tea extract was comparatively large was between 500 and 1000 μg/ml. Their growth inhibitory effects were doubled or more when examined using sensitive meat extract medium. The inhibitory activity of GC and EGC was stronger than (+)-catechin (C) and (–)-epicatechin (EC). Also EGCG was more active than (–)-epicatechin gallate (ECG). These facts indicated that the presence of the three hydroxy moieties at 3′, 4′ and 5′ on the B ring in the catechin and epicatechin molecules strengthens the inhibitory activity. The bactericidal activity of EGCG was stronger than the extract of Japanese green tea. Only a five to ten minutes exposure of the cariogenic bacteria to the tea polyphenols resulted in large reduction of CFU. Kawamura *et al.* (1989) reported the MICs against *S. mutans*. MIC values of catechin fraction A (CF-A), catechin fraction B (CF-B), the mixture of CF-A and CF-B (CF-mix) and EGCG were 100–400, 50–100, 100–200, and 50–100 μg/ml respectively. On the bactericidal activity, 20 mg/ml of CF-B decreased the number of *S. mutans* from 10^7 to 10^2 after 3 min. A cup of green tea (100 ml) usually contains 50–100 mg of the polyphenols of which about 60% are ECG and EGCG. The combined concentrations of these two major compounds (which could range from 300 to 500 μg/ml) is higher than those used in the above experiments.

Ishigami *et al.* (1991) reported the inhibitory effects of tea polyphenols on insoluble glucan synthesis catalyzed by GTase from *S. Mutans*. Crude catechins and crude theaflavins at 10 mg/ml showed appreciable inhibition in the synthesis of insoluble glucan by the inhibition rates of 93% and 90% respectively. Among the components, theaflavins had potent inhibitory activities at a concentration of 10 mM against GTase by the inhibition rates of 97%–98%. (–)EC had moderate inhibitory activity at 10 mM, while ECG and EGCG at the same concentrations had increased inhibitory activities. Sakanaka *et al.* (1990) investigated the effects of the above seven catechins on water-soluble and insoluble glucan synthesis by three GTases from *S. sobrinus* 6715DP, *S. mutans* MT8148 including both cell-free GTase and cell-associated GTase respectively. The results showed that ECG, GCG and EGCG (250 or 500 μg/ml) strongly inhibited glucan synthesis by any of the GTases used. However, other catechins examined were not as inhibitory as the above compounds. The glucan synthesis was inhibited almost proportionally to ECG and EGCG concentrations and this inhibition was independent of the concentration of sucrose. This result may indicate that ECG and EGCG bind GTases to inactivate them irreversibly. Both ECG and EGCG at concentrations more than 25 to 30 μg/ml almost completely inhibited glucan synthesis by the enzymes from 6715DP and MT8148-I (cell-associated GTase catalyzes the synthesis of insoluble glucan). However, for complete inhibition of the synthesis of soluble glucan by MT8148-S enzyme, 60 or > 60 μg/ml of ECG and EGCG were necessary. Again EGCG (50 μg/ml) and ECG (100 μg/ml) almost completely inhibited adherence of the bacterial cells of strain MT8148 to the glass surface. ECG and EGCG both have a galloyl moiety in their molecules. It was found that gallic acid itself showed no inhibitory effect. Therefore, the inhibitory effect shown by these galloyl compounds may be concluded to be due to the chemical groups and their configuration other than the galloyl moiety.

Hattori *et al.* (1990) investigated the effects of instant green tea infusion, instant black tea infusion, various crude tea polyphenolic compounds (designated polyphenon 30, polyphenon 100 and polyphenon B), crude theaflavins, ten kinds of catechins, and four kinds of theaflavins on the activities of GTase from *S. mutans* OMZ 176 (serotype d; alias *S. sobrinus*). All preparations including infusions, polyphenons and crude theaflavins inhibited total- and insoluble-glucan formations strongly at 10 mg/ml, but moderately at 1 mg/ml except for the case of crude theaflavins which had potent inhibition even at the low concentration (1 mg/ml). The tea infusions and polyphenons (30, 100 and B) showed no inhibition on soluble-glucan formation at 1 mg/ml. However, the black tea infusion, polyphenon 100, polyphenon B, and polyphenon-protein complex, and a solution of hydrolyzable soybean protein added to a solution of polyphenon B, significantly stimulated the soluble-glucan formation as compared to a control. Polyphenon-protein complex did not show any inhibitory effect on total- and insoluble-glucan formations. This phenomenon indicated that the presence of any excess polyphenols capable of complexing with protein was important. The effects of catechins on the activity of GTase were similar to the above results. ECG, GCG and EGCG all containing a galloyl group at 3-OH showed strong inhibition on both total- and insoluble-glucan formations (35–50% inhibition at 1 mM and 70–95% inhibition at 10 mM). Free theaflavin and the mono- and digallates had potent inhibition on total- and insoluble glucan formations (45–65% inhibition at 1 mM and 95–100% inhibition at 10 mM). No appreciable stimulatory effect was observed among free theaflavin and its gallates. The authors indicated that the reduction of total-glucan formation relative to a control by individual tea polyphenols is mostly due to the reduction of insoluble-glucan formation. Under similar conditions, the following % inhibitions were observed for gallic acid and its esters at a concentration of 0.1 mg/ml: gallic acid (0), ethyl gallate (0), propyl gallate (9.8%), octyl gallate (83%), lauryl gallate (11.1%) and pentagalloyl glucose (100%). These findings revealed that the relationship between a galloyl group and its inhibitory potency against GTases is quite complex.

Otake *et al.* (1991) also demonstrated the inhibitory effect of the extract of the crude tea polyphenolic compounds (designated Sunphenon^R) from Japanese green tea by another method. The adsorption of Sunphenon-pretreated cells of *S. mutans* to human saliva-coated hydroxyapatite discs (S-HA) or the adsorption of *S. mutans* to Sunphenon-treated S-HV discs were investigated. Both experiments showed the inhibitory effects of Sunphenon on the adsorption of *S. mutans* in a dose-dependent manner. At a concentration of 100 μg/ml the percent inhibition was 35.6–83.1%. Nakahara *et al.* (1993) found a major fraction purified from Oolong tea showing strong inhibition on the water-insoluble glucan-synthesizing enzyme, GTase-I, of *S. sobrinus* 6715. This fraction (designated OTF10) was a novel polymeric polyphenol compound that had a molecular weight of approximately 2000 and differed from other tea polyphenols. The authors compared the inhibitory effects of various tea extracts on S. sobrinus GTase-I, yeast α-glucosidase and salivary α-amylase. The results showed that OTF10 was the strongest inhibitor of GTase-I and α-glucosidase with ID$_{50}$ of 0.002 mg/ml and 0.003 mg/ml respectively but not α-amylase; theaflavin was also a strong inhibitor of GTase-I with ID$_{50}$ of 0.008 mg/ml. The inhibition by (+)-catechin, (–)-epicatechin, (–)-epigallocatechin, (–)-epicatechin gallate and (–)-epigallocatechin gallate on GTase-I was not so impressive. It is known that Oolong tea extract (OTE) and black tea extract contain significant quantities of unknown polyphenols that are not detected in green tea extract (GTE). Green tea leaves contain 14.7% polyphenols (mainly condensed and hydrolyzable monomeric polyphenols), while Oolong tea leaves contain 16.0% polyphenols, including 9.7% monomeric polyphenols and 6.4% unknown polymeric polyphenols. OTE exhibited the most prominent inhibitory action among various tea extracts. Green tea, oolong tea and black tea are manufactured from leaves of the same plant species, *Camellia sinensis*. Unlike green tea, Oolong and black teas are semi-fermented and fermented products respectively during the manufacturing process. The fermentation and heating of tea leaves result in polymerization of monomeric polyphenolic compounds, such as catechins (Hashimoto *et al.* 1989). The results indicate that conformational changes due to polymerization of catechins may be critically important for exerting an inhibitory effect on GTases of mutans streptococci. The inhibition of OTF10 and OTF on GTases was specific and not all GTases prepared from mutans streptococci were inhibited by these substances. For example, *S. sobrinus* GTase-I and *S. mutans* cell-free GTase synthesizing soluble glucan were most susceptible to the inhibitory action of OTF10, while *S. sobrinus* GTase-Sa and *S. mutans* cell-associated GTase were moderately inhibited; no inhibition of S.sobrinus GTase-Sb was observed. Inhibition of specific GTases resulted in decreased adherence of the growing cells of these organisms. The inhibitory effect of OTF10 on cellular adherence was significantly stronger than that of OTE.

Kubo *et al.* (1993), Muroi *et al.* (1993) investigated the inhibitory effects of mate tea and green tea flavor components against the growth of *S. mutans*. Each ten volatile compounds were purified from mate tea and green tea respectively. The 10 main compounds from mate tea were linalool, α-ionone, β-ionone, α-terpineol, octanoic acid, geraniol, 1-octanol, nerolidol, gerany-lacetone and eugenol. The ratios of these compounds in other yerba mates obtained from different locations varied but basically were very similar (Kawakami *et al.* 1991). The same compounds have been reported to be flavor components in *Camellia sinensis* teas such as green tea, black tea and oolong tea, although their ratios differ. The antimicrobial activity of these individual components and caffeine, ursolic acid, and chlorogenic acid was tested against 13 selected microorganisms including 5 Gram-positive bacteria, 3 Gram-negative bacteria and 5 fungi. All of the volatiles tested exhibited moderate to weak activity with MIC values from 12.5–1600 μg/ml. Nerolidol was the most potent compound against *S. mutans* with MIC of 25 μg/ml. This concentration was also MBC (minimum bactericidal concentration) of nerolidol against *S. mutans*. It was found that nerolidol and indole (minute quantities in mate tea flavors) showed additive effect against *S. mutans* resulting in the reduction of MIC from 25 to 12.5 μg/ml. The flavor compounds of green tea tested contained linalool, δ-cadinene, nerolidol, α-terpineol, *cis*-jasmone, β-ionone, 1-octanol, β-caryophyllene, indole, and geraniol. MICs of these compounds against S. mutans ATCC 25175 were 25–1600 μg/ml and MBCs were 200–1600 μg/ml except β-caryophyllene, which did not show any activity. The combination study of some compounds with indole (The most abundant nitrogen-containing substance in green tea flavor) had been examined. It was found that indole exhibited synergistic effects with sesquiterpeneh hydrocarbons (δ-cadinene and β-caryophyllene); their MBC increased 128-fold to 256-fold. The MIC for δ-cadinene decreased from 800 μg/ml to 6.25 μg/ml; the MIC for β-caryophyllene decreased from >1600 μg/ml to 6.25 μg/ml. The additive effects were shown between indole and linalool, geraniol or nerolidol. The synergistic and additive effects were also shown in MBC values.

As mentioned above tea polyphenols and tea flavor components from various tea products showed growth inhibition, bactericidal, anti-GTases and anti-adherence activities against mutans streptococci, the main causative agents of tooth-plaque and caries, although the antimicrobial activity is not as strong as antibiotics. However, the use of antibiotics such as penicillin, erythromycin and tetracycline are accompanied by a potential risk of undesirable and unacceptable side effects. Since tea has been widely consumed by people as a daily beverage, extracts, purified polyphenols or purified flavor compounds from teas may be safe, or risk-free, for use in oral care products for caries control.

2.2. Animal Experiments

Earlier reports on relationship between tea consumption and the reduction of dental caries in human and experimental animals, mostly concentrated on the content of fluorides in tea. Shyu *et al.* (1977) reported that certain Taiwan teas reduced caries activity by 56% and 72% in rats, and that water containing 1 ppm of fluoride reduced caries activity by 73%. On the contrary, Gershon-Cohen *et al.* (1954) showed that tea infusion containing 20 ppm fluoride did not significantly reduce caries. However, Ramsey *et al.* (1975) reported that tea had a cariostatic effect in children. Rosen *et al.* (1984) evaluated the anticariogenic activity of four kinds of teas varying in fluoride (0.38–0.70 ppm) and tannin (0.027–0.079 g/100 ml) concentrations in rat experiments. Tea infusions of Dragonwell, Young hyson, and Panfired, which were Chinese green teas, and Darjeeling, which was a black tea from India, were used in the experiments. The results indicated there was a direct correlation between fluoride in tea and the inhibition of sulcal caries in rats, whereas no relationship was observed between tannin and this type of lesion. But the authors claimed that as the control rats did not develop significant levels of buccal-lingual caries, it could be erroneous to conclude that tannin had no effect on caries inhibition, especially since it had been shown that tannic acid can inhibit glucosyltransferase activity (Paolino *et al.* 1980). The low levels of fluoride in all of the teas used in this study suggested that tea might contain other substances that inhibit caries.

Otake *et al.* (1991) and Sakanaka, *et al.* (1992) reported the effects of green tea polyphenols on caries development in conventional rats. The Japanese green tea product, Sunphenon manufactured by the same company, with little difference in the percentage of catechins was used in both reports. It was composed mainly of (+)-catechin (2.9%), (−)-epicatechin (6.8%), (+)-gallocatechin (12.8%), (−)-epigallo-catechin (16.5%), (−)-epicatechin gallate (6.6%), (−)-gallocatechin gallate (8.5%), and (−)-epigallocatechin gallate (21.3%). The rats were fed on a cariogenic diet and Sunphenon was blended in the powdered diet or dissolved in the drinking water with 0.1%, 0.2%, and 0.5% concentrations. The results showed that the sum of enamel lesions, lesions reaching the dentinoenamel junction and advanced dentin lesions were significantly reduced by the addition of tea polyphenols to the diet or drinking water. Furthermore, the reduction of serious caries was more significant than that of total caries in all the test groups. The effects were independent of the concentration of Sunphenon and there were no significant differences in the percentages of cariogenic streptococci to total Gram positive streptococci between the control group and each test group. However, there was a strong trend indicating that the percentages became lower in the test groups at the final day (40th day) than the 15th day. Further, when Sunphenon was added to both diet and drinking water, synergetic effects were not seen. No side effects were seen in both experiments.

Ooshima *et al.* (1993) reported an animal experiment with Oolong tea extracts. There were two preparations of Oolong tea extracts, one was crude OT-E containing 40% (w/w) polyphenolic compounds consisting of largely monomeric compounds and some unknown polymeric compounds and the other was OT-6 containing 50% (w/w) polymeric polyphenols and no monomeric polyphenols. The animals used were specific pathogen-free (SPF) rats infected with mutans streptococci and fed with a cariogenic diet. OT-E or OT-6 was incorporated in both diets and drinking water from 1 day prior to infection with mutans streptococci (17 days of age) through to the end of the experiment (73 day of age). As reported, the administration of Oolong tea extracts resulted in significant reductions in the plaque index and total caries score in SPF rats infected with *S. mutans* MT8148R or with *S. sobrinus* 6715. The inhibition was not dependent on the concentration of OT-E in the diet, especially in rats infected with *S. mutans* MT8148R. The maximum inhibition of caries was obtained when 1 mg OTE/g diet was administered. However, a dose-dependent effect was obtained in rats infected with *S. sobrinus* and administered OTE. In these experiments, no significant difference in weight gains and no notable side effects were observed in any of the groups.

As described before, green tea leaves and Oolong tea leaves contain different percentages of tea polyphenols especially Oolong tea which contains some unknown polymeric polyphenols. However, both showed significant inhibition activity on caries in animal experiments and both could be useful for controlling dental caries in humans. It was also reported that various tea components such as tannin, catechin, caffeine and tocopherol possessed properties of enhancing acid resistance of dental enamel, especially in combination with fluoride. Tannin-fluoride (Ta-F) was the most effective combination for enhancing acid resistance of dental enamel (Yu, *et al.* 1995). The effect of Ta-F mixture (0.5% tannic acid, 450 ppm fluoride, pH 5.9) on dental enamel had been investigated by using scanning electron microscopy (SEM), electron probe microanalysis (EPMA) and X-ray diffraction (XRD), compared with the effect of acidulated phosphate fluoride (APF, 0.015 M phosphoric acid, 450 ppm fluoride, pH5.3). Under the SEM, a large number of spherical globules (1–5 μm in diameter) were observed on the enamel surface treated with Ta-F. They had a good range and formed a single layer coating on the enamel, whereas on the APF-treated enamel only very small spherical globules (0.1–0.5 μm in diameter) were seen (Yu, *et al.* 1993). These results further support the preventive effect of tea and tea products against caries.

2.3. Clinical Study

Elvin-Lewis *et al.* (1986) reported anticariogenic effects of tea among children in Dallas. One hundred and six children were surveyed for the relationship of their tea and other beverage drinking habits with caries. They were divided into two groups depending upon whether they drank tea in addition to other beverages. Differences between these two groups both for DMFT and plaque scores were significant (P less than 0.05). These studies indicated that consumption of tea in children was beneficial to their dental health by reducing both plaque production and caries risk. Recently Ooshima *et al.* (1994) investigated the inhibitory effect of Oolong tea extract (OTE) on plaque deposition in 35 human volunteers. OTE contained 40% (w/w) polyphenolic compounds. The participants rinsed their mouths with 0.5 mg/ml OTE solution in 0.2% ethanol. Each mouthrinsing consisted of 5 consecutive 10-second treatments with 20 ml of OTE solution, i.e., a total volume of 100 ml OTE. This procedure was done 9 times a day, e.g., before and after meals, before and after between-meal periods, and before sleep for four days. The 4-day study was repeated again with the same procedures and employing the same subjects 1 week after the first trial. In the second trial, 0.2% ethanol solution was used for mouth rinsing. The plaque depositions were assessed by the method of Quigley *et al.* (1962). The results showed that mouth rinsing with 0.5 mg/ml OTE in 0.2% ethanol significantly reduced plaque deposition when compared with mouth rinsing with 0.2% ethanol (p < 0.001). All the plaque indices, except for 1 subject, decreased markedly by mouth rinsing with

OTE. However, there were no significant changes in the CFU of either total streptococci or mutans streptococci in saliva between the two groups. No side effects were recognized during the experimental period. This findings suggest that the OTE preparation could be useful for controlling dental plaque formation and subsequent dental caries development in humans.

3. ANTI-VIRUS ACTIVITY OF TEA

3.1. Antiviral Activity of Tea Extracts

As early as 1949 Green (1949) reported that extracts of black tea inhibited the multiplication of influenza virus A in embryonated eggs. Since then there have been a number of reports dealing with antiviral activities of tea extracts. Nakayama *et al.* (1990) reported the inhibition effects of black tea extracts against influenza virus. Different concentrations of black tea extract were mixed with both influenza viruses A and B respectively for 3 min or 60 min before adsorption to MDCK (Madin-Darby canine kidney) cells. Plaque assays were used to check the antiviral activity. At a concentration of 0.1 μl/ml black tea extract mixed with influenza virus A and B separately for 60 min gave 88% and 80% inhibition of plaque forming units (pfu) respectively. If the contact time was as low as 3 min, it also showed some inhibition of pfu. When MDCK cells were pretreated with black tea extract and washed to remove residual tea extract, and then challenged with influenza A virus, 50 μl/ml gave a 85% plaque reduction. If the black tea extracts were added to MDCK cells after virus adsorption there was no pfu reduction indicating that black tea extract did not inhibit the replication of influenza virus. These results showed that the black tea extract inhibited the adsorption of influenza viruses A and B to the cells but not their replication within the cells. Zhang *et al.* (1993) investigated the inhibitory effects of extracts of black tea, blue tea and dark tea against rotavirus (Wa strain) in MA104 cells by a cytopathic effect (cpe) method. A series of concentrations of extracts of three teas were mixed with rotavirus 100 $TCID_{50}$ for 90 min at 37°C before adsorption to the cells. The results showed that 40 μg/ml of both extracts of black tea and blue tea and 400 μg/ml of extract of dark tea inhibited the cpe formation significantly. 4000 μg/ml of all three extracts were toxic to MA104 cells.

3.2. Antiviral Activity of Tea Polyphenols

Purified tea polyphenols including (−)catechin (C), (−)epicatechin (EC), (−)gallocatechin (GC), (−)epigallocatechin (EGC), (−)catechin gallate (CG), (−)epicatechin gallate (ECG), (−)gallocatechin gallate (GCG), and (−)epigallocatechin gallate (EGCG) were tested against influenza virus A in MDCK cells by a plaque neutralization method (Noda *et al.*, 1994). According to the inhibition concentrations, by which virus infection was totally blocked, the potency of these compounds could be ordered as: GCG, EGCG (4 μM) > CG, ECG (64 μM) > GC, EGC (256 μM) > C, EC (8192 μM). There were no inhibition differences between catechin counterparts.

The antiviral effects of EGCG and theaflavin digallate (TF3) purified from green tea and black tea respectively were further studied (Nakayama *et al.* 1993). EGCG and TF3 were mixed with influenza virus A (A/Yamagata/120/86 H1N1) and B (B/USSR/100/83) for either 5 min or 60 min respectively before being exposed to the cells. 1.5 μM of both compounds inhibited almost 100% of the pfu of both viruses after 60 min treatment. 5 min contact of EGCG or TF3 with the viruses also effectively inhibited the virus infectivity. To understand whether polyphenols are effective if added after virus adsorption to MDCK cells, influenza A viruses were exposed to the cells at 4°C for 30 min, then both polyphenols were added to virus adsorbed cells for 15 min and the cells were washed twice with MEM and cultured. Although the effective dose of polyphenols was higher, inhibition of pfu did occur. However, when polyphenols were added 30 min or more after adsorption of the viruses to MDCK cells at 37°C, pfu was not inhibited. When MDCK cells were pretreated with EGCG or TF3 and washed to remove residual compounds, and then challenged with the virus, pfu formation was not inhibited even at a concentration of 100 μM. The concentrations of EGCG and TF3 greater than 200 μM and 100 μM respectively were toxic to the cells.

The above results suggest that tea polyphenols may bind to surface glycoproteins of the influenza virus. The capacities of EGCG, TF3 and anti-A virus (H1N1) antibody to bind to the A virus were compared by electron microscopy. 1 mM of EGCG and TF3 agglutinated virus particles the same as the antibody did during short-time contact. Viruses pretreated with EGCG (1 mM) or with the antibody failed to bind to MDCK cells. The haemagglutination of influenza viruses was also inhibited by EGCG and TF3. The inhibition of haemagglutination by TF3 was stronger than by EGCG.

These results indicate that tea extracts and tea polyphenols can inhibit the infectivity of influenza virus to MDCK cells by blocking its adsorption and entry into the cells, but not its multiplication inside the cells. They show HA antigen binding properties similar to that of antibody. Beverage concentrations of tea contain at least 500 μM EGCG. Taken together, it appears that polyphenols are responsible for the anti-influenza virus activity of tea extracts. Tea might be useful as a prophylactic agent against influenza virus infection.

The inhibition effects of EGCG and TF3 (theaflavin digallate) were investigated against rotavirus and enterovirus in cell cultures by a cytopathic effect (cpe) method (Mukoyama *et al.* 1991, Zhang *et al.* 1995). Human rotavirus (HRV) strains Wa (serotype 1), TMC2 (serotype 2), YO (serotype 3), Hochi (serotype 4) and 69M (serotype 8); poliovirus 1 strain Sabin 1; coxsackie virus A16 strain G-10; coxsackie virus B3 strain Nancy and ECHO virus 11 strain Gregory were used in the experiments. When the viruses were mixed with serially diluted EGCG or TF3 for 1 hr and then added to cells, 50% inhibition of cpe by EGCG occurred at about 8 μg/ml for Wa and 69M, at about 32 μg/ml for TMC2 and Hochi, and about 62.5 μg/ml for YO; 100% inhibition of cpe was noted at a concentration of 125 μg/ml. The IC$_{50}$ values of EGCG for enterovirus were 15 μg/ml (Sabin 1) and 125 μg/ml (G-10, Nancy and Gregory) respectively. 100% inhibition of cpe was noted at 250 μg/ml for all enteroviruses examined. The IC$_{50}$ values of TF3 were 15 μg/ml for Sabin 1, 32 μg/ml for Wa and 62.5 μg/ml for TMC2 and Nancy. 100% inhibition of cpe was noted at 250 μg/ml for all viruses examined. Their antiviral effects were maximally induced when directly added to virus, and their pre- and post- treatment of the cells produced much weaker antiviral activities. The antiviral activity of EGCG and TF3 seems to be attributable to interference with virus adsorption.

Serial concentrations of catechin were mixed with respiratory syncytial virus (RSV Long strain) and herpes simplex virus type 1 (HSV-1) respectively at room temperature for 1 hr prior to inoculating HEp 2 cells (Kaul *et al.* 1985). After 2 hrs of adsorption the cells were washed and overlaid with medium. Catechin (12.5–200 μM) caused a concentration-dependent inhibition of pfu formation by RSV and HSV-1. As for influenza virus, catechin did not inhibit virus replication if catechin was added after virus adsorption. Catechin also did not inhibit the infectivities of parainfluenza virus type 3, poliovirus type 1 and hepatitis A virus (Biziagos *et al.* 1987). There was some discrepancy about the inhibitory effects of tea polyphenols against poliovirus type 1 and the reason was unknown.

3.3. Antiviral Reverse Transcriptase and DNA Polymerase Activities of Tea Polyphenols

(–)Epigallocatechin gallate (EGCG), (–)epigallocatechin (EGC) and four theaflavins have been investigated for their inhibitory effects against various DNA polymerases (DNAP), reverse transcriptases of human immunodeficiency virus type 1 (HIV-1 RT), Moloney murine leukemia virus (MoMLV RT), terminal deoxynucleotidyltransferase (TDT), and *E. coli* RNA polymerase (RNAP), (Nakane *et al.* 1990, 1991; Ono *et al.* 1991). The results showed that HIV-1 RT was the most sensitive enzyme tested toward the inhibitory effects of EGCG and ECG with IC_{50} values of 0.012 and 0.017 μg/ml respectively and MoMLV RT was the most sensitive enzyme tested toward the inhibitory effects of the theaflavins with IC_{50} values of 0.04–0.7 μg/ml. The mode of inhibition was analyzed kinetically by changing the concentrations of either the template-primer or the triphosphate substrate in the presence of various concentrations of EGCG and ECG. For all enzymes tested except *E. coli* RNAP, competitive inhibition was shown by the template-primer and a noncompetitive inhibition was shown by the triphosphate substrate. For HIV-1 RT mixed type inhibition was shown by the template-primer. For *E. coli* RNAP the mode of inhibition changed from noncompetitive to competitive by increasing the concentration of (dC)n (template-primer) and showed a competitive inhibition with respect to the substrate. A similar mode of inhibition was observed with the theaflavin digallate. The Ki values of EGCG were 2.8 nM for HIV-1 RT, 116 nM for DNAP α, 71.1 nM for DNAP β, 286 nM for DNAP γ and 176 nM for *E. coli* RNAP. The Ki values of ECG were 7.2 nM for HIV-1 RT, 181 nM for DNAP α, 23.7 nM for DNAP β, 298 nM for DNAP γ and 323 nM for *E. coli* RNAP. The Ki values of theaflavin, theaflavin monogallate A and B and theaflavin digallate for HIV-1 RT were 490, 32, 23 and 23 nM respectively. The results show that the galloyl group enhances the inhibitory potency of the theaflavins and only a small difference in inhibitory effects was observed among the three gallic acid esters.

Tao *et al.* (1992) reported the inhibition effects of EGCG, ECG and EGC –(–)epigallocatechin on HIV-1 RT, herpes simplex virus type-1 DNAP (HSV-1 DNAP), duck hepatitis B virus replication complex reverse transcriptase (DHBV RCs RT) and cow thymus DNAP α (CT DNAP α). The IC_{50}s for HIV-1 RT were 6.6 nM for EGCG, 84 nM for ECG and 7200 nM for EGC. It showed that EGCG was a very strong inhibitor against HIV-1 RT activity. The IC_{50} of EGCG, ECG and EGC for HSV-1 DNAP were 440 nM, 1400 nM and >1000 μM respectively. DHBV RCs RT was insensitive toward these compounds. The IC_{50} against CT DNAP α were 2.5 μM for EGCG, 1.6 μM for ECG and 1000 μM for EGC. The inhibition kinetic study with EGCG and HIV-1 RT showed a noncompetitive inhibition with the triphosphate substrate and a mixed type inhibition with the template-primer.

It was found that bovine serum albumin (BSA) or Triton X-100 could counteract the inhibitory effects of tea polyphenols against the above enzymes (Moore *et al.* 1992; Nakane *et al.* 1990; Tao *et al.* 1992). Moore suggested that these catechins bind with no apparent selectivity and that the observed inhibition of HIV-1 RT was non-specific in nature. It was also reported that EGCG and ECG did not show inhibitory effects against HIV-1 in cell cultures.

REFERENCES

Ahn, Y.J., Sakanaka, S., Kim, M.J., Kawamura, T., Fujisawa, T., Mitsuoka, T. (1990) Effect of green tea extract on growth of intestinal bacteria. *Microb. Ecol. Hlth. Dis.*, 3, 335–338.

Ahn, Y.J., Kawamura, T., Kim, M., Yamamoto, T., Mitsuoka, T. (1991) Tea polyphenols: Selective growth inhibitors of *Clostridium spp.*. *Agric. Biol. Chem.*, 55, 1425–1426.

Anonymous (1923) Using tea to fight typhoid. *Tea Coffee J.*, July, p. 129.

Das, D.N. (1962) Studies on the antibiotic activity of tea. *J. Indian Chem. Soc.*, 33, 849–854.

Biziagos, E., Crance, J.M., Passagot, J., Deloince, R. (1987) Effect of antiviral substances on hepatitis A virus replication in vitro. *J. Med. Virol.*, 22, 57–66.

Elvin-Lewis, M., Steelman, R. (1986) The anticariogenic effects of tea drinking among Dallas children. *J. Dent. Res.*, 65, 198.

Eugster, A.K. (1981) Inhibitory growth effect of agar plate added tea powder to pathogenic bacteria. *J. Animal Med.*, 14, 6–10.

Finegold, S.M., Attebery, H.R., Sutter, V.R. (1974) Effect of diet on human fecal flora: comparison of Japanese and American diets. *American J. Clin. Nutr.*, 27, 1456–1469.

Finegold, S.M., Flora, D.J., Attebery, H.R., Sutter, V.L. (1975) Fecal bacteriology of colonic polyp patients and control patients. *Cancer Res.*, 35, 3407–3417.

Flament, I. (1991) Coffee, cocoa, and tea. In H. Maarse (ed.), *Volatile Compounds in Foods and Beverages*, Dekker, New York, pp. 617–669.

Fukai, K., Ishigami, T., Hara, Y. (1991) Antibacterial activity of tea polyphenols against phytopathogenic bacteria. *Agric. Biol. Chem.*, 55, 1895–1897.

Gershon-Cohen, J., McClendon, J.F. (1954) Fluoride in tea and caries in rats. *Nature*, 173, 304–305.

Gorbach, S.L., Nahas, L., Lerner, P.I., Weinstein, L. (1967) Studies of intestinal microflora. 1. Effects of diet, age, and periodic sampling on numbers of fecal microorganisms in man. *Gastroenterology*, 53, 845–855.

Green, R.H. (1949) Inhibition of multiplication of influenza virus by extracts of tea. *Proceed. Soc. Exp. Biol. Med.*, 71, 84–85.

Hamada, S., Fukuchi, J., Sato, S., Kokubun, M., Miyazawa, C., Kusunoki, K. (1996) Effect of Po-lei tea on the growth of human oral bacteria. *Proceed. Internat. Symp. Tea Cult. Hlth. Sci.*, Kakegawa, October 1996, 135–139.

Hamada, S., Koga, T., Ooshima, T. (1984) Virulence factors of *Streptococcus mutans* and dental caries prevention. *J. Dent. Res.*, 63, 407–411.

Hara, Y., Watanabe, M. (1989) Antibacterial activity of tea polyphenols against *Clostridium botulinum* (Studies on antibacterial effects of tea polyphenols part II). *J. Jpn. Soc. Food Sci. Technol.*, 36, 951–955.

Hara, Y., Ishigami, T. (1989) Antibacterial activities of tea polyphenols against foodborne pathogenic bacteria (Studies on antibacterial effects of tea polyphenols part III). *J. Jpn. Soc. Food Sci. Technol.*, 36, 996–999.

Hashimoto, F., Nonaka, G., Nishioka, I. (1989) Tannins and related compounds. XC. 8-C-Ascorbyl (–)-epigallocatechin 3-O-gallate and novel dimeric flavan-3-ols, oolonghomobis-flavans A and B, from oolong tea. *Chem. Pharm. Bull.* (Tokyo), 37, 3255–3263.

Hattori, M., Kusumoto, I.T., Namba, T., Ishigami, T., Hara, Y. (1990) Effect of tea polyphenols on glucan synthesis by glucosyltransferase from *Streptococcus mutans*. *Chem. Pharm. Bull.*, 38, 717–720.

Hentges, D.J. (1983) Role of the intestinal microflora in host defense against infection. In D.J. Hentges, (ed.), *Human Intestinal Microflora in Health and Disease*, Academic, New York, pp. 311–331.

Ikigai, H., Toda, M., Okubo, S., Hara, Y., Shimamura, T. (1990) Relationship between the anti-hemolysin activity and the structure of catechins and theaflavins. *Jpn. J. Bacteriol.*, 45, 913–919.

Ikigai, H., Nakae, T., Hara, Y., Shimamura, T. (1993) Bactericidal catechins damage the lipid bilayer. *Biochim. Biophys. Acta*, 1147, 132–136.

Ishigami, T., Hara, Y., Goto, K., Kanaya, S., Hara, H., Terada, A. (1996) The effect of tea catechins on fecal flora. *Proceed. Internat. Symp. Tea Cult. Hlth. Sci.*, Kakegawa, October 1996, 140–144.

Ishigami, T., Watanabe, M., Fukai, K., Hara, Y. (1991) Antibacterial activity of tea polyphenols against foodborne, cariogenic and phytopathogenic bacteria. *Proceed. Internat. Symp. Tea Sci.*, Shizuoka, August 1991, 248–252.

Kaul, T.N., Middleton, E. Jr., Ogra, P.L. Antiviral effect of flavonoids on human viruses. *J. Med. Virol.*, 15, 71–79.

Kawakami, M., Kobayashi, A. (1991) Volatile constituents of green mate and roasted mate. *J. Agric. Food Chem.*, 39, 418–421.

Kawamura, J., Takeo, T. (1989) Antibacterial activity of tea catechin to *Streptococcus mutans*. *J. Jpn. Soc. Food Sci. Technol.*, 36, 463–467.

Kodama, K., Sagesaka, Y., Goto, M. (1991) Antimicrobial activities of catechins against plant pathogenic microorganisms. *Proceed. Internat. Symp. Tea Sci.*, Shizuoka, August 1991, 294–298.

Koga, T., Okahashi, N., Asakawa, H., Hamada, S. (1986) Adherence of *Streptococcus mutans* to tooth surfaces. In Hamada, S., Michalek, S.M., Kiyono, H., Menaker, L. and McGhee J.R. (eds.), *Molecular Microbiology and Immunobiology of Streptococcus Mutans*, Elsevier Science Publishers, Amsterdam, pp. 111–120.

Kubo, I., Muroi, H., Himejima, M. (1992) Antimicrobial activity of green tea flavor components and their combination effects. *J. Agric. Food Chem.*, 40, 245–248.

Kubo, I., Muroi, H., Himejima, M. (1993) Antibacterial activity against *Streptococcus mutans* of mate tea flavor components. *J. Agric. Food Chem.*, 41, 107–111.

Moore, P.S., Pizza, C. (1992) Observations on the inhibition of HIV-1 reverse transcriptase by catechins. *Biochem. J.*, 288, 717–719.

Mukoyama, A., Ushijima, H., Nishimura, S., Koike, H., Toda, M., Hara, Y., Shimamura, T. (1991) Inhibition of rotavirus and enterovirus infection by tea extracts. *Jpn. J. Med. Sci. Biol.*, 44, 181–186.

Muroi, H., Kubo, I. (1993) Combination effects of antibacterial compounds in green tea flavor against *Streptococcus mutans. J. Agric. Food Chem.*, 41, 1102–1105.

Nakahara, K., Kawabata, S., Ono, H., Ogura, K., Tanaka, T., Ooshima, T., Hamada, S. (1993) Inhibitory effect of Oolong tea polyphenols on glucosyltransferases of mutans streptococci. *Appl. Envir. Microbiol.*, 59, 968–973.

Nakane, H., Hara, Y., Ono, K. (1991) Differential inhibition of HIV-reverse transcriptase and various DNA polymerases by theaflavins. *Proceed. Internat. Symp. Tea Sci.*, Shizuoka, August, 282–286.

Nakane, H., Ono, K. (1990) Differential inhibitory effects of some catechin derivatives on the activities of human immunodeficiency virus reverse transcriptase and cellular deoxyribonucleic and ribonucleic acid polymerases. *Biochemistry*, 29, 2841–2845.

Nakayama, M., Suzuki, K., Toda, M., Okibo, S., Hara, Y., Shimamura, T. (1993) Inhibition of the infectivity of influenza virus by tea polyphenols. *Antiviral Res.*, 21, 289–299.

Nakayama, M., Toda, M., Okubo, S., Shimamura, T. (1990) Inhibition of influenza virus infection by tea. *Letters Appl. Microbiol.*, 11, 38–40.

Noda, M., Toda, M., Endo, W., Hara, Y., Shimamura, T. (1994) The anti-influenza virus activities of catechins. *Jpn. J. Bacteriol.*, 49, 117.

Okubo, S., Ikigai, H., Toda, M., Shimamura, T. (1989) The anti-haemolysin activity of tea and coffee. *Lett. Appl. Microbiol.*, 9, 65–66.

Okubo, S., Toda, M., Hara, Y., Shimamura, T. (1991) Antifungal and fungicidal activities of tea extract and catechin against *Trichophyton. Jpn. J. Bacteriol.*, 46, 509–514.

Okubo, T., Ishihara, N., Oura, A., Serit, M., Kim, M., Yamamoto, T., Mitsuoka, T. (1992) *In vitro* effects of tea polyphenol intake on human intestinal microflora and metabolism. *Biosci. Biotech. Biochem.*, 56, 588–591.

Onisi, M., Ozaki, F., Yoshino, F., Murakami, Y. (1981) An experimental evidence of caries preventive activity of non-fluoride component in tea. *J. Dent. Hlth.*, 31, 158.

Onisi, M., Shimura, N., Nakamura, C., Sato, M. (1981) A field test on the caries preventive effect of tea drinking. *J. Dent. Hlth.*, 31, 13.

Ono, K., Nakane, H. (1991) Catechins as a novel class of inhibitors for HIV-reverse transcriptase and DNA polymerases. *Proceed. Internat. Symp. Tea Sci.*, Shizuoka, August, 277–281.

Ooshima, T., Minami, T., Aono, W., Tamura, Y., Hamada, S. (1994) Reduction of dental plaque deposition in humans by Oolong tea extract. *Caries Res.*, 28, 146–149.

Ooshima, T., Minami, T., Aono, W., Izumitani, A., Sobue, S., Fujiwara, T., *et al.* (1993) Oolong tea polyphenols inhibit experimental dental caries in SPF rats infected with mutans streptococci. *Caries Res.*, 27, 124–129.

Otake, S., Makimura, M., Kuroki, T., Nishihara, Y., Hirasawa, M. (1991) Anticaries effects of polyphenolic compounds from Japanese green tea. *Caries Res.*, 25, 438–443.

Paolino, V.J. (1982) Anti-caries activity of cocoa. *J. Dent. Res.*, 62, 171.

Quigley, G.A., Hein, J.W. (1962) Comparative cleansing efficiency of manual and power brushing. *J. Am. Dent. Assoc.*, 65, 26–29.

Ramsey, A.C., Hardwick, J.L., Tamacas, J.C. (1975) Fluoride intakes and caries increments in relation to tea consumption. *Caries Res.*, 9, 312.

Rosen, S., Elvin-Lewis, M., Beck, F.M., Beck, E.X. (1984) Anticariogenic effects of tea in rats. *J. Dent. Res.*, 63, 658–660.

Ryu E. (1980) Prophylactic effect of tea on pathogenic micro-organism infection to human and animals (1) Growth inhibition and bacteriocidal effect of tea on food poisoning and other pathogenic enterobacterium *in vitro*. *Int. J. Zoon.*, **7**, 164–170.

Ryu, E., Boenden, D.C., Wendall, D. (1982) The inhibition of growth of selected bacteria by incorporating powdered tea in the medium. *Int. J. Zoon.*, **9**, 73–76.

Sakanaka, S., Kim, M., Taniguchi, M., Yamamoto, T. (1989) Antibacterial substances in Japanese green tea extract against *Streptococcus mutans*, a cariogenic bacterium. *Agric. Biol. Chem.*, **53**, 2307–2311.

Sakanaka, S., Sato, T., Kim, M., Yamamoto, T. (1990) Inhibitory effects of green tea polyphenols on glucan synthesis and cellular adherence of cariogenic streptococci. *Agric. Biol. Chem.*, **54**, 2925–2929.

Sakanaka, S., Shimura, N., Aizawa, M., Kim, M., Yamamoto, T. (1992) Preventive effect of green tea polyphenols against dental caries in conventional rats. *Biosci. Biotech. Biochem.*, **56**, 592–594.

Shetty, M., Subbannayya, K., Shivananda, P.G. (1994) Antibacterial activity of tea (*Camellia Sinensis*) and coffee (*Coffee Arabica*) with special reference to *Salmonella Typhimurium*. *J. Com. Dis.*, **26**, 147–150.

Shyu, K., Meng, C., Sun. J. (1977) The anticariogenic effect of Taiwan tea. *Chinese Med. J.*, **24**, 55–61.

Tao, P.Z., Zhang, T., Zhou, P., Wang, S.Q., Chen, S.J., Jiang, J.Y., *et. al.* (1992) The inhibitory effects of catechin derivatives on the activities of human immunodeficiency virus reverse transcriptase and DNA polymerases. *Acta Academiae Medicinae Sinicae*, **14**, 334–338.

Toda, M., Okubo, S., Hiyoshi, R., Shimamura, T. (1989a) The bactericidal activity of tea and coffee. *Lett. Appl. Microbiol.*, **8**, 123–125.

Toda, M., Okubo, S., Ohnishi, R., Shimamura, T. (1989b) Antibacterial and bactericidal activities of Japanese green tea. *Jpn. J. Bacteriol.*, **44**, 669–672.

Toda, M., Okubo, S., Ohnishi, R., Shimamura, T. (1990) The antibacteria and anti-toxin effects of tea catechins and structure analoges. *Jpn. J. Bacteriol.*, **45**, 561–566.

Toda, M., Okubo, S., Hara, Y., Shimamura, T. (1991) Antibacterial and bactericidal activities of tea extracts and catechins against methicillin resistant *Staphylococcus aureus*. *Jpn. J. Bacteriol.*, **46**, 839–845.

Toda, M., Okubo, S., Ikigai, H., Suzuki, T., Suzuki, Y., Shimamura, T. (1991) The protective activity of tea against *Vibrio cholerae 01*. *Proceed. Internat. Symp. Tea Sci.*, Shizuoka, August 1991, 287–290.

Toda, M., Okubo, S., Ikigal, H., Suzuki, T., Suzuki, Y., Shimamura, T. (1991) The protective activity of tea against infection by *Vibrio cholerae 01*. *J. Appl. Bacteriol.*, **70**, 109–112.

Toda, M., Okubo, S., Ikigai, H., Suzuki, T., Suzuki, Y., Hara, Y., *et al.* (1992) The protective activity of tea catechins against experimental infection by *Vibrio cholerae 01*. *Microbiol. Immunol.*, **36**, 999–1001.

Yamazaki, K. (1996) Enhancing effect of Japanese green tea extract on the growth-inhibitory activity of antibiotics against clinically isolated MRSA strains. *Jpn. J. Chemoth. Soc.*, **44**, 477–482.

Yu, H., Oho, T., Xu, L.X. (1995) Effects of several tea components on acid resistance of human tooth enamel. *J. Dent.*, **23**, 101–105.

Yu, H., Xu, L.X., Oho, T., Morioko, T. (1993) The effect of a tannin-fluoride mixture on human dental enamel. *Caries Res.*, **27**, 161–168.

Zhang, G.Y., He, L.N., Li, Y.Y., Wu, Y.L., Sun, Y.P. (1993) An experimental study on antiviral activities of black tea, blue tea and dark tea against rotavirus. *Virologica Sinica*, **8**, 151–153.

Zhang, G.Y., Wang, Z.H., Sun, Y.P. (1995) Inhibitory effects of (–)-epigallocatechin gallate and theaflavin-3,3'-o-digallate on human rotavirus. *Proceed. '95 Internat. Tea-Quality-Human Hlth. Symp.*, Shanghai, November, 1995, 80–82.

10. ANTICARCINOGENIC ACTIVITY OF TEA

SHU-JUN CHENG

Cancer Institute, Chinese Academy of Medical Sciences and Peking Union Medical College, Panjiayuan, Beijing 100021, China

There is sufficient evidence that dietary and nutritional factors are of great importance in the causation of human cancer, and the imbalance between dietary carcinogens and anticarcinogens may play an important role in the development of human cancer (Cheng and Ho, 1988). Epidemiological studies have shown that the incidence of certain kinds of human cancer is relatively low in communities where fresh vegetables and fruits are frequently consumed (Block *et al.* 1992). It is generally recognized that naturally occurring substances in the food consumed are one of the most promising sources of inhibitors of carcinogenesis. The search for natural edible products which might have a preventive effect on human cancer has thus attracted the attention of many scientists. Since 1984, we have studied over 100 varieties of fresh vegetables, fruits and beverages for their antimutagenicity (Luo *et al.* 1987). Green tea, among others, was found to be highly antimutagenic. The research advances from our collaborative groups as well as other laboratories on the antimutagenesis and anticarcinogenesis effects of tea and tea polyphenolic components are summarized in this chapter.

Tea originated in China and is the most popular beverage in the world. According to the different manufacturing processes, teas can be classified into three types: non-fermented tea (green tea); fermented tea (black tea); and semi-fermented tea (oolong tea). Tea polyphenols are the main composition of fresh tea leaves, and account for 17~30% of dry tea weight. Tea polyphenols include flavanols, flavandiols, phenolic acids, flavonoids (Chen, 1989). The most important components of green tea polyphenols are various catechins, such as (–)-epigallocatechin gallate (EGCG), (–)-epicatechin gallate (ECG), (–)epigallocatechin (EGC), and (–)-epicatechin (EC). Tea polyphenols are oxidized to thearubigins and theaflavins during the processing of black tea known as fermentation.

The IARC Working Group on the evaluation of carcinogenic risks to humans has reviewed a wide range of studies on carcinogenicity of tea both in animals and humans. The Working Group considers that there is inadequate evidence for carcinogenicity of tea drinking in humans, and there is inadequate evidence for carcinogenicity of tea in experimental animals. The overall evaluation from the Working Group is that tea is not classifiable as carcinogenic to humans (Groups 3) (WHO International Agency for Research on Cancer, 1991).

1. INHIBITION OF MUTAGENICITY

There were a few studies which reported that the components of green and black tea were mutagenic to bacteria (Nagao *et al.* 1979; Ariza *et al.* 1988; Alejandre-Duran *et al.* 1987). The mutagenicity of tea was strongly inhibited by catalase. The results indicated that hydrogen peroxide was produced in tea solution, playing an essential role in its mutagenicity. On the other hand, there is increasing evidence to show the antimutagenic and antigenotoxic activities of tea and their ingredients in bacteria and mammalian cells. It has been reported (Jain *et al.* 1989) that tea extract and catechins such as EC, EGCG and EGC decreased the mutagenic effect of N-methyl-N'-nitro-N-nitrosoganidine (MNNG) in the stomachs of rats using *E. coli* WP2 and *S. typhimurium* TA_{100}. The water extracts of green tea and their major polyphenol fraction, including EGCG, EGC, ECG and EC, were investigated for antimutagenicity in our laboratory. Water extracts of green tea showed a strong inhibitory effect on the backward mutation induced by aflatoxin B_1(AFB_1) and benzo (α)pyrene (Bap) in *S. typhimurium* TA_{100} and TA_{98}. The green tea epicatechin compounds (GTEC) decreased the frequency of sister chromatid exchanges (SCE) and chromosomal aberrations in V_{79} cells treated with AFB_1 (Cheng *et al.* 1986) (Table 1 and Table 2). In 6-thioguanine-resistant mutation assay, it was observed that GTEC significantly inhibited gene forward mutation, in a dose-dependent manner, in V_{97} cells treated with AFB_1 or Bap (Cheng *et al.* 1986) (Table 3). Fried fish extract (FFE) and its related compound 2-amino-3,4-dimethylimidazo(4,5-f)quinoline (MeIQ)

Table 1 Inhibitory Effect of GTEC on Sister Chromatid Exchanges Induced by AFB_1 in V_{79} Cells.

GTEC (µg/ml)	AFB_1 (M)	SCE (Mean ± SE)
0	0	6.7 ± 0.4
0	3×10^{-8}	14.8 ± 0.8*
20	3×10^{-8}	11.6 ± 0.8*
50	3×10^{-8}	11.9 ± 0.8*
100	3×10^{-8}	9.3 ± 0.5*
200	3×10^{-8}	7.8 ± 0.6*

*P < 0.05

Table 2 Inhibitory Effect of GTEC on Chromosomal Aberrations Induced by AFB_1 in V_{79} Cells.

GTEC (µg/ml)	AFB_1(M)	Cells with chromosomal aberration (%)
0	0	4
0	10^{-6}	20*
100	10^{-6}	12
200	10^{-6}	14
400	10^{-6}	6*

*P < 0.05

Table 3 Inhibitory Effect of GTEC on 6-Thioguanine Resistant Mutation Induced by AFB_1 or Bap in V_{79} Cells.

GTEC ($\mu g/ml$)	AFB_1 or Bap (M)	Resistant mutant 10^6 survivors
	AFB_1 0	0
0	AFB_1 10^{-6}	48.3*
100	AFB_1 10^{-6}	52.6
200	AFB_1 10^{-6}	11.3*
400	ABF_1 10^{-6}	7.5*
0	0	1.4
0	Bap 8×10^{-6}	55.1*
50	Bap 8×10^{-6}	35.4*
100	Bap 8×10^{-6}	7.8*
200	Bap 8×10^{-6}	0*

*$P < 0.05$

showed potent genotoxicity to bacteria and mammalian cells. GTEC strongly inhibited the increase of SCE and micronuclei induced by FFE or MeIQ in V_{79} or IAR_{20} cells, and MeIQ-induced backward mutation in *S. typhimurium* (Liu *et al.* 1989; Liu *et al.* 1990). An inhibitory effect of GTEC on unscheduled DNA synthesis induced by MeIQ or FFE was also found in primary rat hepatocytes (Liu *et al.* 1990) (Table 4). In an *in vivo* experiment, green tea water extract or GTEC inhibited micronuclei induced by 1,2-dimethylhydrazine (DMH) in the colon crypt cells of mice (Zhao *et al.* 1992). Ito *et al.* (1989) reported that rats given the hot water extract from green tea 24h before they were injected with AFB_1 displayed considerably suppressed AFB_1-induced chromosomal aberrations in their bone marrow cells.

Table 4 Inhibitory Effect of GTEC on Unscheduled DNA Synthesis Induced by MeIQ or FFE in Primary Rat Hepatocytes.

MeIQ or FFE		GTEC ($\mu g/ml$)	Grains/nucleus (Mean $\pm$ SE)
MeIQ	0	0	0.58 $\pm$ 0.50
(M)	10^{-4}	0	3.82 $\pm$ 0.74*
	10^{-4}	10	3.45 $\pm$ 0.79
	10^{-4}	20	2.61 $\pm$ 0.53*
	10^{-4}	40	1.24 $\pm$ 0.41*
FFE	0	0	0.25 $\pm$ 0.49
(mg/ml)	125	0	3.36 $\pm$ 0.69*
	125	10	2.68 $\pm$ 0.61
	125	20	1.50 $\pm$ 0.44*
	125	40	1.41 $\pm$ 0.58*

*$P < 0.01$

Table 5 Inhibitory Effect of GTEC on Transformation Induced by MCA/TPA in BALB/3T3 cells.

GTEC + MCA treatment for 3 day* (μg/ml)		GTEC + TPA in each medium change** (μg/ml)		Plating efficiency (%)	No. of transformant/ 10^5 survivors
Exp. I	0 0	0 0		41.4	0
	0 0	0 0.1		40.4	0
	0 0.5	0 0		24.8	81
	0 0.5	0 0.1		23.6	42.3***
	2.5 0.5	2.5 0.1		31.6	9.6***
	0 0.5	2.5 0.1		29.4	17.0***
Exp. II	0 0	0 0		52.5	0
	0 0.5	0 0.1		28.5	63.2***
	5.0 0.5	5.0 0.1		53.2	5.6***
	0 0.5	5.0 0.1		46.5	8.6***

*GTEC was added 6h before MCA;
**After MCA treatment for 3 days;
***P < 0.01

2.   INHIBITION OF PROMOTION AND CELL TRANSFORMATION

It is generally believed that malignant transformation is a multistep process apparently caused by carcinogens and subject to the continuous influence of promoters, such as 12-o-tetradecanoylphorbol-13-acetate (TPA) (Berenblum, 1978; Hecker, 1971). Antipro-moters in natural edible products might have preventive effects on human cancers. A significant inhibition of malignant transformation by GTEC was observed in BALB/3T3 cells treated first with 3-methylcholanthrene (MCA) and then with TPA (Table 5). GTEC (5 μg/ml) reduced the transformation frequency of BALB/3T3 cells from 63.2/10^5 survivors to 5.6~8.6/10^5 survivors, showing inhibitory effects on both initiation and promotion (Cheng *et al.* 1989). In another transformation assay, EGCG, the main component of tea polyphenols, was used. We observed that EGCG showed the same inhibitory effect on the cell transformation as GTEC (Table 6) (Tong *et al.*

Table 6 Inhibitory Effect of EGCG on Transformation Induced by MCA/TPA in BALB/3T3 Cells.

MCA treatment for 3 days (μg/ml)	EGCG + TPA treatment in each medium change* (μg/ml)		Plating efficiency (%)	No. of foci/ plate	No. of transformant/ 10^5 survivors
0	0	0	40.7	0	0
0.5	0	0	31.2	0.2	6.39
0.5	0	0.1	41.8	1.7	40.87**
0.5	2.5	0.1	41.8	0.6	14.42**
0.5	0.5	0.1	38.5	1.0	25.87

*After MCA treatment for 3 days; **P < 0.001

Table 7   Inhibitory Effect of GTEC on TPA-induced Edema in Mouse Ear.

GTEC (mg)	TPA (μg)	Mean weight of ear plug ± SE (mg)
0	0	6.36 ± 0.22
0	0.4	13.42 ± 0.56*
2.0	0.4	7.02 ± 0.37*
1.0	0.4	9.00 ± 0.86*
0.5	0.4	10.40 ± 0.70*

*$P < 0.05$

Table 8   Inhibitory Effect of EGCG on TPA-induced Epidermal Hyperplasia in Mouse Skin.

Group	EGCG (mg)	TPA (μg)	Thickness of epidermis Mean ± SE (μm)
I	0	0	22.30 ± 0.77*
	0	1	48.05 ± 1.57*
II	0	1	42.31 ± 1.60*
	2	1	32.76 ± 1.26*
III	0	1	40.83 ± 2.50*
	5	1	23.71 ± 1.74*

*$P < 0.001$ for group I and III, $P < 0.01$ for group II, TPA was applied topically alone or with EGCG every other day for a total of 3 applications. Mice were sacrificed at day 7. The thickness of the dorsal epidermis was determined in histological section by a micrometer.

1992). These experimental results also indicate that EGCG is probably the main component in green tea polyphenols which inhibits TPA-induced promotion. In mouse skin tests, topical application of GTEC or EGCG significantly inhibited both edema and hyperplasia induced by TPA (Table 7 and Table 8) (Tong et al. 1992). Nakamura et al. (1991) reported antitumor promoting effects of four kinds of tea extracts in mouse epidermal JB6 cells. The preparation of hot water extracts of green-, black-, Pu-erh, and Oolong teas decreased the TPA-induced anchorage independent colony induction by 70%, 67%, 32% and 46% respectively. They considered that the green tea fraction containing catechins, such as (–)-epigallocatechin gallate had the highest activity, as much as 68.8% of total activity. It has been reported (Huang et al. 1992; Wang et al. 1992) that topical application of green tea polyphenols or green tea in drinking water inhibited both initiation and promotion of mouse skin carcinogenesis. In X-ray-induced BALB/3T3 cell transformation assay, it was found that GTEC significantly inhibited the transformation (Table 9) (Cheng et al. 1991).

Table 9   Inhibitory Effect of GTEC on Transformation Induced by X-rays in BALB/3T3 Cells.

GTEC (μg/ml)	X-ray (Gy)	Plating efficiency (%)	No. of foci/plate	No. of transformants/ 10^5 survivors
0	0	52.5	0	0
0	4	2.4	1.20	333.33*
5.0	4	41.6	0.11	1.76*

*$P < 0.001$

3. INHIBITION OF CIGARETTE SMOKE-INDUCED EFFECT

Many reports have consistently estimated that cigarette smoking is responsible for between 80% and 90% of all lung cancer deaths in the United States, and lung cancer is the dominate form of neoplastic disease in the population (Shopland et al. 1991; Peto, R. et al. 1992). It was suggested (Pryor 1986) that the carcinogenic effect of cigarette smoking appears to be associated with the carcinogens as well as the radicals contained in the smoke and tar. Therefore, we investigated the possible effect of green tea polyphenols on cigarette smoke-related mutagenesis and carcinogenesis. A comparative study on the effect of various antioxidants on the mutagenicity of cigarette smoke condensate (CSC) was carried out in S. typhymurium TA$_{98}$. The result revealed that GTEC had a stronger inhibitory effect than ellagic acid, butylated hydroxyanisol, ascorbic acid, β- carotene and tocopherol (Cheng et al. 1991). CSC-induced chromosomal aberration was inhibited by GTEC treatment in V$_{79}$ cells (Han et al. 1989). It was suggested (Trosko, et al. 1982) that gap-junction-mediated intercellular communication plays an important role in carcinogenesis. In an assay of scrape loading/Lucifer Yellow dye transfer, CSC inhibited intercellular communication. GTEC were found to prevent CSC-induced inhibition of intercellular communication in BALB/3T3 cells (Cheng 1996). In an in vivo assay, topical application of CSC increased epidermal thickness of mouse skin about twofold. Treatment with GTEC reduced the CSC-induced epidermal thickness to normal level (Cheng et al. 1991) (Table 10). In an in vivo-in vitro transformation of rat tracheal epithelial cells, pretreatment by intraperitoneal injection of EGCG significantly inhibited the transformation induced by Bap in rat tracheal epithelial cells (Feng et al. 1992) (Table 11). It was also found that GTEC treatment significantly decreased Bap-induced transformation frequency of BALB/3T3 cells (Cheng et al. 1989) (Table 12). It has been reported (Xu et al. 1992) that the tobacco-specific nitrosamine 4-(methylnitrosamino)-1-(3-pyridyl)-1-butanone (NNK) treatment induced lung adenomas in A/J mice. Green tea and its major component EGCG showed a protective effect against NNK-induced lung tumorigenesis in A/J mice. The study indicated that polyphenol EGCG appeared to be a major active component for this inhibitory activity. Wang et al. (1992a) observed that green tea and black tea infusion inhibited NNK-induced pulmonary adenomas in A/J mice. Wang et al. (1992b) also reported the inhibitory effect of green tea infusion on Bap-induced lung tumorigenesis. In a human epidemiological study (Xue et al. 1992),

Table 10   Inhibitory Effect of GTEC on CSC-induced Hyperplasia in Mice.

CSC (mg)	GTEC (mg)	Mean thickness of epidermis $\pm$ SE (μm)
0	0	16.30 $\pm$ 1.25
12.5	0	32.16 $\pm$ 2.81*
12.5	0	28.70 $\pm$ 1.26*
12.5	30	16.35 $\pm$ 0.99*
12.5	0	28.90 $\pm$ 1.19**
12.5	15	19.76 $\pm$ 0.71**

* or ** P < 0.01

Table 11   Inhibitory Effect of EGCG on Bap-induced Transformation in Rat Tracheal Epithelial Cells.

Bap (mg/kg.bw)	EGCG (mg/kg.bw)	Plating efficiency (%)	Transformation frequency (%)
0	0	1.26	0.76
25	0	0.51	4.79*
25	300	0.56	1.09*
25	600	0.58	0.64*

*P < 0.01

Table 12   Inhibitory Effect of GTEC on Bap-induced Transformation in BALB/3T3 Cells.

GTEC ($\mu g/ml$)	Bap ($\mu g/ml$)	Plating efficiency (%)	No. of transformation/ 10^5 survivors
0	0	64.5	2
0	2	6.3	222*
5	2	7.0	124*
25	2	13.4	65*

GTEC was added into culture medium 6 h before Bap. *P < 0.05

the frequency of micronuclei in peripheral-blood lymphocytes in 220 healthy cigarette smokers was detected by a micronucleous test. On the basis of analysis of matched-pair data, the results show that cigarette smoking significantly increases the frequency of micronuclei in lymphocytes as compared with healthy non-smokers, and green tea consumption decreases the frequency of micronuclei in smokers.

4.   INHIBITION OF CARCINOGENESIS

In recent years, many experimental studies have shown that green tea or black tea extract and polyphenolic components inhibited carcinogen-induced tumors of different organs in rodents. Esophageal papillomas and carcinomas of the esophagus and forestomach were induced in mice by the *in vivo* formation of nitrososarcosine from its precursors, sarcosine and $NaNO_2$. Administration of water-soluble green tea extract significantly decreased the frequency of induced papillomas and carcinomas (Lin *et al.* 1990) (Table 13). It has been reported (Xu and Hanm, 1990) that oral feeding of Chinese tea infusions inhibited esophageal tumors induced by the *in vivo* formation of N-nitrosomethylbenzylamine from its precursors methylbenzylamine and $NaNO_2$ in rats. It was observed (Yamane *et al.* 1995) that EGCG inhibited glandular stomach carcinogenesis induced by MNNG in rats. Fujita *et al.* (1989) found that EGCG inhibited duodenal tumors in mice treated with N-ethyl-N'-nitrosoguanidine. Large intestinal cancers were induced in mice by subcutaneous injection of 1,2-dimethylhydrazine. Oral feeding of GTEC or EGCG significantly inhibited the

Table 13 Inhibitory Effect of Water Soluble Green Tea Extract (GTE) on Tumor Induction in Mice Caused by the *in vivo* Formation of Nitrososarcosine from its Precursors.

	Chemicals (g/kg BW)	No. of mice	Incidence of esophageal papilloma (%)	Incidence of carcinoma (%)		
				Esophagus	Forestomach	Total
Exp. I	A. Sarcosine (2.0) + NaNO$_2$ (0.3)	47	33(70.2)*	3	16	19(40.4)*
	B. GTE + sarcosine (2.0) + NaNO$_2$ (0.3)	45	20(44.4)*	0	8	8(17.8)*
Exp. II	A.Sarcosine (1.0) + NaNO$_2$ (0.15)	37	22(59.5)*	1	11	12(32.4)**
	B. GTE + Sarcosine (1.0) + NaNO$_2$ (0.15)	34	10(29.4)*	0	2	2(5.9)**

*P < 0.05; **P < 0.01.

Mice (group B) were given GTE (5 mg/day) by stomach intubation. From the second week [I] or third week [II] on, mice in both groups A and B were given sarcosine and NaNO$_2$ by intubation (5 h after GTE). The animals were sacrificed at 14 weeks [I] or 18 weeks [II] and tumors were examined histologically.

incidence of large intestinal tumor (Yin *et al.* 1994) (Table 14). Hagiwara *et al.* (1991) also found that azoxymethane-induced colon carcinogenesis in rats was inhibited by 0.01~0.1% tea polyphenolic compounds in drinking water. In rat hepatocarcinogenesis, gamma-glutamyl transpeptidase(GGT)-positive foci were induced by diethylni-

Table 14 Inhibitory Effect of GTEC or EGCG on DMH-induced Large Intestinal Tumors in Mice.

Group	Mean No. of tumors in each experimental mouse ± SD
1. DMH + water	4.389 ± 4.709*
2. DMH + GTEC1	0.667 ± 1.312*
3. DMH + GTEC2	1.410 ± 2.993*
4. DMH + EGCG	0.795 + 1.343*
5. GTEC3	0
6. EDTA control	0

GTEC1. 1 mg/mouse per day, GTEC2. 2 mg/mouse per day,
GTEC3. 3 mg/mouse per day, EGCG. 2 mg/mouse per day,
*P < 0.05

Table 15   Inhibitory Effect of GTEC on DEN and 2AAF-induced GGT-positive
Liver Foci in Rats.

Group	Number density Mean ± SD	Volume density Mean ± SD
GDA	234.56 ± 74.81*	7.91 ± 4.25*
DA	417.05 ± 204.42*	12.75 ± 3.73*
A	20.99 ± 10.04	0.70 ± 0.30*
D	0	0

G: GTEC; D: DEN; A: 2AFF *P < 0.05

Table 16   Inhibitory Effect of GTEC on DEN and PB-induced GGT-positive
Liver Foci in Rats.

Group	Number density Mean ± SD	Volume density Mean ± SD
HDGP	175.88 ± 70.34*	2.61 ± 0.97*
HDP	286.71 ± 99.27*	4.22 ± 0.98*
HD	189.51 ± 71.32	2.84 ± 1.41
H	0	0
N	0	0

G: GTEC; H: Hepatectomy; D: DEN; P: PB
N: Normal, *P < 0.01

trosamine (DEN) and 2-acetylaminofluorene (2AAF) treatment (initiation assay), or by
DEN and phenobarbital (PB) treatment (promotion assay). Administration of GTEC
significantly inhibited the development of GGT-positive foci in both experiments,
indicating anti-initiation and anti-promotion effects on hepatocarcinogenesis (Ding et
al. 1990; Mao et al. 1993) (Table 15 and Table 16). Chen et al. (1987) reported that
green tea or black tea in the basal diet (5% concentration) inhibited AFB_1-induced
GGT-positive hepatocyte foci in rats. Li et al. (1991) also observed that 2.5% green tea
in the diet showed an inhibitory effect against DEN-induced hepatocarcinogenesis in
rats. The inhibitory effects of tea and tea polyphenols on carcinogenesis were also
demonstrated in other organs such as pancreatic cancer in the golden hamster (Harada
et al. 1991), small intestinal tumor in rats (Hirose et al. 1991), mammary tumor in rats
(Hirose et al. 1991, 1994). It has been reported (Isemura et al. 1995) that green tea
infusion inhibited in vitro invasion and in vivo metastasis of mouse lung carcinoma cells.

5.   MECHANISMS OF ANTICARCINOGENESIS

The series of above-mentioned investigations have demonstrated that tea and tea
polyphenolic components possess antimutagenicity and anticarcinogenicity. The
mechanisms related to the biological activities are very complicated. The experimental
studies completed in our and other laboratories suggest that the mechanisms are
probably related to the following functions of tea polyphenols.

5.1. Effect on Enzymes

It is known that the conjugation of electrophilic carcinogens with glutathione is an important mechanism of detoxification. The reaction is catalyzed by glutathione-s-transferase (GST). GTEC fed to mice increased the activity of liver GST by 36% and that of liver superoxide dismutase (SOD) by 25% (Cheng *et al.* 1991). Rats treated with carbon tetrachloride showed severe liver function damage, such as increased serum glutamate pyruvate transaminase, decreased serum albumin/globulin ratio, and low GST activity in liver tissue. However, GTEC treatment could strongly prevent these carbon tetrachloride-induced toxic effects (Cheng *et al.* 1991). In the study of DMH-induced large intestinal cancer, the activity of SOD was detected in red cells and in the plasma of mice. It was found that the SOD activity in mice treated with GTEC was much higher than in mice treated with DMH (Yin *et al.* 1994). It has been reported (Khan *et al.* 1992) that green tea polyphenols in drinking water for 30 days significantly increased the activities of glutathione, peroxidase, catalase, and quinone reductase in small bowel, liver, and lungs, and GST in small bowel, and liver of female SKH-1 mice. Induction of epidermal ornithine decarboxylase (ODC) is a biochemical parameter closely related to tumor promotion. Topical application of GTEC significantly inhibited TPA-induced ODC activity of mouse skin (Cheng *et al.* 1989).

5.2. Scavenging Effect on Free Radicals

It has become clear that free radicals are involved in many of the biological processes that occur during carcinogenesis (Amesm, 1983). Tea polyphenolic components have been discovered to possess excellent antioxidant activity (Cheng *et al.* 1988; Namiki *et al.* 1986; Korverm, 1995). It has been found (Zhao *et al.* 1989) that TPA can initiate the formation of superoxide anion radicals and hydroxyl radicals in human leukocyte cells in culture. Measurement of free radicals by electron spin resonance showed that when GTEC was added to the system, the formation of free radicals was inhibited. The inhibitory effect on the TPA-induced free radicals by GTEC was stronger than either ascorbic acid or tocopherol. In IAR$_{20}$ liver cell assay, GTEC as well as SOD were found to inhibit the frequency of SCE induced by free radicals generated by hypoxanthine and xanthine oxidase in the culture medium (Cheng *et al.* 1989) (Table 17). Wang *et al.* (1989) reported that green tea polyphenols resulted in a dose-dependent disappearance of the electron spin resonance signal due to the radicals in perylene solution. Wu *et al.* (1993) observed that singlet oxygen produced by illuminated photosensitizer Rose Bengal was detected by means of bleaching of N,N-dimethyl-nistrosoaniline. The singlet oxygen could be scavenged by green tea polyphenolic components, and the scavenging effects were obviously at the beginning of photosensitized oxidation.

5.3. Effect on Promoter-Induced Biological Activities and Gene Expression

It has become clear that tumor promoters play a critical role during carcinogenesis. A wide range of studies demonstrated that tea and tea polyphenolic components

Table 17 Inhibitory Effect of GTEC and SOD on SCE Induced by Hypoxanthine and Xanthine Oxidase in IAR$_{20}$ Cells.

	Chemicals (μg/ml)	Hypoxanthine (7 μg/ml)	Xanthine Oxidase (15 μg/ml)	SCE (Mean ± SE)
GTEC	0	0	0	11.3 ± 0.8
	0	+	0	10.4 ± 0.6
	0	0	+	12.0 ± 0.9
	0	+	+	25.1 ± 1.5*
	2.5	+	+	17.4 ± 0.9*
	5.0	+	+	16.7 ± 1.1*
	10.0	+	+	15.3 ± 1.1*
	20.0	+	+	12.6 ± 0.5
SOD	10.0	+	+	16.8 ± 1.1*
	20.0	+	+	16.0 ± 0.8*

*$P < 0.001$

inhibited TPA-induced inflammation, hyperplasia, cell transformation and tumors. These inhibitory effects might be related to the alterations of gene expressions. Our slot blot analysis (Cheng *et al.* 1993) showed that TPA treatment induced the expression of *c-fos* gene in cultured BALB/3T3 cells, and the level of *c-fos* mRNA reached a maximum at 30 min after exposure to TPA. Pretreatment of GTEC decreased the TPA-induced overexpression of *c-fos* threefold at this time. It was also observed that TPA enhanced overexpression of *c-myc* gene mRNA about twofold 2h after the treatment, and GTEC reduced the TPA-altered level of *c-myc* mRNA to control level. The TPA-induced overexpression of ornithine decarboxylase gene mRNA was also inhibited by GTEC treatment. In an *in vivo* experiment, it was found that GTEC inhibited protein kinase C gene overexpression that was induced by croton oil in rat liver. Lin (1995) reported that EGF receptor gene was overexpressed in human A-431 epidermal carcinoma cells. The growth of the cells was significantly inhibited by tea polyphenols in a dose-dependent manner. The autophosphorylation of EGF receptor and phosphorylation of extracellular signal-related kinase (ERK-1 and ERK-2) were remarkably inhibited by tea polyphenols.

5.4. Effect on Intercellular Communication

It was believed that gap junction-mediated intercellular communication plays an important role in cell growth and carcinogenesis (Trosko *et al.* 1982). In Lucifer Yellow dye transfer assay, GTEC prevented both the inhibition of intercellular communication (IC) by paraquat, glucose oxidase, and phenobarbital in mouse hepatocytes, and the inhibition of IC by TPA (Ruch *et al.* 1989). It was also observed that the TPA-induced inhibitory effect on IC was blocked in BALB/3T3 cells in the presence of GTEC (Feng, *et al.* 1992).

5.5.   Effect on Cell-Mediated Immunity

The anticarcinogenic effect of tea polyphenolic components may be involved in the enhancement of cell-mediated immunity. Hu and Zhang (1991) investigated the *in vitro* immunological effects of catechins. They found that catechin itself was weakly mitogenic but strongly synergized with ConA in inducing lymphocyte proliferation. The NK cell activity was also significantly increased when assayed in the presence of catechin. Catechin could also synergize with rIL-2 in the induction of LAK cell activity. Luo *et al.* (1995) reported that after immunizing DBA/2 mice with syngenic $L_{5178}Y$ lymphoma, catechin given to mice by gastric intubation for 7–10 days could markedly enhance T cell proliferation. The generation of tumor specific CTL and the NK cell activity were also significantly increased. Catechin could increase IL-2 mRNA expression levels and IL-2 secretion by T cells. The enhancing effect of catechin on tumor immune response seems to be related to its ability to abrogate the immunosuppressive activity of accessory cells on T cells and NK cells.

5.6.   Other Biologic Activities

It has been reported (Suganuma *et al.* 1995) that treatment of stomach cancer cell line KATO III with EGCG for 24 h increased the cells of G2/M phase from 8.2% to 21.6%. EGCG inhibited release of tumor necrosis factor (TNF-α), an endogenous tumor promoter, suggesting reduction of the amount of an endogenous tumor promoter in tissues. Fan *et al.* (1992) reported that GTEC inhibited cell growth of Hela and KB cell line in culture, and significantly reduced the [^{3}H]thymidine incorporation rate into HeLa and KB cells in a dose-dependent manner. It has been shown (Zhen. *et al.* 1991) that GTEC is a highly active nucleoside transport inhibitor. GTEC significantly inhibited radiolabelled thymidine and uridine transport in mouse leukemia L1210 cells; blocked the rescue effect of exogenous nucleosides and therefore enhanced the cytotoxicity of cytarabine and methotrexate in L_{1210} and human hepatoma BEL-7402 cells. Kinae *et al.* (1995) reported that tea water infusion inhibited the formation of heterocyclic amines during cooking of meat. It has been observed (Tao *et al.* 1992) that catechin derivatives (ECG, EGCG, EGC) and green tea extract inhibited the activities of cloned human immunodeficiency virus type-I reverse transcriptase, duck hepatitis B virus replication complexes reverse transcriptase, herpes simplex virus 1 DNA polymerase and cow thymus DNA polymerase α.

6.   CONCLUSIONS

It is believed that dietary factors play an important role in human carcinogenesis. Doll and Peto (1981) estimated that diet can account for up to 70% of all avoidable cancers. Many mutagens and carcinogens that have been isolated from foods and environmental samples are known to be directly or indirectly responsible for the occurrence of human cancer. It is extremely important to reduce or completely remove these mutagens or carcinogens from our environment in order to reduce the incidence of human cancer. However, this is an extremely difficult task. Alternatively humans can protect themselves from tumor occurrence by increasing the intake of anti-carcinogens.

Tea and polyphenolic components could affect carcinogenesis through a number of mechanisms. Of particular interest among the biological effects is the finding of a very strong antipromotion activity of epicatechin compounds in both *in vitro* and *in vivo* studies. Many experimental and epidemiological studies (Cheng *et al.* 1982; Day, 1982; Weisburger *et al.* 1982) indicate that cancer development is a multistep process, and promoting action may be an important determinant of risk for human cancer. The promotional events in carcinogenesis appear to be reversible, and inhibition of the promotion may prove to be of particular importance in the prevention of human cancer. As mentioned above in this chapter, tea polyphenolic compounds showed a strong inhibitory effect on promoter-induced inflammation, hyperplasia, and tumors. As potential chemopreventive agents, tea polyphenolic compounds may be useful for blocking or reversing the progression of premalignant cells. However, further studies, especially human clinical trials, should be done before any conclusion can be reached.

Many mutagens and carcinogens may exert their effects by inducing the generation of free radicals, which in turn may play a major role in tumor initiation and promotion and in the development of human heart disease and the aging process. The carcinogenic effect of cigarette smoke appears to be related to the carcinogens as well as the radicals contained in smoke and tar (Pryor, 1986). It has been reported (Shopland, 1995) that smoke from a single cigarette is composed of over 4000 different constituents, including nearly 60 agents that are known carcinogens, tumor promoters, or tumor initiators. Cigarette smoking is causally related to a host of fatal disease, including several major sites of cancer, such as lung, larynx, oral cavity and esophagus. Our experimental results have shown that mutagenic effect, inhibition of intercellular communication and skin hyperplasia induced by CSC were prevented by GTEC treatment. Bap and NNK are two well-known carcinogens contained in cigarette smoke. Tea and tea polyphenolic compounds inhibit Bap or NNK-induced lung tumors in rodents. Tea catechins, as efficient free radical scavengers and strong antioxidants, may protect against the noxious effects produced by cigarette smoke. It may be wise to encourage cigarette smokers to use tea as a daily natural beverage.

Although anticarcinogenic activities of tea and its components have been investigated in animal experiments, the epidemiological evidence of a protective role of tea consumption and cancer incidence in human populations is controversial. The death rate from cancer, especially stomach cancer, is reportedly much lower in inhabitants of the midwest areas of Shizuoka region in Japan than in the general Japanese population, and green tea is a major product and the main beverage in this area (Oguni and Nasu, 1987). Recently, Kohlmeier *et al.* (1997) have reviewed epidemiological literature related to tea consumption and human cancer incidence. They considered that the evidence to demonstrate a protective role of tea consumption in human cancer incidence is weak. If benefits exist, they may be restricted to high consumption levels in populations at high risk of the cancer in question. Further epidemiological studies on the correlation between tea drinking and cancer incidence are necessary. However, a wide range of research data discussed in this chapter strongly suggest that tea and tea polyphenolic components, as a natural antioxidant, may be useful for the prevention of human cancers caused by cigarette smoking, radiation and some chemical carcinogens.

REFERENCES

Alejandre-Duran, E., Alonso-Moraga, A., and Pueyo, C. (1987) Implication of active oxygen species in the direct-acting mutagenicity of tea. *Mutation Res.*, **188**, 251–257.

Ames, B.N. (1983) Dietary carcinogens and anticarcinogens, oxygen radicals and degenerative disease. *Sciences*, **221**, 1256–1264.

Ariza, R.R., Dorado, G., Barbancho, M., and Pueyo, C. (1988) Studies of the causes of direct-acting mutagenicity in coffee and tea using the Ara test in *Salmonella typhimurium. Mutation Res.*, **201**, 89–96.

Berenblum, I. (1978) Established principles and unsolved problems in carcinogenesis. *J. Natl. Cancer., Inst.*, **60**, 723–726.

Block, G., Patterson, B., and Subar, A. (1992) Fruits, vegetables and cancer prevention: a review of the epidemiological evidence. *Nutr. Cancer*, **18**, 1–29.

Chen, Z.M. (1989) Tea production in China and therapeutic effect of tea. *Food Science and Industry*, **22**, 28–43.

Chen, Z.Y., Yan, R.Q., Qin, G.Z., and Qin, L.L. (1987) Effect of six edible plants on the development of AFB_1-induced γ-glutamyltranspeptidase positive hepatocyte foci in rats. *Chin. J. Oncology*, **9**, 109–111. (in Chinese)

Cheng, S.J., Sala, M., Li, M.H., and Chroulinkov. I. (1982) Esophageal cancer in Linxian County, China: A possible etiology and mechanism (initiation and promotion). In: Hecker, E., Fuseng, N.E., Kunz, W., Marks, F., and Thielman H.N., (eds.) C ocarcinogenesis and Biological Effects of Tumor Promoters, Raven Press. pp. 167–174.

Cheng, S.J., Ho, C.T., Lou, H.Z., Bao, Y.D., Jian, Y.Z., Li, M.H., et al. (1986) A preliminary study on the antimutagenicity of green tea antioxdants. *Acta Biological experimental Sincia*, **19**, 427–431. (in Chinese)

Cheng, S.J. and Ho, C.T. (1988) Mutagens, Carcinogens, and Inhibitors in Chinese Food. *Food Review International*, **4**, 353–374.

Cheng, S.J., Wang, Z.Y., and Ho, C.T. (1988) Studies on antimutagenic and anticarcinogenic properties of green tea antioxidant. In Wang J.W. (eds.) *Current Medicine in China*, Peoples Health Press, pp. 165–172.

Cheng, S. J., Ho, C. T., Huang, M.T., Wang, Z. Y., Liu, S.L., Gan, Y. N., et al., Li, S.Q. (1989) Inhibitory effect of green tea extracts on promotion and related action of TPA. *Acta. Acad. Med. Sin.*, **11**, 259–264. (in Chinese)

Cheng, S.J., Din, L., Zhen, Y.S., Lin, P.Z., Zhu, Y.J., Chen, Y.Q., et al. (1991) Progress in studies on the antimutagenicity and anticarcinogencity of green tea epicatechins. *Chin. Med. Sci. J.*, **6**, 233–238.

Cheng, S.J. (1996) Study on antimutagenicity and anticarcinogenicity of green tea epicatechins A natural free radical scavanger. In: Packer, L., Traber, M.G., Xin W. (eds.) *Proceeding of the International Symposium on Natural Antioxidants, Molecular Mechanisms and Health Effects*. AOCS Press. pp. 392–396.

Day, N.E. (1982) Epidemiological evidence of promoting effects- the example of breast cancer. In: Heker, E., Fusenig, N.E., Kunz, W., Marks, F., Thielman, H.W. (eds.) *Cocarcinogenesis and Biological Effects of Tumor Promoters*. Raven Press. pp. 183–199.

Din, L., Xing, Y.Q., Chen, J.Y., Cheng, S.J. and Sun, Y.H. (1990) Inhibition effect of epicatechin on phenobarbitol-induced proliferation of precancerous liver cells. *Chin. J. Pathol.*, **19**, 261–263. (in Chinese)

Doll, R. and Peto, R. (1981) The cause of cancer: Quantitative estimates of avoidable risks of cancer in the United States today. *J. Natl. Cancer. Inst.*, **66**, 1191–1308.

Fan, X.J., Wang, U.X., and Feng, S.Y. (1992) *In vitro* effect of catechin on cell growth. *Chin. J. Oncology*, **14**, 190–192. (in Chinese)

Feng, J., Cheng, S.J., Li, X.Q., Guo, S.P., Xu, R.X. and Zhou, Y.L. (1992) The use of scrape-loading/dye transfer assay for detection of antipromoters. *Carcinogenesis. Teratogenesis and Mutagenesis*, **4**, 45–48. (in Chinese)

Feng, J.N., Xu, J.K., and Cheng, S.J. (1992) Inhibitory effect of epigallocatechin gallate on transformation. *Carcinogensis, Mutagenesis, Teratogenesis*, **4**, 5–7. (in Chinese)

Fujita, Y., Yamane, T., Tanaka, M., Kuwata, K., Okuzumi, J., Takahashi, T., *et al.* (1989) Inhibitory effect of (–) epigallocatechin gallate on carcinogenesis with N-ethyl-N'-nitro-N-nitrosoguanidine in mouse duodenum. *Jpn. J. Cancer*, **80**, 503–505.

Hagiwava, N., Tateishi, M., Kim, M., Yamane, T., and Takahashi, T. (1991) Inhibition of azoxymethane-induced colon carinogenesis in rats by tea polyphenols. *Proceeding of the International Symposium on Tea Science*. Shizuoka, Japan. Kurofune Printing Co. Ltd. pp. 190–194.

Han, X.L., Cheng, S.J. and Li, M.H. (1989) Effect of green tea extract on the genotoxicity of cigarette smoke condensate. *Heredity and Disease*, **6**, 81–83 (in Chinese).

Harada, N., Takabayashi, F., Oguni, I. and Hara, Y. (1991) Anti-promotion effect of green tea extracts on pancreatic cancer in golden hamster induced by N-nitro-bis(2-oxoplopyl)amine. *Proceedings of the International Symposium on Tea Science*. Japan, Kurofune printing Co. Ltd. pp. 200–204.

Hecker, E. (1971) Isolation and characterization of the cocarcinogenic principle from croton oil. *Methods Cancer Res.*, **6**, 439–484.

Hirose, M., Hosiya, T., Takahashi, S., Hara, Y. and Ito, N. (1991) Inhibition of carcinogenesis by green tea catechins in rats. *Proceedings of the International Symposium on Tea Science*. Japan, Kurofune Printing Co. Ltd. pp. 210–214.

Hirose, M., Hosiya, T., Akagi, K., Futakuchi, M. and Ito, N. (1994) Inhibition of mammary gland carcinogenesis by green tea catechins and other naturally occurring antioxidants in female Sprague-Dawley rats pretreated with 7,12-dimethyl-benz(α)anthracene, *Cancer Lett.*, **83**, 149–156.

Hu, X.Z. and Zhang, Y.H. (1991) The enhancing effect of green tea catechin on cell mediated immunity in mice. *Chin. J. Microbiology and Immunology*, **11**, 97–99 (in Chinese).

Huang, M.T., Ho, C.T., Wang, Z.Y., Ferraro, T., Finnegan-Olive, T., Lou, Y.R., *et al.* (1992) Inhibitory effect of topical application of a green tea polyphenol fraction on tumor initiation and promotion in mouse skin. *Carcinogensis*, **13**, 947–954.

Isemura, M., Sazuka, M., Noro, T., Nakamura, Y. and Hara, Y. (1995) Inhibitory effects of green tea infusion on *in vitro* invasion and in vivo metastasis of mouse lung cancer cells. *International Conference on Food Factors*: *Chemistry and Cancer Prevention*, Japan, Abstract, pp. 118.

Ito, Y., Ohnishi, S., and Fujie, K. (1989) chromosome aberration induced by aflatoxin B_1 in rat bone marrow cells *in vivo* and their suppression by green tea. *Mutation Res.*, **222**, 253–261.

Jain, A.K., Shimoi, K., Nakamura, Y., Kada, T., Hara, Y., and Tomota, I. (1989) Crude tea extracts decrease the mutagenic activity of N-methyl-N'-nitro-nitrosoguanidine *in vitro* and in intragastric tract of rats. *Mutation Res.*, **210**, 1–8.

Khan, S.G., Katiyar, S.K., Agawal, R., and Mukhtar (1992) Enhancement of antioxidant and phase II Enzymes by oral feeding of green tea polyphenols in drinking water to SKH-I hairless mice: Possible role in cancer chemoprevention. *Cancer Res.*, **52**, 4050–4052.

Kinae, N., Furugori, M., Takemura, H., Iwazaki, M., Shimoi, K., and Wakabayashi, K. (1995) Inhibitory effect of tea extracts on the formation of heterocyclic amines during cooking of hamburger. *International Conference on Food Factors: Chemistry and Cancer Prevention.* Japan, Abstracts, pp. 109.

Kohlmeier, L., Weterings, K.G.C., Steck, S., and Kok, F.J. (1997) Tea and cancer prevention: an evaluation of the epidemiological literature. *Nutrition and Cancer.*, **27**, 1–13.

Korver, O. (1995) Tea components and cancer prevention, *International Conference on Food Factors: Chemistry and Cancer Prevention.* Japan, Abstracts, pp. 95.

Li, Y., Yan, R.Q., Qin, G.Z., Duan, X.X., and Qin, L.L. (1991) Comparative study on the inhibitory effect of green tea, coffee and levamisole on the hepatocarcinogenic action of diethylnitrosamine. *Chin. J. Oncology*, **13**, 193–195 (in Chinese).

Lin, J.K. (1995) Anticarcinogenesis of tea polyphenols. *International Conference on Food Factors: Chemistry and Cancer Prevention.* Japan. Abstract, pp. 99.

Lin, P.Z., Cheng, S.J., Zhang, J.S., and Oguni, I. (1990) Green tea extract inhibits carcinogen-induced tumors of the forestomach and esophagus in mice. *Acta. Acad. Med. Sin.* **12**, 156 (in Chinese).

Liu, X.L., Cheng, S.J., and Li, M.H. (1989) Preliminary study on mutagenicity of MeIQ and extracts of fried fish and antimutagenicity of some dietary factor. *Acta. Acad. Med. Sin.*, **11**, 97–101 (in Chinese).

Liu, X.L., Cheng, S.J., Li, M.H. (1990) Genotoxicity of fried fish extract, MeIQ and inhibition by green tea antioxidant. *Chin. J. Oncology*, **12**, 170–173 (in Chinese).

Luo, H.Z., Cheng, S.J., Jang, Y.Z., Han, N.Y., Li, X.Q., Yie, S.Y., *et al.* (1987). Preliminary study on antimutagenesis of vegetables and fruits. *Chin. J. Oncol.*, **9**, 328–332 (in Chinese).

Luo, L.Q., Zhang, Y.H. and Hong, R. (1995) The enhancing effect of green tea catechin on *in vivo* tumor immunity in mice. *Chin.. J. Immunol.*, **11**, 294–297 (in Chinese).

Mao, R., Din, L., Chen, J.Y., and Cheng S.J. (1993) The inhibitory effects of epicatechin complex on diethylnitrosamine induced initiation of hepatocarcinogenesis in rats. *Chin. J. Prev. Med.*, **27**, 201–204 (in Chinese).

Nagao, M., Takahashi, Y., Yamanaka, H., Sugimura, T. (1979) Mutagens in coffee and tea, *Mutation Res.*, **68**, 101–106.

Nakamura, Y., Harada, S., Kawase, I., Matsuda, M. and Tomita, I. (1991) Inhibitory effect of tea in gradients on the in vitro tumor promotion of mouse epidermal JB6 cells. *Proceedings of the International Symposium on Tea Sciences.* Shizuoka, Japan. Kurofune Printing Co, Ltd. pp. 205–209.

Namiki, M. and Osawa, T. (1986) Antioxidants/antimutagens in foods. In D.M. Shankel, P.E. Hartman, T. Kada, and A. Hollaender (eds.), *Antimutagenesis and Anticarcinogenesis Mechanisms*, Plenum Press, New York, pp. 131–142.

Oguni, I. and Nasu, K. (1987) Epidemiological and physiological studies on the antitumor activity of the fresh green tea leaf. *Proceeding of International Tea-Quality-Human Health Symposium.* Hangzhou, China. pp. 222–226.

Peto, R., Lopez, A.D., Boreham, J., Thun, M. and Heath, C. (1992) Mortality from smoking in developed countries 1950–2000, indirect estimation from national vital statistics. *Lancet*, **339**, 1268–1278.

Pryor, W.A. (1986) Cancer and free radicals. In D.M. Shankel, P.E. Hartman, T. Kada, and A. Hollaender (eds.), *Autimutagenesis and Anticarcinogenesis Mechanisms,* Pleum Press, New York, pp. 45–59.

Ruch, R.J., Cheng, S.J. and Klaunig, J.E. (1989) Prevention of cytotoxicity and inhibition of intercelluar communication by antioxidant catechins isolated from Chinese green tea. *Caninogenesis*, **10**, 1003–1008.

Shopland, D.R. (1995) Tobacco use and its contribution to early cancer mortality with a special emphasis on cigarette smoking. *Environmental Health Perspective* 103: Supplement 8, 131–142.

Shopland, D.R., Eyre, H.J. and Pechacek, T.F. (1991) Smoking-attributable cancer mortality in 1991: Is lung cancer now the leading cause of death among smokers in the United States? *J. Natl. Cancer Inst.*, **83**, 1142–1148.

Suganuma, M., Okabe, S., Oniyama, M., Sueoka, N., Kozu, T., Komori, A., *et al.* (1995) Mechanisms of EGCG and green tea in inhibition of carcinogenesis. *International conference on Food Factors: Chemistry and Cancer Prevention*, Japan, Abstract, pp. 119

Tao, P.Z., Zhang, T., Zhou, P., Wang, S.Q., Cheng, S.J., Jian, J.I., *et al.* (1992) The inhibitory effects of catechin derivatives on the activities of human immunodeficiency virus reverse transcriptase and DNA polymerase. *Acta. Acad. Med. Sin.*, **14**, 332–338 (in Chinese).

Tong, T., Cheng, S.J., Li, S.Q., Bai, J.F. and Hara, Y. (1992) Inhibitory effect of (–)-epigallocatechin gallate on TPA-induced activities. *Carcinogenesis. Mutagenesis. Teratogenesis*, **4**, 1–4 (in Chinese).

Trosko, J.E., Yotti, L.P., Warren, S.T., Tsushimoro, G. and Chang, C.C. (1982) Inhibition of cell-cell communication by tumor promoters. In E. Hecker, N.E. Fusenig, W. Kunz, F. Marks, H.W. Thielman (eds.) *Cocanogenesis and Biological Effects of Tumor Promoters*. Raven Press, New York, pp. 565–585.

Wang, Z.Y., Cheng, S.J., Zhou, Z.C., Athar, M., Khan, W.A., Bickers, D.R., *et al.* (1989) Antimutagenic activity of green tea polyphenols. *Mutation Res.*, **223**, 273–285.

Wang, Z.Y., Agarwl, R., Khan, W.A. and Mukhtar, H. (1992b) Protection against benzo(a) pyrene-and N-nitrosodiethylamine-induced lung and forestomach tumorigenesis in A/J mice by water extracts of green tea and licorice. *Carcinogenesis*, **13**, 1491–1494.

Wang, Z.Y., Hong, J.Y., Hunag, M.T., Reuhl, K.R., Conny, A.H., and Yang, C.S. (1992a). Inhibition of N-nitrosodiethylamine-and 4-(methylnitrosamino)-1-(3-pyridyl)-1butanone-induced tumorigenesis in A/J mice by green tea and black tea. *Cancer Res.*, **52**, 1943–1947.

Wang, Z.Y., Huang, M.T., Ferraro, T., Wong, C.Q., Lou, Y.R., Latropoulos, M., *et al.* (1992) Inhibitory effect of green tea in the drinking water on tumorigenesis by ultraviolet light and 12-o-tetradecanoylphobol-13-acetate in the SKH-1 mice. *Cancer Res.*, **52**, 1162–1170.

Weisburger, J.H., Reddy, B.S., Cohen, Z.A., Hill, P. and Wynder, E.L. (1982) Mechanisms of promotion in nutritional carcinogenesis. In: E. Hecker, N.E. Fusenig, W. Kunz. F. Marks, H. W. Thielman (eds.) *Cocarinpgensis and Biological Effects of Tumor Promoters*, Raven Press. pp. 175–182.

WHO International Agency For Research On Cancer (1991) Coffee, tea, mate, methylxanthines and methylglyoxal. *IARC Monographs on the Evaluation of carcinogenic Risks to Humans*, **51**, 207–271.

Wu, Y.D., Qiu, Q., Yu, Z.H., Qian, S.S. and Gao, C.M. (1993) Comparison of green tea extracts with other antioxidants on scavenging singlet oxygen. *Basic Medicine and Clinic*, **13**, 63–66 (in Chinese).

Xu, Y., Ho, C.T., Amin, S.G., Han, C., and Chang, F.L. (1992) Inhibition of tabecco-specific nitrosamine-induced lung tumorigenesis in A/J mice by green tea and its major polyphenol as antioxidants. *Cancer Res.*, **52**, 3875–3879.

Xu, K.X., Wang, S., Ma, G.J., Zhou. P., Wu, P.Q., Zhang, R.F. *et al.* (1992) Micronucleus formation in peripheral-blood lymphocytes from smokers and the influence of alcohol-and tea-drinking habits. *Int. J. Cancer*, **50**, 702–705.

Xu, Y. and Han, C. (1990) The effect of Chinese tea on the occurrence of esophasgeal tumors induced by N-nitrosomethylbenzylamine formed *in vivo*. *Biomed. Environ. Sci.*, **3**, 35–42.

Yamane, T., Nakatani, H., Matsumoto, H., Iwata, Y., and Takahashi, T. (1995) Inhibitory effects and toxicity of green tea components for the prevention of gastro-intestinal carcinogenesis. *International Conference on Food Factors: Chemistry and Cancer Prevention*, Japan, Abstracts, pp. 118.

Yin, P.Z., Zhao, J.Y., Cheng, S.J., Hava, Y., Zhu, Q.F., and Liu, Z.G. (1994) Experimental studies of the inhibitory effects of green tea catechin on mice large intestinal cancers induced by 1,2-dimethylhydrazine. *Cancer Lett.*, **79**, 33–38.

Zhao, B.L., Li, X.J., He, R.G., Cheng, S.J., and Xin, W.J. (1989) Scavenging effect of extracts of green tea and natural antioxidants on active oxygen radicals. *Cell Biophys.*, **14**, 175–185.

Zhao, J.Y., zhu, Q.F., Cheng, S.J., Ling, Y. and Fu, S.L. (1992) The antagonistic effects of green tea extract on micronuclei and apoptosis induced by 1,2-dimethylhydrazine (1,2-DMH) in the colon crypt cells of mice. *Acta. Nutrimenta. Sincia.*, **14**, 255–258 (in Chinese).

Zhen, Y.S., Can, S.S., Xue, Y.C. and Wu, S.Y. (1991) Green tea extract inhibits nucleoside transport and potentiates the antitumor effect of antimetabolites. *Chin. Med. Sci. J.*, **6**, 1–5.

11. ANTITUMOR ACTIVITY OF TEA PRODUCTS

YONG-SU ZHEN

Institute of Medicinal Biotechnology, Chinese Academy of Medical Sciences and Peking Union Medical College, Beijing 100050, China

Cancer is a serious disease that causes global concern. As one of the most popular beverages in the world, tea has been found to display a great variety of biological activities. Therefore, the possible effect of tea on cancer has drawn much attention in biomedical research. Epidemiological and experimental investigations have provided substantial evidence that tea may exert a blocking effect on carcinogenesis, the process of cancer development, and tea may be useful for the prevention of cancer. In addition to the blocking effect on carcinogenesis, tea and its components have shown antitumor activity, namely, active against existing cancer cells and the growth of tumors. Tea infusion, tea extracts and some tea components are reported to be active against tumors in experimental studies. Among the tea constituents tea polyphenols have been intensively investigated for antitumor activity. Those include (−)-epigallocatechin gallate (EGCG), (−)-epicatechin gallate (ECG), (−)-epigallocatechin (EGC), and (−)-epicatechin (EC).

1.   CYTOTOXICITY TO CANCER CELLS

There are many reports that tea extract and its components including polyphenol catechins and caffeine show cytotoxicity to cancer cells as observed in *in vitro* assays. Tea products inhibit the proliferation of the cells, reduce cell viability, induce cell apoptosis, and block cell cycle.

1.1   Inhibiting Cell Proliferation

Tea extracts and tea polyphenols inhibit the proliferation of cancer cells. As reported (Yan *et al*. 1989), a preparation of green tea extract inhibited the growth of SGC-7901 cells, a gastric cell line, at 0.5 mg/ml. Over a 20-hour exposure, the increase in cell number was markedly suppressed. The mitosis rate in treated cells was also reduced. Zhen *et al*. (1991) reported that green tea polyphenols inhibited the proliferation of L1210 mouse leukemia cells. The 50% inhibition concentration (IC$_{50}$) was 19 μM. In a comparison study, Fan *et al*. (1992) found that the 3T3 cells were less sensitive to the effect of green tea catechins than HeLa and KB cells, indicating that the difference in sensitivity may be related to the degree of malignancy. Green tea catechins suppress DNA biosynthesis in cancer cells. By radiolabeled thymidine incorporation assay, the inhibition rate by 10 μg/ml of green tea catechins reached 68.7% in HeLa cells. Nishida *et al.* (1994) reported that EGCG, a polyphenolic main constituent of green tea, inhibited the growth and the secretion of alpha-fetoprotein in human hepatoma-derived PLC/PRF/5 cells. The inhibitory effect was dose-dependent over the concentration range of 0.5 μM to 100 μM. AFP secretion reduced to approximately 60% of the control at 0.5 μM of EGCG. Lin (1995) found that tea polyphenols inhibited dose-dependently the growth of human A-431 epidermal carcinoma cells, in which the EGF receptor gene was overexpressed. Tea polyphenols notably inhibited the autophosphorylation of EGF receptor and the phosphorylation of extracellular signal-related kinase.

Valcic *et al.* (1996) compared the effect of caffeine and six green tea catechins including GC, EG, EGC, ECG, EGCG and catechin on four human tumor cell lines by measurement of cell proliferation. In terms of IC_{50} values (concentration for 50% inhibition), EGCG, GC and EGC were more potent than the other three catechins and caffeine. Notably, EGCG was the most potent among the seven green tea components against three out of the four cell lines (i.e. MCF-7 breast cancer, HT-29 colon cancer and UACC-375 melanoma). Chung *et al.* (1998) investigated the influence of green tea extract and its components on the growth of human squamous carcinoma A431, prostatic carcinoma PC3, and DU145 cells. Tea extract suppressed the growth of all three cell lines of which A431 cells are the most sensitive to treatment. All four green tea epicatechins, EGCG, ECG, EGC and EC, inhibited the growth of PC3 and DU145 cells; and the inhibition is dose dependent. Regarding A431 cells, EC stimulated growth while the effect of EGCG, ECG and EGC is biphasic; growth stimulation at low concentration and growth suppression at high concentration. As measuring by [^{3}H]thymidine incorporation assay (Yang *et al.* 1998), EGCG and EGC displayed growth inhibition against lung cancer cell lines H661 and H1299, with IC_{50} values of 22 μM, but were less effective against lung cancer H441 cell line and colon cancer HT-29 cell line with IC_{50} values 2- to 3-fold higher. ECG had lower activity and EC was even less effective. Moreover, preparations of green tea polyphenols and theaflavins had higher activities than extract of green tea and decaffeinated green tea. Stammler and Volm (1997) reported that the growth of S180/dox (doxorubicin-resistant murine sarcoma 180 cells) was reduced at concentrations higher than 10 μg/ml of EGCG and 5 μg/ml of EGC, and completely stopped at concentrations higher than 50 μg/ml and 40 μg/ml, respectively. There is a difference between the EGCG growth inhibitory effect on cancer cells and that on normal cells. In a comparison study (Chen *et al.* 1998), the inhibitory effect of EGCG on the growth of SV40 virally transformed WI38 human fibroblasts (WI38VA) was stronger than that of normal WI38 cells, the IC_{50} values being 10 μM and 120 μM, respectively. Similar differential growth inhibition was also observed between the human colorectal cancer cell line (Caco-2) and breast cancer cell line (Hs578T) and their respective normal counterparts. In addition, EGCG did not affect the serum-induced expression of *c-fos* and *c-myc* genes in normal WI38 cells; however, EGCG significantly enhanced their expression in transformed WI38VA cells. It is suggested that differential modulation by EGCG of certain genes, such as *c-fos* and *c-myc*, may cause differential effects on the growth and death of cancer cells.

1.2. Decreasing Cell Survival

The viability of tumor cells and the surviving fraction of tumor cell population can be determined by clonogenic assay. Zhen *et al.* (1991) reported that green tea polyphenols inhibited colony formation of human hepatoma BEL-7402 cells with an IC_{50} value of 21 μM. For comparison, MTX, an antitumor antimetabolite, at the concentration of 0.02 μM completely killed hepatoma BEL-7402 cells with a 100% colony inhibition rate. Also examined by clonogenic assay (Fan *et al.* 1992), IC_{50} values of green tea polyphenols for HeLa, KB and 3T3 cells were 3.68 μg/ml, 5.36 μg/ml and 11.8 μg/ml, respectively. Examination by microscope revealed cell colonies treated with green tea polyphenols were smaller and the cell density was much lower compared with those of untreated control.

The inhibitory effects of different tea catechins against HeLa cells vary remarkably in clonogenic assay (Han, 1997). At a concentration of 50 μg/ml, the colony inhibition rates by EGCG, ECG, EGC and EC were 2.8%, 26.5%, 38.7% and 6.3%, respectively. Green tea catechins significantly inhibit the growth ability of HeLa cells in soft agar. At 100 μg/ml, the inhibition rates by EGCG, ECG, EGC and EC reached 71.6%, 99.9%, 100% and 68.6%, respectively.

Recently, telomerase activity has been considered an important indicator that differentiates cancer cells from normal cells. Telomerase is an enzyme that adds repeated telomere sequences to the ends of chromosomes. It helps maintain both the telomere length and infinite proliferation of cancer cells. Interestingly, Naasani *et al.* (1998) have found that EGCG inhibits telomerase in a cell-free system (cell extract) as well as in living cells. In the presence of non-toxic concentrations of EGCG, the continued growth of two human cancer cell lines, U937 monoblastoid leukemia cells and HT-29 colon cancer cells, showed life span limitations accompanied with telomere shortening, chromosomal abnormalities, and expression of the senescence-associated beta-galactosidase.

1.3. Inducing Programmed Cell Death

In a meeting report by Boone and Wattenburg (1994), EGCG was found to induce apoptosis (programmed cell death) in six leukemia cell lines, as shown by DNA laddering of multiples of an approximately 200-base pair subunit. Hibasami *et al.* (1996) reported that the exposure of human lymphoid leukemia Molt 4B cells to EGC or EGCG led to both growth inhibition and induction of apoptosis in a manner of concentration- and time-dependent. Zhao *et al.* (1997) reported the apoptosis effect of tea polyphenols on a human promyelocytic leukemia line, HL-60 cells. When HL-60 cells were exposed to 250 μg/ml of tea polyphenols for 5 hours, DNA ladder was found in agarose gel electrophoresis and apoptotic bodies were observed by transmission electronic microscopy. The changes were obvious at the concentrations of 500 μg/ml and 1000 μg/ml; however, DNA laddering was not shown when the concentration raised to 2000 μg/ml. Induction of apoptosis by ECG, EGC, or EGCG was also observed in human lymphoid leukemia 4B cells (Achiwa *et al.* 1997).

Using a polyphenolic fraction of green tea (GTP), Mukhtar *et al.* (1997) found that treatment of human epidermoid carcinoma A431 cells with GTP (80 μg/ml for 48 h) resulted in formation of inter-nucleosomal fragments. Apoptosis was evident even 10 h after treatment of cells at 20 μg/ml of GTP. Among the four polyphenols present in GTP, EGCG and ECG are more effective than EGC whereas EC is ineffective. Tan *et al.* (1998) observed the effects of green tea aqueous extract on the human colon cancer LoVo cell line. The growth of LoVo cells was inhibited by 67.6% at a concentration of 5 mg/ml with an IC_{50} of 3.03 mg/ml. Induction of apoptosis was found at concentrations of 3–5 mg/ml. Paschka *et al.* (1998) tested the inhibitory effect of green tea components on prostate cancer cell lines LNCaP, PC-3 and DU145. EGCG proved to be the most potent polyphenol in terms of inhibiting cell growth and the inhibition was found to occur via apoptotic cell death. Yang *et al.* (1998) reported that the exposure of lung cancer line H661 cells to EGCG, EGC or theaflavins for 24 hours led to apoptosis. With 30 μM of EGCG, EGC or theaflavins, the apoptosis index was 23, 26 or 8%, respectively; and with 100 μM of these compounds, the apoptosis index was 82, 76 or 78%, respectively. In a study on the effect of green tea catechin extract and its main component EGCG on human stomach cancer KATO III cells, Hibasami *et al.* (1998a) reported that the induction of apoptosis is concentration- and time-dependent. In addition, the black tea theaflavin extract also induced apoptosis concentration- and time-dependently (Hibasami *et al.* 1998b).

Suganuma *et al.* (1998) reported the synergistic effects of EC and EGCG on induction of apoptosis in the human lung cancer cells PC-9. Both EGCG and ECG induced apoptosis of PC-9 cells but EC did not. When examined by ^{3}H-EGCG assay, the incorporation of ^{3}H-EGCG was inhibited by unlabeled EGCG, ECG or EGC dose-dependently; however, EC enhanced ^{3}H-EGCG incorporation. Notably, the combination of EGCG with EC synergistically induced apoptosis of PC-9 cells. Chen *et al.* (1998) have observed that there was some difference between the effects of EGCG on SV40 virally transformed WI38 human fibroblasts (WI38VA) and that on normal WI38 cells. GCG at a concentration range of 40–200 μM induced a significant amount of apoptosis in WI38VA cells, but not in WI38 cells. As shown, EGCG caused a dose-dependent formation of H_2O_2 in cells. The addition of H_2O_2 to cells induced apoptosis similar to that induced by EGCG; and the EGCG-induced apoptosis was completely inhibited by adding exogenous catalase. The mechanistic study indicates that the production of H_2O_2 may mediate tea polyphenol-induced apoptosis.

The induction of apoptosis also occurs *in vivo*. In a UVB-induction tumor model of mice (Lu *et al.* 1997), oral administration of green or black tea increased the apoptosis indices in nonmalignant neoplasms by 46% and 35%, and those in malignant neoplasms by 61% and 51%, respectively.

1.4. Blocking Cell Cycle Progression

Tea and tea products may block the cell cycle progression and change the phase distribution of the cell cycle. Using green tea extract, Yan *et al.* (1990) found that the transition from G_1 to S was blocked, showing a significant increase of G_1 and a decrease of S phase. Suganuma *et al.* (1995) reported that Kato III gastric cancer cells treated with EGCG showed an increase of the cells in G2/M phase from 8.2% to 21.6%. Han reported (1997) that the cell cycle phase distribution of Ehrlich carcinoma cells was changed by exposure to green tea polyphenols. In untreated Ehrlich carcinoma cells, G_1 and S phases consisted of 61% and 27% respectively. At 50 μg/ml of EGCG, G_1 phase increased to 81% and S phase decreased to 7%. At 10 μg/ml of ECG, G_1 reached 83% and S phase was down to 5%. In comparison, the inhibitory effect of EGCG and ECG on Ehrlich carcinoma cells was stronger than that on mouse bone marrow cells.

Okabe et al. (1997) investigated the mechanism of growth inhibition of cancer cells by tea polyphenols. The study found that ECG, EGCG and EGC inhibited the growth of PC-9 cells, a human lung cancer cell line, with IC_{50} values of 78 μM, 140 μM, and 275 μM, respectively; whereas EC did not show significant growth inhibition. Flow cytometric analysis revealed that treatment with 50 μM and 100 μM EGCG increased the percentages of cells in the G_2-M phase from 13.8% to 25.6% and 24.1%, respectively. EGCG at 100 μM weakly decreased the mitotic index from 11.16% to 7.25%. EGCG induced the G_2-M arrest in human stomach cancer cells at concentrations of 60 to 100 μM; however, EGC, EC and ECG did not show significant G_2-M arrest, indicating that each tea polyphenol may induce a different pattern of G_1, S and G_2-M phase cells. The study of the intracellular localization of [^{3}H]EGCG by microautoradiography has found that [^{3}H]EGCG was incorporated into cytosol and nuclei, as well as being present in the membrane. Notably, the incorporation of [^{3}H]EGCG was inhibited by excess amounts of unlabeled EGCG as well as ECG.

2. EFFECTS ON BENIGN TUMORS AND RELATED LESIONS

2.1. Skin Papilloma

Skin tumorigenesis has been widely used as model system for the study of tumor prevention. The formation of a papilloma, a benign tumor in nature, is a critical stage in tumorigenesis. In further development a papilloma may undergo malignant conversion. Prior to papilloma formation, a series of related skin lesions exists. In this system the size of papillomas can be used to evaluate the inhibitory effect of tea and its constituents on tumor growth. Wang et al. (1992a) observed the effect of green tea on ultraviolet induced skin tumorigenesis. A preparation of 1.25% green tea extract (1.25 g of tea leaves in 100 ml of water) was used for drinking water for mice. Skin tumorigenesis was induced by ultraviolet light, UVB, in mice previously initiated with DMBA, 7,12-dimethylbenz(a)anthracene. The results showed that administration of green tea not only decreased the number of induced tumors in mice, but tumor size was also markedly decreased. In three separate experiments, 1.25% green tea extract as the sole source of drinking water decreased tumor size by 75%, 77% and 84%, respectively, indicating inhibition of tumor growth. In a study of the inhibitory effect of green tea on the growth of skin papillomas, Wang et al. (1992b) reported that both oral administration of tea extract and intraperitoneal administration of EGCG suppressed the growth of skin papillomas. Partial tumor regression or >90% inhibition of tumor growth occurred in 5 experiments and marked inhibition of tumor growth (46–89%) in 5 additional experiments. Complete tumor regression occurred in 14 of 346 papillomas in mice of the treated group, but in none of the 220 papillomas in control mice.

Katiyar et al. (1992) investigated the effect of green tea polyphenols on skin tumorigenesis in mice which was initiated by DMBA and promoted by TPA. The green tea preparation was applied topically. The animals pretreated with green tea polyphenols showed substantially lower tumor burden, as compared with the animals without pretreatment. In terms of average volume, tumors of the treated group were much smaller than that of untreated controls. Studying the effect of tea products on skin carcinogenesis induced by ultraviolet B light, Wang et al. found (1994) that tea in drinking fluid showed an inhibitory effect on the growth of skin tumors in mice. In addition to reducing the number of mice with tumors and the number of tumors per mouse, the volume per tumor and tumor volume per mouse were also decreased. Black tea, green tea, decaffeinated black tea, and decaffeinated green tea in drinking fluid were all effective. The administration of drinking fluid containing 1.25% tea may inhibit the tumor volume per mouse by over 90%.

Investigation with a model of UVB light-induced skin carcinogenesis, Huang et al. (1997) found that caffeine, a constituent of tea, has an inhibitory effect. Oral administration of caffeine alone had a substantial inhibitory effect on UVB-induced carcinogenesis. According to the report, the tumor volume per mouse in 0.036% caffeine-treated group and water control group was 82 mm^3 and 351 mm^3, respectively. Decaffeinated black tea or green tea was less effective; however, adding caffeine to decaffeinated teas restored the inhibitory effects of these teas. Topical application of a green tea polyphenol fraction after each UVB exposure also inhibited UVB-induced carcinogenesis. Tumor volume per mouse was 123 mm^3 in the treated group and 712 mm^3 in the control group. In their study on protection against induction of mouse skin papillomas, Katiyar et al. found that topical application of a green tea polyphenolic fraction resulted in protection against carcinogenesis. The application of green tea polyphenol reduced tumor size by up to 90%; in terms of tumor volume per mouse, the reduction reached 74%. As reported by Liu et al. (1997), oral administration of green tea or black tea (4–9 mg tea solids/ml) to UVB-treated mice inhibited the formation and size of nonmalignant tumors including papillomas and keratoacanthomas. In one of these studies in which green tea or black tea (6 mg tea solids/ml) was administered for 20 weeks, the volume of nonmalignant tumors per mouse was decreased by 69 and 86%, respectively. In a study of esophageal tumorigenesis induced by N-nitrosomethylbenzy-lamine in rats, Yang et al. (1992) found that the administration of 0.6% decaffeinated green tea or black tea extracts as the sole source of drinking water not only decreased the papilloma multiplicity but also reduced the size of esophageal papillomas.

2.2. Lesions Related to Carcinogenesis

In the process of carcinogenesis, a number of pathological lesions exist prior to the formation of tumors, benign or malignant. It is of interest to evaluate the effects of tea and tea products on the development of these preexisting changes during carcinogenesis. In a skin carcinogenesis model in mice initiated by DMBA and promoted by TPA, Katiyar *et al.* (1992) found that prior application of green tea polyphenols to mouse skin inhibited the TPA-induced skin edema and epidermal hyperplasia by 30–40%. By histopathological examination, the application of TPA resulted in a significant increase in mean epidermal thickness (74 μm) and mean vertical thickness of epidermal cell layers (7.6). In solvent-treated animals, the corresponding values were 2.5 μm and 2.8, respectively. Notably, the pre-application of green tea polyphenols significantly suppressed TPA-induced changes, reducing those parameters to 42 μm and 4.4, respectively. TPA also caused mixed cell inflammation reaction in dermis, that was comprised of mostly neutrophils with some mononuclear cells, and these changes in dermis were also inhibited by pre-application of green tea polyphenols.

During skin tumorigenesis induced by ultraviolet light and TPA, skin lesions exist and the formation of these skin lesions was inhibited in a dose-dependent manner by oral administration of green tea in drinking water for 2 weeks prior to and during application of UVB (Wang *et al.* 1992a). This inhibitory effect was characterized by a decreased area and intensity of red color of the lesions in green tea-treated mice. Treatment with green tea extract did not change the body weight of mice. It is of interest that tea or tea products may provide protection against photodamage of skin. Topical application of 8-methoxypsoralen followed by irradiation with ultraviolet A light in mice can caused skin photodamage that includes erythema and edema after 2–3 days and hyperplasia/hyperkeratosis after 1–2 weeks; evidently, the skin was protected from photodamage by topical application of green tea polyphenol fraction (Zhao *et al.* 1997). As compared with the photodamage control group, topical application of 0.4 mg/cm^2 of a green tea polyphenol fraction resulted in 78% ($P <$ 0.001) and 98.6% ($P <$ 0.001) inhibition of the net increase in dorsal skin thickness and lesion severity index, respectively.

Zhao *et al.* also reported (1998) the photoprotection effect in humans. In human volunteers, pretreatment of the skin with 0.2 mg/cm^2 of standardized green tea extracts (SGTE) and standardized black tea extracts (SBTE) 30 min prior to ultraviolet B light irradiation resulted in dose-dependent protection against acute erythema formation. In addition, topical application of SBTE 5 min after UBV irradiation reduced UBV-induced inflammation. Mitchell and Liebler (1998) have investigated the contribution of DNA photoprotection by EGCG. EGCG dispersed in a hydrophilic cream vehicle was topically applied to dorsal epidermis of mice and then the animals were exposed to UVB for 60 min. Thymine dimer levels in DNA were analyzed immediately after exposure. They found that topical application of EGCG inhibited dimer formation in a dose dependent manner and significant inhibition was seen in a 5% EGCG dispersion, which inhibited thymine dimer formation by 34% as compared with vehicle controls.

Tea and tea products have been shown to suppress the proliferation of bronchial epithelial cells during carcinogenesis. As reported by Kim *et al.* (1996), in a model of NNK-induced lung carcinogenesis in mice, hyperproliferation of bronchiolar epithelia was observed from day 3 to day 7 after NNK treatment. At day 14, proliferation of bronchiolar cells returned to basal level. Administration of decaffeinated green tea (0.6% as the sole source of drinking fluid) starting on day 1 decreased the hyperproliferation by 50% for the period of day 3 to day 7. The researchers pointed out that this activity may be important for the chemopreventive action of tea against lung carcinogenesis. In a further study, Kim *et al.* (1997) found that green tea polyphenol preparation (GTPP) and its major constituent EGCG were effective in the inhibition of hyperproliferation of bronchial epithelial cells. In an experiment, 0.3% GTPP and 0.1% EGCG as the sole source of drinking fluid were shown to reduce the NNK-induced hyperproliferation of bronchial epithelial cells, respectively, by 44% and 34%. Yang *et al.* (1997) reported the effects of black tea theaflavins. Theaflavins (0.1%) were administered to mice through drinking water, starting 24 hours after using a single dose of NNK, and the proliferation index in the bronchiolar epithelial cells was measured. The NNK-induced highest proliferation rate of bronchiolar cells, observed on day 5, was inhibited by 30% by treatment with theaflavins.

Both green and black teas have been shown to inhibit lung carcinogenesis in laboratory animals and the antiproliferative effects of tea may be responsible for these anti-carcinogenic actions (Yang *et al.* 1998). Liu *et al.* (1997) reported that hyperproliferation of gastric glandular epithelial cells can be suppressed by administration of decaffeinated green tea. A model with high salt-induced gastric epithelial injury and hyperproliferation in rats was used. Cellular injury and hyperproliferation were found after intragastric intubation of NaCl (1 g/kg). The hyperproliferation was characterized by a 5-fold increase in the bromodeoxyuridine labeled index. In rats treated with 0.6% decaffeinated green tea as the sole drinking fluid for two weeks before NaCl administration, the induced proliferation of fundic epithelial cells was inhibited by 57% and 31% in two experiments. As suggested by the authors, the anti-proliferative effect of tea may play a role in the inhibition of gastric carcinogenesis.

2.3. Malignant Conversion

Carcinogenesis in mouse skin, and possibly in other tissues, is a multistage process composed of initiation, promotion, and progression. Among these stages, the tumor progression stage in which nonmalignant tumors convert into malignant ones is of greatest concern since malignant tumors are capable of metastasis, eventually causing death. Tea and tea products have been found to show an inhibitory effect to some degree on the growth of benign tumors, e.g. papillomas. It is interesting to examine the effect of tea on malignant conversion. Katiyar *et al.* (1993) used a skin carcinogenesis model in mice to examine the effect of green tea on malignant conversion of chemically induced benign skin papillomas to squamous cell carcinomas. Benign skin papillomas were induced by DMBA and promoted by TPA. Enhanced malignant conversion of the papillomas to carcinomas was achieved by topical application of benzoyl peroxide (BPO) or 4-nitroquinoline-N-oxide (4-NQO), whereas spontaneous malignant conversion was associated with topical application of acetone. A polyphenolic fraction isolated from green tea (6 mg/mouse) was topically applied 30 min prior to skin application of BPO, 4-NQO or acetone. As reported, the application of green tea polyphenolic fraction resulted in reducing BPO- and 4-NQO-enhanced malignant conversion, showing 31%

and 29% protection, respectively, in terms of the percentage of mice with carcinomas, and 35% and 43% protection, respectively, in terms of the number of carcinomas per mouse. The BPO- and 4-NQO-enhanced rate of malignant conversion was found to be decreased significantly by topical application of green tea polyphenolic fraction; however, such effects were less profound in the case of spontaneous malignant conversion. In a study on skin carcinogenesis in DMBA plus TPA-treated mice, Katiyar et al. (1997) reported that in three separate experiments topical application of GTP, a polyphenolic fraction isolated from green tea, resulted in significant protection against the malignant conversion of papillomas. These protective effects were evident in terms of mice with carcinomas (35–41%), carcinomas per mouse (47–55%) and percent malignant conversion of papillomas to carcinomas (47–58%). As pointed out by Yang et al. (1998), black tea polyphenol preparations decreased NNK-induced hyperproliferation and also inhibited the progression of pulmonary adenomas to adenocarcinomas in mice.

3. EFFECTS ON MALIGNANT TUMORS

There are many reports on experimental studies of the effects of tea and its products on the growth of tumors in mice or in rats. Most of the model systems used in the studies were transplantable tumors. In addition, some investigations have been performed on spontaneous tumors or chemically induced tumors. As in the case of studying antitumor drugs, the criteria for evaluation of therapeutic efficacy include inhibition of tumor growth, increase of life span, and suppression of metastases.

3.1. Effect on Transplanted Tumors

As reported by Ye et al. (1984), tea infusion was effective against Ehrlich carcinoma in mice. Ehrlich ascites carcinoma cells were inoculated subcutaneously in mice and 5% green tea infusion was administered intragastrically, twice daily for 8 days. Comparing tumor weights, the growth of the tumor was inhibited by 41.4%. In an experiment comparing surviving time of the animals, the life span was increased by 55%. The preliminary study showed that green tea may exert an inhibitory effect on transplanted tumor in mice. Oguni et al. (1988) reported the antitumor activity of green tea extract in which caffeine and dyes had been eliminated. Sarcoma 180 cells were transplanted subcutaneously in mice and the green tea extract was administered orally once a day for 4 days. Tumor weight was compared with that of the control on day 21 after tumor transplantation. The growth of sarcoma 180 in mice was suppressed by 50% and 59.8%, respectively, at dosage levels of 400 mg/kg/day and 800 mg/kg/day. Hara et al. (1989) reported the inhibitory effect of green tea catechin on the growth of transplanted tumor in mice. Yan et al. (1990) reported that the growth of Ehrlich ascites carcinoma, ascites hepatoma and sarcoma 180 in mice was suppressed by oral administration of green tea extract at a dose of 50 mg/kg, resulting in inhibition rates of 45%, 57% and 55%, respectively.

As reported by Yang et al. (1991), the growth of Ehrlich carcinoma (solid tumor) and sarcoma 180 in mice was suppressed by the administration of tea polyphenols.

Dose levels of 40 mg/kg, 80 mg/kg, and 120 mg/kg inhibited the growth of Ehrlich carcinoma by 66.8%, 50.0%, and 48.6%, respectively. It seems that tea polyphenols did exert an inhibitory effect on tumor growth; however, the effectiveness was not quite correlated with the doses. Yan et al. (1992) observed the inhibitory effect of various dose levels of green tea extract on the growth of transplanted tumors in mice. As reported, intraperitoneal administration of green tea extract at dosage levels of 10, 50, and 100 mg/kg resulted in growth inhibition of sarcoma 180 by 31%, 54%, and 42%, respectively. Evidently, more marked inhibition was found at the dose of 50 mg/kg; however, increasing the dose to 100 mg/kg resulted in a lower inhibition rate. Similar patterns of the dose effect relationship were found in experiments with Ehrlich carcinoma and mouse hepatoma. From the above-mentioned results, it appears that green tea extract exhibits a moderate inhibitory effect on the growth of transplanted tumors; however, the inhibition rate could not go much higher even though the dose was increased. As reported by Taniguchi et al. (1992), oral administration of 0.1% green tea catechins was found to reduce the growth of B16 melanoma transplanted into the footpads of mice. So far, most of the reports indicated that green tea extract or its catechins have moderate inhibitory effect on the growth of transplanted tumors in animals. On the other hand, Sazuka et al. (1995) reported that a green tea infusion did not prove effective for inhibiting the growth of local tumor after subcutaneously transplanted Lewis lung carcinoma cells.

The effect of tea or tea products has been tested in nude mice transplanted with human cancer. Liao et al. (1995) reported the growth inhibitory effect of EGCG on human prostate and breast cancers in nude mice. Cells from human prostate cancer cell lines, PC-3 or NCaP 104-R, were inoculated subcutaneously into nude mice. Intraperitoneal injection of green tea EGCG (50 mg/kg) inhibited tumor growth and reduced the tumor size rapidly. The inhibition of PC-3 tumor growth seemed to be EGCG specific; notably, PC-3 tumor growth was not inhibited by intraperitoneal injection of EC, EGC, or EGC. In a comparison of the structures of these compounds, the galloyl group of EGCG appears to be necessary since EGC is inactive; and it is striking that the addition of one hydroxyl group to the dihydroxybenzene makes the compound (EGCG) active. As reported, the growth of tumor after inoculation with human mammary cancer MCF-7 cells was also inhibited by intraperitoneal injection of EGCG. Gotoh et al. (1998) reported that the development of human prostatic carcinoma transplanted in nude mice was inhibited by green tea extract. PC-3 cells, a human prostatic cancer cell line, were inoculated orthotopically into the prostate of BALB/c nude mice. Green tea extract was administered for 4 weeks, namely, 2 weeks prior to and 2 weeks after the inoculation. As a result, prostatic cancer was found in 40% in green tea extract-treated group, and in 80% in the water control group, respectively.

3.2. Effect on Induced Tumors

There are many reports on the effectiveness of tea or tea products in the prevention of cancer using experimental models. In those investigations the number of tumors is reduced in carcinogenesis by administration of tea or tea products. This reduction of tumor number may be related to effects on the early stage of carcinogenesis. It is often not clear what effect tea and tea products have on the development of well established malignant tumors. Therefore, it is of interest to evaluate the effect of tea on induced malignant tumors. Wang *et al.* (1994) reported that tea preparations were found to decrease the number of carcinomas in skin carcinogenesis in mice. The experimental model used was ultraviolet B light-induced carcinogenesis in DMBA-initiated mice. Tea preparations in water were used as the sole source of drinking fluid. As reported, the administration of 1.25% of black or green tea reduced the percentage of mice with carcinomas by 88 or 79%, respectively, and the number of carcinomas per mouse reduced by 88 or 79%, respectively. Administration of 1.25% of decaffeinated black or decaffeinated green tea reduced the percentage of mice with carcinomas by 9 or 49%, and the number of carcinomas per mouse by 77 or 72%, respectively. Huang *et al.* (1997) have studied the effects of tea, decaffeinated tea, and caffeine on UVB light-induced skin carcinogenesis. Oral administration of 0.9% lyophilized green or black tea decreased the average number of carcinomas (squamous cell carcinomas and carcinomas *in situ*) by 80 and 41%, respectively. Oral administration of caffeine alone also exerted a substantial inhibitory effect on UVB-induced carcinogenesis. The experimental results indicate that green tea or decaffeinated green tea plus caffeine is the most effective regimen for decreasing the percent of mice with carcinomas.

Hirose *et al.* (1991) investigated the effect of green tea catechins on DMBA-induced rat mammary carcinogenesis. Dietary intake of 0.1% green tea catechins did not have a statistically significant effect on the number of mammary tumors including adenocarcinomas and fibroadenomas, but the survival time of the green tea catechins-treated animals was significantly longer than that of the controls. In the experiment, mortality due to hemorrhage from large tumors was significantly lower in the catechins-treated group. Yan *et al.* (1993) reported that oral administration of green tea extract inhibited the growth of chemically induced gastric carcinomas, which were identified as highly differentiated adenocarcinoma in rats. In addition to the reduction of tumor incidence, the size of the tumors was inhibited by 88% in animals treated with green tea extract. In a study of chemically induced esophageal carcinogenesis, Wang *et al.* (1995) found that oral administration of 0.9% regular green tea and decaffeinated green tea significantly decreased the esophageal tumor volumes by 57% and 35%, respectively. In an experiment in which the esophageal specimens were examined histopathologically, 40% of the rats developed carcinomas; and oral administration of regular or decaffeinated green tea significantly decreased the carcinoma incidence and multiplicity. Yang *et al.* (1997) investigated the effect of theaflavins, constituents of black tea, on 4-(methylnitrosamino)-1-(3-pyridyl)-buta-none (NNK)-induced lung carcinogenesis. As reported, in the group which received theaflavins (0.1%) through drinking water, the tumor multiplicity and tumor volume were significantly lowered, by 23.5% and 33.6% respectively.

3.3. Effect on Invasion and Metastasis

Invasion and metastasis are the subjects of wide concern in cancer therapy. It is of interest to investigate the activity of tea and tea products against invasion and metastasis. Bracke *et al.* (1984) reported that (+)-catechin inhibited the invasion of MO4 fibrosarcoma cells into heart fragments of the chick embryo. The inhibition reached the maximum at 500 μM of (+)-catechin. Pretreatment of the host tissue with (+)-catechin appeared to be a prerequisite for the inhibitory action. The researchers hypothesized that the collagen-stabilizing effect of (+)-catechin is the key to the explanation of its anti-invasion properties. Isemura *et al.* (1995) found that green tea infusion showed an inhibitory effect on the *in vitro* invasion and *in vivo* metastasis of mouse lung carcinoma cells. Bracke *et al.* (1987) investigated the effect of (+)-catechin on the cell-matrix interaction in two cell types, MO4 and M5076 cells. As shown, MO4 fetal transformed mouse cells adhered and spread when cultured on laminin-coated coverslips or on amnion basement membrane, while M5076 mouse reticulum sarcoma cells adhered poorly to these substrates. Pretreatment of these substrates by (+)-catechin abrogated the effect of laminin on cell morphology and adhesion. Consequently, (+)-catechin inhibited the invasion of MO4 cells but not of M5076 cells into embryonic chick heart *in vitro*. It was suggested that the anti-invasion activity of (+)-catechin is due to its interference with the adhesion of MO4 cells to laminin.

Taniguchi *et al.* (1992) observed that oral administration of a green tea preparation containing EGCG (85%), EC (10%), and ECG (5%) inhibited lung metastasis of B16 melanoma and Lewis lung carcinoma in mice. B16-F10 melanoma cells were injected into the tail vein of mice to form artificial metastasis. They found that the average number of lung metastatic nodules reached >300 in untreated controls; however, administration of green tea preparation, 0.05% and 0.1% EGCG in drinking water, decreased the number of lung metastatic nodules to 107 and 76, respectively. In an experiment on spontaneous metastasis, in which melanoma cells were inoculated into the foot pad and then the tumor-bearing leg was amputated 9 days after inoculation, the average number of lung metastatic nodules was also significantly reduced by oral administration of EGCG solution. Sazuka *et al.* (1995) reported the inhibitory effects of green tea components on the invasion and metastasis of cancer cells. The percentage of cells that penetrated into the lower side of the filter was significantly reduced by EGCG and ECG but not by catechin at the concentration of 50 μM as examined by Matrigel invasion assay. In another experiment on spontaneous metastasis, oral administration of green tea infusion reduced the number of pulmonary metastatic nodules of Lewis lung carcinoma in mice.

4. MODULATING EFFECT

In recent years, the possible application of biochemical modulators in cancer chemotherapy has become a highly attractive topic. Generally speaking, biochemical modulator acts on a relevant molecular target or a key step in a vital biological process to modulate the effect of other drugs. In the field of cancer therapy, the modulating effect of a biochemical modulator may include the augmentation of antitumor effects and the reduction of toxic effects of chemotherapeutic drugs. Recent studies have shown that tea may act as a biochemical modulator when used in combination with chemotherapeutic drugs.

4.1. Augmentation of Antitumor Effect

Zhen *et al.* (1991) found that green tea polyphenolic extract (GTE) inhibited nucleoside transport and potentiated the antitumor effect of antimetabolites. As determined by nucleoside transport assay, GTE inhibited radiolabeled thymidine and uridine transport into leukemia L1210 cells with IC_{50} values of 3.2 μM and 8.0 μM, respectively. In a clonogenic assay, GTE blocked the rescue effect of exogenous nucleoside and enhanced the cytotoxicity of cytarabine and methotrexate against leukemia L1210 cells and human hepatoma BEL-7402 cells. In animal experiments, GTE markedly potentiated the inhibitory effect of cytarabine against mouse leukemias L1210 and P388. As shown, cytarabine or GTE alone at the tested dose level showed no antitumor effect; however, the combination of cytarabine and GTE increased the survival time of mice by 124% for L1210 and by 133% for P388, respectively. Zhen *et al.* (1983, 1992) previously reported that dipyridamole, a nucleoside transport inhibitor, can be used to block nucleoside salvage and potentiate the effect of antimetabolites. A combination of a nucleoside transport inhibitor with an antimetabolite such as acivicin, 5-fluorouracil and methotrexate is potentially useful in cancer chemotherapy. The strategy of using biochemical modulators from nucleoside transport inhibitors has been applied to new drug discovery. Zhen *et al.* (1994, 1996) and Su *et al.* (1994, 1995) found that several nucleoside transport inhibitors including green tea polyphenol extract, salvianolic acid A and some antibiotics may act as biochemical modulators which can augment the antitumor effect of chemotherapeutic drugs and reverse multidrug resistance in cancer cells.

Sadzuka *et al.* (1998) investigated the effect of green tea and tea components on the antitumor activity of doxorubicin (adriamycin). Oral administration of green tea enhanced 2.5-fold the inhibitory effects of doxorubicin on the growth of Ehrlich ascites carcinoma in mice. Furthermore, the potentiation of doxorubicin antitumor activity by green tea was observed in M5076 ovarian sarcoma, which has low sensitivity to doxorubicin.

In addition to green tea polyphenols, theanine may also act as a biochemical modulator. As previously determined (Graham, 1992), theanine, an amino acid specifically found in green tea, is contained in 1–2% of the dry weight of tea leaves and is one of the taste components. Sadzuka *et al.* (1996) reported the effects of theanine on the antitumor activity of adriamycin from the view point of biochemical modulation. In an *in vitro* assay, theanine at 100 nM inhibited adriamycin efflux from Ehrlich carcinoma cells by 30.5% and maintained a higher concentration of adriamycin in carcinoma cells. In animal experiments, Ehrlich ascites carcinoma cells were inoculated subcutaneously in mice and adriamycin or theanine was given through intraperitoneal injection. Adriamycin (2 mg/kg, ×4) alone and the combination of adriamycin and theanine (10 mg/kg) reduced the tumor weight by 25% and 54%, respectively. The result shows that theanine enhances the inhibitory effect of adriamycin on tumor growth. Sugiyama *et al.* (1997) reported the enhancing effect of theanine on the inhibition of metastasis by adriamycin. In M5076 ovarian sarcoma–bearing mice, theanine increased adriamycin concentration in the tumor and significantly enhanced the inhibitory effect of adriamycin on tumor growth compared with adriamycin alone. The combination of theanine with adriamycin inhibited liver metastasis of M5076 cells. Sugiyama *et al.* (1998) further reported the enhancing effect of theanine on antitumor activity of adriamycin in P388 leukemia bearing mice.

Caffeine also displays a modulating effect on the action of antitumor drugs. Sadzuka *et al.* (1993, 1995) reported that caffeine enhanced the antitumor activity of adriamycin against Ehrlich carcinoma in mice by increasing the concentration of adriamycin in the tumor.

4.2. Modulation of Multidrug Resistance

In cancer chemotherapy, the development of drug resistance is considered to be the most significant obstacle to curing cancer. The multidrug resistance phenomenon is of importance because it may play a significant role in clinical resistance to certain antitumor drugs. As is well known, tumor cells selected for resistance to a particular drug of natural origin such as adriamycin and vincristine display cross-resistance to these and other agents of unrelated chemical structure. Therefore, those agents which can modulate and reverse multidrug resistance may be useful in cancer chemotherapy. As reported by Stammler and Volm (1997), green tea polyphenols EGCG and EGC have modulating effects on the activity of doxorubicin (adriamycin) in drug-resistant cells. In their experiments, two doxorubicin-resistant cell lines, mouse sarcoma 180/dox and human colon carcinoma SW620/dox, were used for growth inhibition assays. In S180/dox cells, the number of cells treated by doxorubicin alone, EGCG alone, and doxorubicin plus EGCG was 99%, 56%, and 34% of the control, respectively; and the number of cells treated by doxorubicin, EGC, and doxorubicin plus EGC was 84%, 44%, and 10%, respectively. Similar results were obtained with SW620/dox cells. Evidently, both EGCG and EGC showed a sensitizing effect on the doxorubicin-resistance cell lines. This effect may be related to the inhibition of protein kinase C by EGCG and EGC, leading to reduced expression of some drug resistance related proteins. Theanine may also exert a modulating effect on drug resistance in cancer cells. Sugiyama and Sadzuka (1998) investigated the action of theanine on sensitive P388 leukemia and resistant P388/ADR leukemia that overexpressed P-glycoprotein. For P388 leukemia bearing mice, adriamycin reduced the tumor growth and theanine enhanced the antitumor activity of adriamycin. For P388/ADR leukemia–bearing mice, adriamycin failed to inhibit the tumor growth; however, the combination of theanine and adriamycin significantly reduced the P388/ADR tumor growth. In addition, theanine was found to inhibit the efflux of adriamycin from P388/ADR cells.

4.3. Reduction of Toxic Effects

There are reports that tea and tea products may reduce the toxic effects of exogenous agents. Sugiyama and Ueda (1991) found protective effects of tea on toxic side effects of cisplatin. As reported, cisplatin (3 mg/kg, ×9) was given intraperitoneally and tea or tea constituents were administered orally to mice. The cisplatin-treated animals showed a significant increase of BUN, serum creatinine, SGOT and SGPT, and showed a significant decrease in kidney weight, urinary volume, WBC and platelet counts, and body weight. These indicate the toxic effects of cisplatin. However, treatment with tea extract (600 mg/kg) or ECG (50 mg/kg), reduced these toxic effects. Moreover, treatment with tea extract or ECG did not reduce the antitumor activity of cisplatin against sarcoma 180 in mice. Sugiyama *et al.* (1997) found that theanine reduces adriamycin concentration in normal tissue of mice, while it increases the adriamycin concentration in the tumor. It is suggested that theanine might reduce the side effects of adriamycin.

5. CONCLUSION

Tea consumption is highly popular around the world and has a long history. Biomedical research has shown that tea consists of highly diverse constituents and displays a great variety of bioactivities. There is evidence that tea may be beneficial to human health and potentially useful for prevention and treatment of various diseases.

Cancer is a severe disease of global concern. The topic of antitumor activity of tea and tea products has drawn very much attention, particularly in the last two decades. Tea extract and its separate components have been found to be active, to a certain degree, against cancer. Tea extract and certain components inhibit the proliferation of cancer cells, block cell cycle progression and induce programmed cell death. In animal experiments, tea extracts and certain components suppress the growth of various tumors; furthermore, they modulate the effect of antitumor drugs. The above-mentioned investigations have provided important clues leading to the conclusion that tea, particularly its related active components, may play a positive role in cancer therapy. In comparison with those antitumor drugs that are currently in clinical use, the antitumor activity of tea or its components is rather moderate. Tea extract or its component might not be used alone as a chemotherapeutic agent for cancer, however, tea has been found to exhibit a modulating effect that may enhance the antitumor activity of other drugs and reverse multidrug resistance in cancer cells. In terms of biochemical modulation, tea extract or its component is a promising agent in cancer therapy.

REFERENCES

Achiwa, Y., Hibasami, H., Katsuzaki, H., Imai, K., and Komiya, T. (1997) Inhibitory effects of persimmon (*Diospyros kaki*) extract and related polyphenol compounds on growth of human lymphoid leukemia cells. *Biosci. Biotechnol. Biochem.*, 61, 1099–1101.

Boone, C.W. and Wattenburg, L.W. (1994) Current strategies of cancer prevention: 13th Sapporo Cancer Seminar. *Cancer Res.*, 54, 3315–3318.

Bracke, M.E., Van Cauwenberge, R.M. and Mareel, M.M. (1984) (+)-catechin inhibits the invasion of malignant fibrosarcoma cells into chick heart *in vitro*. *Clin. Exp. Metastasis*, 2, 161–170.

Bracke, M.E., Castronovo, V., Van Cauwenberge, R.M., Coopman, P., Vakaet, L. Jr., Strojny, P., et al. (1987) The anti-invasive flavonoid (+)-catechin binds to laminin and abrogates the effect of laminin on cell morphology and adhesion. *Exp. Cell Res.*, **173**, 193–205.

Chen, Z.P., Schell, J.B., Ho, C.T., and Chen, K.Y. (1998) Green tea epigallocatechin gallate shows a pronounced growth inhibitory effect on cancerous cells but not on their normal counterparts. *Cancer Lett.*, **129**, 173–179.

Chung, L.Y., Cheung, T.C. and Kwok, T.T. (1998) The influence of green tea epicatechin isomers on the growth of human cancer cells. *Proc. Am. Assoc. Cancer Res.*, **39**, 392.

Fan, Y.J., Wang, Y.X. and Feng, S.Y. (1992) *In vitro* effect of catechin on cell growth. *Chin. J. Cancer*, **14**, 90–92.

Gotoh, A., Matsumoto, H., Wada, Y., Gohji, K., Okada, H., Arakawa, S., et al. (1998) Inhibitory effects of green tea extract on human prostatic cancer cell line. *Proc. Am. Assoc. Cancer Res.*, **39**, 392.

Graham, H.N. (1992) Green tea composition and polyphenol chemistry. *Prev. Med.*, **21**, 334–350.

Han, C. (1997) Screening of anticarcinogenic ingredients in tea polyphenols. *Cancer Lett.*, **114**, 153–158.

Hara, Y., Matsuzaki, S., and Nakamura, K. (1989) Antitumor activity of green tea catechin. *J. Jpn. Soc. Nutr. Food Sci.*, **42**, 39–45.

Hibasami, H., Achiwa, Y., Fujikawa, T. and Komiya, T. (1996) Induction of programmed cell death (apoptosis) in human lymphoid leukemia cells by catechin compounds. *Anticancer Res.*, **16**, 1943–1946.

Hibasami, H., Komiya, T., Achiwa, Y., Ohnishi, K., Kojima, T., Nakanishi, K., et al. (1998a) Induction of apoptosis in human stomach cancer cells by green tea catechins. *Oncol. Rep.*, **5**, 527–529.

Hibasami, H., Komiya, T., Achiwa, Y., Ohnishi, K., Kojima, T., Nakanishi, K., et al. (1998b) Black tea theaflavins induced programmed cell death in cultured human stomach cancer cells. *Int. J. Mol. Med.*, **1**, 725–727.

Hirose, M., Hoshiya, T., Takahashi, S., Hara, Y. and Ito, N. (1991) Inhibition of carcinogenesis by green tea catechins in rats. *Proceedings of the International Symposium on Tea Science, Shizuoka, Japan,* 210–214.

Huang, M.T., Xie, J.G., Wang, Z.Y., Ho, C.T., Lou, Y.R., Wang, C.X., et al. (1997) Effects of tea, decaffeinated tea, and caffeine on UVB-induced complete carcinogenesis in SKH-1 mice: demonstration of caffeine as a biologically important constituent of tea. *Cancer Res.*, **57**, 2623–2629.

Isemura, M., Sazuka, M., Noro, T., Nakamura, and Y., Hara, Y. (1995) Inhibitory effects of green tea infusion on *in vitro* invasion and *in vivo* metastasis of mouse lung cancer cells. *International Conference on Food Factors: Chemistry and Cancer Prevention*, Japan, Abstract, p. 118.

Katiyar, S.K., Agarwal, R., Wood, G.S. and Mukhtar, H. (1992) Inhibition of 12-O-tetradecanoylphorbol-13-acetate-caused tumor promotion in 7,12-dimethylbenz(*a*)anthracene-induced SENCAR mouse skin by a polyphenolic fraction isolated from green tea. *Cancer Res.*, **52**, 6890–6897.

Katiyar, S.K., Argawal, R., and Mukhtar, H. (1993) Protection against malignant conversion of chemically induced benign skin papillomas to squamous cell carcinomas in SENCAR mice by a polyphenolic fraction isolated from green tea. *Cancer Res.*, **53**, 5409–5412.

Katiyar, S.K., Mohan, R.R., Agarwal, R. and Mukhtar, H. (1997) Protection against induction of mouse skin papillomas and high risk of conversion to malignancy by green tea polyphenols. *Carcinogenesis*, **18**, 497–502.

Kim, S., Yang, G.Y., and Yang, C.S. (1996) Decaffeinated green tea inhibited 4-(methylnitrosamino)-1-(3-pyridyl)-1-butanone (NNK)-induced early pulmonary hyperproliferation in A/J mice. *Proc. Am. Assoc. Cancer Res.*, 37, 274.

Kim, S., Yang, G.Y., and Yang, C.S. (1997) Assessment of the inhibitory effects of green tea polyphenols on the 4-(methylnitrosamino)-1-(3-pyridyl)-1-butanone (NNK)-induced early pulmonary hyperproliferation in A/J mice. *Proc. Am. Assoc. Cancer Res.*, 38, 367.

Liao, S., Umekita, Y., Guo, J., Koontis, J.M., and Hiipakka, R.A. (1995) Growth inhibition and regression of human prostate and breast tumors in athymic mice by tea epigallocatechin gallate. *Cancer Lett.*, 96, 239–243.

Lin, J.K. (1995) Anticarcinogenesis of tea polyphenols. *International Conference on Food Factors: Chemistry and Cancer Prevention*, Japan, Abstract, p. 99.

Liu, Z.J., Yang, G.Y., Landau, J., Liao, J., Seril, D. and Yang, C.S. (1997) High salt-induced hyperproliferation and its inhibition in gastric glandular epithelium of F344 rats by decaffeinated green tea. *Proc. Am. Assoc. Cancer Res.*, 38, 367.

Lu, Y.P., Lou, Y.R., Xie, J.G., Yen, P., Huang, M.T., and Conney, A.H. (1997) Effect of green or black tea on epidermal carcinogenesis in UVB-treated SKH-1 mice: inhibition by tea administration after discontinuation of UVB treatment and prior to tumor formation. *Proc. Am. Assoc. Cancer Res.*, 38, 367.

Mitchell, A.C. and Liebler, D.C. (1998) Inhibition of DNA photodamage by (–)-epigallocatechin gallate (EGCG). *Proc. Am. Assoc. Cancer Res.*, 39, 391.

Mukhtar, H., Ahmad, N., Feyes, D.K. and Agarwal, R. (1997) Green tea polyphenols induce apoptosis and alter the progression of cell cycle in human epidermoid carcinoma cells A431. *Proc. Am. Assoc. Cancer Res.*, 38, 580.

Naasani, I., Seimiya, H., and Tsuruo, T. (1998) Telomerase inhibition, telomere shortening, and senescence of cancer cells by tea catechins. *Biochem. Biophys. Res. Commun.*, 249, 391–396.

Nishida, H., Omori, M., Fukutomi, Y., Ninomiya, M., Nishiwaki, S., Suganuma, M., *et al.* (1994) Inhibitory effect of (–)-epigallocatechin gallate on spontaneous hepatoma in C3H/HeNCrj mice and human hepatoma-derived PLC/PRF/5 cells. *Jpn. J. Cancer Res.*, 85, 221–225.

Oguni, I., Nasu, K., Yamamoto, S., and Nomura, Y. (1988) On the antitumor activity of green tea leaf. *Agric. Biol. Chem.*, 52, 1879–1880.

Okabe, S., Suganuma, M., Hayashi, M., Sueoka, E., Komori, A., and Fujiki, H. (1997) Mechanisms of growth inhibition of lung cancer cell line, PC-9, by tea polyphenols. *Jpn. J. Cancer Res.*, 88, 639–643.

Paschka, A.G., Butler, R. and Young, C.Y. (1998) Induction of apoptosis in prostate cancer cell lines by the green tea component, (–)-epigallocatechin-3-gallate. *Cancer Lett.*, 130, 1–7.

Sadzuka, Y., Mochizuki, E. and Takino, Y. (1993) Caffeine modulates the antitumor activity and toxic side effects of adriamycin. *Jpn. J. Cancer Res.*, 84, 348–353.

Sadzuka, Y., Mochizuki, E., Iwazaki, A., Hirota, H., and Takino, Y. (1995) Caffeine enhances adriamycin antitumor activity in Ehrlich ascites carcinoma bearing mice. *Biol. Pharm. Bull.*, 18, 156–161.

Sadzuka, Y., Sugiyama, T., Miyagishima, A., Nozawa, Y. and Hirota, S. (1996) The effects of theanine, as a novel biochemical modulator, on the antitumor activity of adriamycin. *Cancer Lett.*, 105, 203–209.

Sadzuka, Y., Sugiyama, T. and Hirota, S. (1998) Modulation of cancer chemotherapy by green tea. *Clin. Cancer Res.*, 4, 153–156.

Sazuka, M., Murakami, S., Isemura, M., Satoh, K. and Nukiwa, T. (1995) Inhibitory effects of green tea infusion on *in vitro* invasion and *in vivo* metastasis of mouse lung carcinoma cells. *Cancer Lett.*, 98, 27–31.

Stammler G., and Volm, M. (1997) Green tea catechins (EGCG and EGC) have modulating effects on the activity of doxorubicin in drug-resistant cell lines. *Anti-Cancer Drugs*, 8, 265–268.

Su, J., Zhen, Y.S., Qi, C.Q. and Chen, W.J. (1994) A fungus-derived novel nucleoside transport inhibitor potentiates the activity of antitumor drugs., *Acta Pharm. Sin.*, 29, 58–661.

Su, J., Zhen, Y.S., Qi, C.Q. and Hu, J.L. (1995) Antibiotic C3368-A, a fungus-derived nucleoside transport inhibitor, potentiates the activity of antitumor drugs. *Cancer Chemother. Pharmacol.*, 36,149–154.

Suganuma, M., Okabe, S., Oniyama, M., Sueoka, N., Kozu, T., Komori, A., *et al.* (1995) Mechanisms of EGCG and green tea in inhibition of carcinogenesis. *International Conference on Food Factors: Chemistry and cancer Prevention*, Japan, Abstract, p. 119.

Suganuma, M., Okabe, S., Kai, Y., Sueoka, E., and Fujiki, H. (1998) Synergistic effects of (–)-epicatechin and (–)-epigallocatechin gallate on induction of apoptosis in human lung cancer cell line, PC-9. *Proc. Am. Assoc. Cancer Res.*, 39, 392.

Sugiyama, K. and Ueda, H. (1991) Protective effects of tea on toxic side effects by anticancer drug, cis-diamminedichloro-platinum in mice. *Proceedings of the International Symposium on Tea Science*, 353–356.

Sugiyama, T., Sadzuka, and, Y., Hirota, S. (1997) Effects of theanine, a tea leaf component, on antitumor activity and inhibition of tumor metastasis of adriamycin. *Proc. Am. Assoc. Cancer Res.*, 38, 611.

Sugiyama, T. and Sadzuka, Y. (1998) Green tea component, theanine, enhances the antitumor activity of adriamycin against sensitive P388 leukemia and resistant P388/ADR. *Proc. Am. Assoc. Res.*, 39, 528.

Taniguchi, S., Fujiki, H., Kobayashi, H., Go, H., Miyado, K., Sadano, H., *et al.* (1992) Effect of (–)-epigallocatechin gallate, the main constituent of green tea, on lung metastasis with mouse B16 melanoma cell line. *Cancer Lett.*, 65, 51–54.

Tan, X.H., Zhang, Y.L., Zhou, D.Y., Jiang, B. and Zhang, L. (1998) Apoptosis of colon cancer cell line, LoVo, induced by green tea aqueous extract *in vitro*. *Chin. J. cancer*, 17, 171–174.

Valcic, S., Timmermann, B.N., Alberts, D.S., Waechter, G.A., Krutzsch, M., Wymer, J., *et al.* (1996) Inhibitory effect of six green catechins and caffeine on the growth of four selected human tumor cell lines. *Anti-Cancer Drugs*, 7, 461–468.

Wang, Z.Y., Huang, M.T., Ferraro, T., Wong, C.Q., Lou, Y.R., Reuhl, K., *et al.* (1992a) Inhibitory effect of green tea in the drinking water on tumorigenesis by ultraviolet light and 12-O-tetradecanoylphorbol-13-acetate in the skin of SKH-1 mice. *Cancer Res.*, 52, 1162–1170.

Wang, Z.Y., Huang, M.T., Ho, C.T., Chang, R., Ma, W., Ferraro, T., *et al.* (1992b) Inhibitory effect of green tea on the growth of established skin papillomas in mice. *Cancer Res.*, 52, 6657–6665.

Wang, Z.Y., Huang, M.T., Lou, Y.R., Xie, J.G., Reuhl, K.R., Newmark, H.L., *et al.* (1994) Inhibitory effect of black tea, green tea, decaffeinated black tea, and decaffeinated green tea on ultraviolet B light-induced skin carcinogenesis in 7,12-dimethyl[*a*]anthracene-initiated SKH-1 mice. *Cancer Res.*, 54, 3428–3435.

Wang, Z.Y., Wang, L.D., Lee, M.J., Ho, C.T., Huang, M.T., Conney, A.H., *et al.* (1995) Inhibition of N-nitrosomethylbenzylamine-induced esophageal tumorigenesis in rats by green and black tea. *Carcinogenesis*, 16, 2143–2148.

Yan, Y.S. (1990) Antitumor activity of Chinese green tea. *Tea Science*, 10, 79–84.

Yan, Y.S., Zhou, Y.Z., You, L.Q., and Tian Z.Q. (1990) Effect of Chinese green tea extracts on the human gastric carcinoma cell *in vitro*. *Chin. J. Prevent. Med.*, 24, 80–82.

Yan, Y.S., Wang, Q.D., Zhou, Y.Z., and Zhao, X.P. (1992) Effect of Chinese green tea extract on the immune function of mice bearing tumor and its antitumor activity. *Chin. J. Prevent. Med.*, 26, 5–7.

Yan, Y.S., Zhou, Y.Z., Su, C.Q., *et al.* (1993) The experiment of tumor-inhibiting effect of green tea extract in animal and human body. *Chin. J. Prevent. Med.*, **27**, 129–131.

Yang, C.S., Hong, J.Y. and Wang, Z.Y. (1992) Inhibition of nitrosamine-induced tumorigenesis by diallyl sulfide and tea. Cited from Yang, C.S. and Wang, Z.Y. (1993) Tea and cancer. *J. U.S. Natl. Cancer Inst.*, **85**, 1038–1049.

Yang, C.S., Yang, G.Y., Landau, J.M., Kim, S. and Liao, J. (1998) Tea and tea polyphenols inhibit cell hyperproliferation, lung tumorigenesis, and tumor progression. *Exp. Lung Res.*, **24**, 629–639.

Yang, G.Y., Liu, Z., Ding, W., Bondoc, F., Kim, F., Seri, D.N., *et al.* (1997) Inhibition of 4-(methylnitrosamino)-1-(3-pyridyl)-butanone (NNK)-induced early bronchiolar cell hyperproliferation and tumorigenesis in A/J mice by theaflavins, constituents of black tea. *Proc. Am. Assoc. Cancer Res.*, **38**, 368.

Yang, G.Y., Liao, J., Kim, K., Yurkow, E.J. and Yang, C.S. (1998) Inhibition of growth and induction of apoptosis in human cancer cell lines by tea polyphenols. *Carcinogenesis*, **19**, 611–616.

Yang, X.Q., Shen, S.R., Luo, S.J., Huang, P.J., Fang, Y.Z., Liu, M.Z., *et al.* (1991) The biological activity and toxicological test of tea polyphenols. *Proceedings of International Symposium on Tea Science, Shizuoka, Japan*, 263–267.

Ye, Z.X., Ye, Q.S., Li, Z.Y., Wu, P.Y., Fang, L., Fu, Q., *et al.* (1984) Effect of green tea on transplanted tumor in mice. *Tumor (Shanghai, China)*, **4**, 128.

Zhao, J.F., Bickers, D.R., and Wang, Z.Y. (1997) Protective effect of green tea on 8-methoxypsoralen plus UVA (PUVA)-induced skin photodamage in SKH-1 mice. *Proc. Am. Assoc. Cancer Res.*, **38**, 366.

Zhao, J.F., Jin, X.H., Mukhtar, H., Bickers, D.R., and Wang, Z.Y. (1998) Photoprotection in human skin by green tea and black tea. *Proc. Am. Assoc. Cancer Res*, **39**, 392.

Zhao, Y., Cao, J., Ma, H. and Liu, J.W. (1997) Apoptosis induced by tea polyphenols in HL-60 cells. *Cancer Lett.*, **121**, 163–167.

Zhen, Y.S., Cao, S.S., Xue, Y.C. and Wu, S.Y. (1991) Green tea extract inhibits nucleoside transport and potentiates the antitumor effect of antimetabolites. *Chin. Med. Sci. J.*, **6**, 1–5.

Zhen, Y.S., Lui, M.S. and Weber, G. (1983) Effects of acivicin and dipyridamole on hepatoma 3924A cells. *Cancer Res.*, **43**, 1616–1619.

Zhen, Y.S., Taniki, T. and Weber, G. (1992) Azidothymidine and dipyridamole as biochemical response modifier: Synergism with methotrexate and 5-fluorouracil in human colon and pancreatic carcinoma cells. *Oncology Res.*, **4**, 73–78.

Zhen, Y.S., Su, J., Xue, Y.C., Qi, C.Q. and Hu, J.L. (1994) Novel nucleoside transport inhibitors of natural origin. *Adv. Exper. Med. Biol.*, **370**, 779–782.

Zhen, Y.S., Su, J. and Xue, Y.C. (1996) Salvianolic acid A inhibits nucleoside transport and potentiates the activity of antitumor drugs. *Proc. Am. Assoc. Cancer Res.*, **37**, 287.

12. THERAPEUTIC USES OF TEA IN TRADITIONAL CHINESE MEDICINE

WEI-BO LU

Institute of Basic Theory, China Academy of Traditional Chinese Medicine, 18 Beixincang, Dongzhimennei, Beijing 100700, China

Tea is one of the most widely used and most popular common drinks in the world. Tea drinking is not only an indispensable part of daily life, but also an important component of the oriental culture, so-called "tea culture". Tea and tea culture originated in China, the earliest literature in history appeared in the *"Poem Classic"* of 1100–476 BC, it stated: "Who said that tea is bitter? It tasted sweet as Jicai" (a kind of vegetable).

1.   THREE APPROACHES TO TEA DRINKING

There are three approaches held by different countries and peoples towards tea and tea drinking. The most widely held viewpoint is: Tea is a drink. It is further developed as instant tea, iced tea, lemon tea, etc. for more convenient drinking and better taste. Chinese people are very fond of tea. Emperor Qian Long of Qing Dynasty (1736–1795) once said: "You cannot live one day without tea" (Luo, 1992). Tibetan people also have similar saying: "You can live three days without milk, but cannot live one day without tea" (Cheng *et al.* 1991). The frequency of tea drinking by English people may amount to six to seven times a day, from morning till late in the night.

What is the benefit of tea drinking? "It may supplement the essential of rice and vegetables, antagonize the grease of meat, best serve the drink of Summer and wash out the drowsiness throughout the night" (Cheng *et al.* 1991). The principles of drinking tea have been described by Lu Yu in his *"Tea Classic"* as: "It is preferably mild in taste, with an appropriate amount, drink less after a meal, don't drink any tea before going to bed". China is also entitled "Kingdom of Tea", where 350 species of tea tree have been cultured and over 1000 brands of tea products manufactured, such as Green tea (Long Jing, Bi Luo Chun), Red tea (Qi Hong, Dian Hong) and Blue tea (Oolong, Tie Guan Yin), etc.

The second approach takes tea drinking as a cultural rite, so-called "Tea Ceremony", which also originated in ancient China, but stressed and developed by Japanese people after Song Dynasty (1259 AD) (Mao, 1996). Tea is used as a media to calm down psychological agitation, to get rid of confused and distracting thinking so as to cultivate mental calmness. It coincided perfectly with Oriental philosophical thinking, as well as the thinking of Buddhist, Taoist and Confucianists, who encouraged the thinking of "Tranquility, no lust". The set ceremonial procedure of preparing tea and tea cake, applying the teacup, teaspoon and teapot, and the steps of drinking tea, decorating the tea room and environment have been formulated and strictly followed (Teng, 1992).

The third approach is for medical purposes. Tea is mainly taken as a medicinal herb, with its property and function being designated as "Bitter, sweet, cool, entering the heart, lung and stomach meridians, it may clear up the head and eyes, relieve the thirsty, assist the digestion, promote diuresis and detoxification".

Lu Yu said in 758 AD: "Tea drinking started from Shen Nong, it became well-known in Zhou Dynasty". Shen Nong (1 century BC) was the earliest pharmacist in ancient China, author of *Shen Nong's Herbal*, in which the discovery of tea was described: "Shen Nong tasted hundreds of herbs and became intoxicated by 72 herbs in one day, and was detoxified by tea". And it was said that when Shen Nong was intoxicated by gold-green colored Gun Shan Zhu and fell unconscious under the tea tree, he was awakened by the dew on the leaves flowing into his mouth. Thus, the detoxifying function of tea became the earliest pharmaceutical action described.

Later on, famous physicians of various dynasties had greater understanding of tea: The Saint of Medicine, Dr. Zhang Zhongjing (150–219 AD) of Han Dynasty in his *"Treatise on Febrile and Miscellaneous Diseases"* recorded his experiences in using tea to treat bloody purulent stool. Dr. Hua Tuo of Han Dynasty in *"On Food"* said: "The bitter tea, when taken for a long period, will do benefit to mental activity"; Dr. Tao Hongjing of Jin Dynasty (456–536 AD) commented: "After drinking for a long time, you may become lighter in body weight and change the bone". Sun Simiao (581–682 AD) in *"Prescriptions Worth a Thousand Gold"* said: Tea "makes one strong with pleasant thinking"; and Dr. Chen Zangqi (8 century AD) in *"Supplement to Materia Medica"* said: " Drugs are the medicine of various diseases, while tea is the medicine of all diseases". It is a concise and excellent summary of Chinese views of taking tea as medicine at that time.

In China, tea was treated at first as a medicine, and gradually, some characteristics have been found that tea not only can relieve thirst, but also is "cold" in nature, it may clear up heat, particularly in summer time or after a long journey. A cup of tea relieves fatigue and serves as good refreshment. It drives away drowsiness and headache. Moreover, its property is mild and tastes slightly bitter. These features of tea make it suitable for a long-term drink. From medicine to beverage to herbal tea, that is the development of tea drink in China.

2. CHINESE HERBAL TEA

2.1. Definition of Herbal Tea

According to the Pharmacopoeia of the People's Republic of China (1963 ed: App.12), herbal tea (health tea) is a kind of herbal medicine mixture with a dosage-form of coarse powder, small cube and granule form, it is prepared with boiling water as infusion or short time decoction, and then taken as tea drink. There are two kinds of herbal tea:

- Tea with medicinal herbs.
- Medicinal herbs without tea, tea drink is the means of administration.

Most herbal tea consists of two or more herbs.

2.2. Indications for Herbal Tea

- Chronic diseases, for long-term medication.
- Acute but mild ailments.
- For aged or weak patients.
- Not for severe and acute cases.

2.3. Advantages of Herbal Tea

- It is convenient to take, just like daily tea drinking, without using any herbal decoction–making devices.
- It is composed of coarse powder or granules, the contact surface area is enlarged, which facilitates the active principle to be dissolved.
- The enzyme is rapidly inactivated by boiling water, which avoids its active principle being disintegrated.
- It is mild in action and taste, and may be used repeatedly, the effect is gentle but consistent, which facilitates long term drinking, and is particularly indicated for chronic diseases, and for weak aged people.
- It is particularly suitable for those herbs and recipes that are not tolerable to high temperature such as those with volatile oil to treat acute diseases, e.g., common cold and influenza.
- It is safe, without any side effects.
- It is flexible, its constituents can be modified, and is adapted to syndrome differentiation in traditional Chinese medicine.

2.4. The Theoretical Basis of Herbal Tea

In traditional Chinese medicine (TCM), it is considered that "Food and drugs have the same origin", the nutrient food and the pharmacologically active drug has no clear demarcation line between them. Many foods are active pharmaceutically, and many (not all) herbal drugs have nutrients, the astragalus porridge, the wolfberry chicken, and the carp decoction are popular examples frequently used by Chinese people, which yields "The Chinese Medicinal Diets" a borderline subject between medicine and dietetics. Tea belongs to the daily food and drink category; therefore, "Tea is medicine" is a natural conclusion.

2.5. The Development of Herbal Tea

The earliest herbal tea appeared in the Tang Dynasty, developed in the Song, Yuan, Ming and Qing Dynasties as well as modern China. Wang Tao in *"Medical Secrets of an Official"* (752 AD) described the preparation, application and indication of herbal tea. He said that *Astragalus membranaceus, Tetrapanax papyriferus* each 1 kg, *Poria cocos, Zingiber officinalis, Pueraria lobata, Morus alba* each 0.5 kg, (total 14 herbs), pounded in a motar, screened twice, boiled to a concentrated liquid, mixed with excipient, made into ring-form cake, dried, heated and taken for tea is indicated for expelling the evil wind. Sun Simiao (581–682 AD) in *"Prescriptions Worth a Thousand Gold"* and *"Supplement to Prescriptions Worth a Thousand Gold"* carried over ten herbal tea recipes to treat non-stroke headache, vomiting, etc.

An official publication called *"The Holy Benevolent Prescriptions"* in Song Dynasty (992 AD) carried 8 herbal tea recipes such as Menthol tea, Allium-Glycine soja fermented tea and Gypsum tea, to treat the common cold and its complicating headache and fever respectively. The royal doctor Hu Sihui (1279–1368) in his book *"Synopsis of Drink and Food"* described "Yumo Tea and Qing Tea" which was appreciated by the royal family at the time.

From Ming Dynasty (1368–1644) herbal tea has been further widely used. The complete collection of prescriptions *"Prescriptions for Universal Relief"* (1406) carried 8 herbal tea formulae; the famous *"Compendium of Materia Medica"* (1590) by Li Shizhen also carried 8 herbal tea recipes such as Imperata tea to treat hematuria.

In Qing Dynasty, the royal palace paid great attention to the herbal tea, it had Qing Palace Elixir Tea which consisted of Oolong tea, Liu An tea etc., as well as Qingre tea including Qingre Liqi tea, Qingre Huashi tea, Qingre Zhike tea, etc. They selected the tea recipes according to different syndromes in traditional Chinese medicine of Syndrome Differentiation.

In addition, there were many famous herbal tea recipes such as Tian Zhong tea, Wu He tea, Wu Shi tea, Wan Ying tea, Jian Qu tea, Gan Lu tea, Gan Mao tea, Jian Fei tea, Oolong tea, Cassia tea, Salvia tea, Crataegus tea, Nelumbo tea, Qing Cao tea in Europe; and Laxation tea, Anti-tussive tea in Russia, etc. As reported, they have good efficacy and have been used until recently.

On the whole, the herbal tea has a long history, it developed before the use of tea as a drink. It is convenient to take, easy to prepare, mild in action and negligible in side-effect, and good in efficacy, particularly in chronic disease patients and weak aged people. It is either effective in preventing and treating diseases or beneficial for preserving health. It may serve as a self protecting health care measure.

So far, it has proven efficacy as non-specific wide spectrum medicine enjoyed by many people combining medication with the pleasant experience of tea drinking. As research proceeds further, many biological effects of herbal tea have been explored such as anti-cancer, lowering the blood lipid and blood sugar, anti-hypertensive, anti-oxidation, anti-aging, anti-fatigue, anti-radiation, preventing caries, etc.

Many active principles such as tea polyphenol, tea pigments, caffeine, theophylline, tannin, vitamins A, B, C, D, K, P, and trace elements (F, Cu Zn, Se) have been isolated from tea. Many studies have been conducted to elucidate the mechanisms. In combining traditional Chinese medicine with modern Western medicine, herbal tea has a bright prospect.

3. THERAPEUTIC USES OF TEA AND TEA PRODUCTS

Since tea can be used with other herbs, the indication of tea and tea products is greatly expanded. In this chapter, it is confined to the therapeutic uses of tea, its components, and its active principles.

3.1. Non-Specific Therapeutic Effect

Relief of thirst is the most fundamental function of tea as a universally cherished drink. "*Supplement of Materia Medica*" stated, "Relieve the thirst and drive away the plague, how precious is the tea!" The tea expert Lu Tong (775–835 AD) described the feeling after drinking tea vividly. He wrote: "One bowl may moisten the mouth and throat. Two bowls breaks the loneliness. Three bowls promotes writing a thousand sentences. Four bowls make one sweat and calm down. Five bowls lightens the muscle and bone. Six bowls opens the way to heaven; and after seven bowls the breeze comes out under the axilla". The above-mentioned pleasant feelings make many people like tea very much, some even craving for it.

In addition, the anti-fatigue action and the use for refreshment make tea a favorite daily drink. As *Shen Nong's Herbal* (1 century BC) described: "Tea is bitter, the drink is beneficial to thinking, and sleeping less". Sun Simiao in "*Prescriptions Worth a Thousand Gold*" mentioned: "Tea makes one strong, pleasant in thinking". Its component caffeine may enhance excitement of the cerebral cortex, stimulate the central nervous system, and thus alleviate and eliminate fatigue.

3.2. Specific Therapeutic Effect

3.2.1. *Lowering blood lipid*

Lipid metabolic disturbance usually manifests as hyperlipidemia, one of the risk factors for cardio-cerebrovascular diseases. Improving lipid metabolism and lowering blood lipid is vital in preventing and treating arteriosclerosis, coronary heart diseases (CHD), and obesity. Many Chinese herbal medicines have been proven to reduce hyperlipidemia. At present, Oolong tea and green tea have been revealed to reduce high blood lipid. They not only lower cholesterol and triglyceride, but also raise the high density lipoprotein cholesterol (HDL-C). Green tea is better than Oolong tea at lowering blood lipid and systolic pressure; while in lowering diastolic pressure and raising HDL-C, Oolong is better than Green tea (Chen, W.Y. *et al.* 1997).

Using green tea or Oolong tea could lower both the blood lipid and the blood pressure, the result would be satisfactory with either. It is suitable for drinking daily for a prolonged term as a treatment for hyperlipidemia. In traditional Chinese medicine, the Syndromes include Qi Deficiency, Phlegm-Dampness, Blood Stasis, and Spleen Deficiency. Herbs that could replenish the Qi, remove the Phlegm-Dampness and blood stasis, and activate the Spleen might be even more effective in treating hyperlipidemia and atherosclerosis. The reported effective recipes are as follows:
Oolong tea or green tea daily, one drink of tea infusion each morning and afternoon, treatment course is one month (Chen *et al.* 1997). Alisma tea: Alisma orientale and Jasmine green tea, infused with boiling water (Mao, 1996). Artemisia Curcuma tea: Artemisia capillaris, Curcuma aromatica, and green tea infused with boiling water (Mao, 1996). Polygonum Crataegus tea: Polygonum multiflorum, Crataegus pinnatifida, and Oolong tea, using a decoction of the former two herbs to make tea (Wu, 1995).

Obesity has already become a common social problem, usually the result of excess nutrition. In some developed countries obesity has become a very common ailment, there is a need to keep slim. A number of "fat reducing" herbal teas have appeared in the market, such as: Qing Palace Elixir Fat-Reducing tea: Folium Nelumbo nucifera, Crataegus pinnatifida, Oolong tea, and Liu-an tea in pulverized form, to be infused with boiling water and drunk (Yang, 1991). Strong & Handsome Fat-Reducing tea: Oolong tea 50% combined with Crataegus pinnatifida, Hordeum vulgare, Citri aurantii, Alismatis orientale, Massa Fermentata medicinalis, and others (Luo, 1992).

3.2.2. *Hypertension*

Hypertension is one of the most common diseases and its pathology is complicated. If it is kept under control, and the blood pressure is stabilized within the normal range, then the patient can not only live and work as usual, but the disease would not progress to the 2nd and 3rd stage, the cardiac and cerebral complications could thus be avoided. Therefore, taking comprehensive measures, including sufficient sleep, reducing work, a rational diet, lower body weight, proper physical exercise, etc. is important. It is a good idea to take TCM hypotensive herbs constantly, particularly hypotensive herbal teas. The advantages far outweigh the disadvantages.

The effective hypotensive Chinese herbs are: Chrysanthemum indicum, Scutellaria baicalensis, Eucommia ulmoides, Paeonia suffroticosa, Coptis chinensis, Ligusticum wallichii, Loranthus parasiticus, Clerodendrum trichotomum, Prunella vulgaris, Apocynum venetum and Aristolochia debilis. Green tea and Oolong tea is not only effective in reducing high blood lipid, but also in lowering high blood pressure (Chen *et al.* 1997). The hypotensive herbal teas are usually used as follows:

Prunella tea: Prunella vulgaris, and green tea infused with 200 ml boiling water (Mao, 1996).
Chrysanthemum tea: Oolong or green tea, Chrysanthemum morifolium, infused with boiling water (Cheng *et al.* 1991).
Three Herbs tea: Plantago asiatica, Siegesbeckia orientalis, Cephalanoplos segetum, and green tea, the decoction is taken orally (Luo, 1992).

Headache occurs frequently in hypertension. According to the view of traditional Chinese medicine, this is mainly caused by Liver Yang (Fire) flaring up; therefore, it should be treated with herbs that calm down Liver Fire. The herbal teas for that purpose are:

Ligusticum tea: Ligusticum wallichii, and green tea, take decoction (Cheng *et al.* 1991).
Rheum & Angelica tea: Rheum palmatum, Angelica dahurica, and green tea, infused with boiling water (Mao, 1996).

3.2.3.   *Lowering blood sugar*

Tea can improve the glucose tolerance test (Ruan, 1987), promote pancreatic juice secretion, and reduce the blood sugar source (Mao, 1996). Tea polysaccharide (TPS) can markedly lower the blood sugar (Wang *et al.* 1995). In traditional Chinese medicine, diabetes mellitus is equivalent to so-called "Xiao Ke", the Syndrome of Yin Deficiency and flaring-up of Fire; therefore, those herbal teas so-called Nourishing Yin and Lowering Fire would be helpful in reducing blood sugar. The herbal teas usually used are as follows:

Coptis & Anemarrhena tea: Coptis chinensis, Anemarrhena aspheloides, and green tea, infuse the tea with the decoction of the former two herbs (Mao, 1996).
Polygonum tea: Polygonum multiflorum, Alismatis orientale, Salvia miltiorrhiza, and green tea, boiling with water, take the decoction (Huang, 1995).
Ophiopogon tea: Ophiopogon japonica, Rehmannia glutinosa, Scrophularia ning-poensis, and green tea, boil with water, take the decoction (Luo, 1992).
Gourd tea: Towel gourd, green tea 5 g, and a little salt, boil with water, take the decoction (Zhou, 1993).

3.2.4.   *Cardiovascular diseases*

Some herbal teas are indicated for cardiovascular diseases, particularly coronary heart diseases (CHD), hypertension, chest pain, etc.
CHD (so-called Qi Stagnation & Blood Stasis Syndrome in TCM):   Leonurus tea: Leonurus heterophyllus, Crataegus pinnatifida, and green tea, infused with boiling water (Huang, 1995).
CHD (so-called Blood Stasis Syndrome in TCM):   Nelumbo tea: Folium Nelumbo nucifera, green bean, and green tea, infused with boiling water (Ruan, 1987).
Chest pain (so-called Blood Stasis Syndrome in TCM): Ligusticum tea: Ligusticum wallichii and green tea, infused with boiling water (Wu, 1995).

3.2.5.   *Anti-aging*

The imbalance between the production of free radicals and scavenging leads to the accumulation of free radicals, which are harmful to the body cell function and cause diseases. The free radical induced lipid peroxidation that plays important role to the pathogenesis of many diseases. Tea polyphenol (TP) is a highly efficient scavenger of free radicals, the clearance reached 98% or beyond. Both the TP and tea pigment can elevate the superoxide dismutase (SOD) activity, lower malondialdehyde (MDA) content, weaken and lessen the lipid peroxidation (LPO) injury, and scavenge the free radicals from the body. Thus protect the body and keep in good health (Du *et al.* 1997; Li, 1997; Yang *et al.* 1997).

The herbal tea used by ancient Chinese mostly consisted of so-called Liver and Kidney tonics, Qi and Blood tonics and green tea (Cheng *et al.* 1991).

Celestial Longevity tea: Panax ginseng, Achyranthes bidentata, Morindae officinalis, Eucommia ulmoides, Lycium barbarum, and black tea, infuse black tea with decoction of the other herbs (Mao, 1996).

Poria tea: Poria cocos, Rehmannia glutinosa praeparata, Chrysanthemum morifolium, Panax ginseng, Biota orientalis, and black tea, infuse black tea with decoction of the other herbs (Mao, 1996).

Rheum tea: Green tea and Rheum palmatum, infuse with boiling water (Cheng *et al.* 1991).

3.2.6. Cancer

The most interesting thing was the epidemiological investigation, which showed that in tea producing region, those constantly drank green tea could prevent the occurrence of cancer. In liver cancer high prevalent area, Qidong County of Jiangsu Province in China, the incidence was 48.7/100,000, where the rate of tea drinking in the population was 15.4%. However, in the nearby Jurong County, where the rate of tea drinking was 61.47%, the incidence of liver cancer was only 14.96/100,000, the difference was significant (Cheng, 1990). In Shizuoka of Japan, the green tea growing area, the mortality of cancer was lower than other areas in Japan (Oguni, 1987). In traditional Chinese medicine, cancer usually manifested as Phlegm Coagulation, Blood Stasis, Toxin Accumulation and lowered body resistance (hypo-immunity). Any herbs that can remove Phlegm and blood stasis, detoxify, and replenish vitality are helpful in treating neoplasms. The herbal tea formula for cancer frequently used are as follows:

Coix tea: Semen Coix lacryma-jobi and green tea (Mao, 1996).   Garlic tea: Green tea, garlic and brown sugar (Wu, 1995).   Glycyrrhiza tea:Green tea and Glycyrrhiza uralensis (Wu, 1995).   Curcuma tea: Green tea, Curcuma aromatica, Glycyrrhiza uralensis, and honey (Wu, 1995).

3.2.7. Nervous system

Stroke (so-called Qi Deficiency & Blood Stasis Syndrome in TCM): Polygonum tea: Polygonum multiflorum, Alismatis orientale, Salvia miltiorrhiza, and green tea, boiling, take decoction daily (Huang, 1995).

Neurasthenia (so-called LIver Qi Stagnation Syndrome in TCM): Albizia tea: Albizia julibrissin, Glycyrrhiza uralensis, Euryale ferox, brown sugar, and black tea, decoct the former 3 herbs, and then add sugar and tea (Huang, 1995).

Hemiplegia (so-called Liver Wind Syndrome in TCM): Gastrodia tea: Gastrodia elata and green tea, infused with boiling water (Wu, 1995).

3.2.8. Digestive diseases

Peptic ulcer (so-called Liver Qi affects Stomach Syndrome in TCM): Mume tea: Flos Mume album and green tea, infused with boiling water Huang, 1995).

Chronic gastritis (so-called Spleen-Stomach Deficiency Cold Syndrome in TCM): Zingiber tea: Zingiber officinalis and green tea, take decoction (Huang. 1995).

Acute and chronic dysentery (so-called Dampness Heat Syndrome in TCM): Green tea, add water, take decoction (Chen et al. 1997).

3.2.9. Respiratory diseases

Bronchial asthma (so-called Heat Syndrome in TCM): Tussilago tea: Tussilago farfara, Asteria tartaricus, and green tea, infuse with boiling water (Wu, 1995).

Cough (bronchitis, common cold, Phlegm Syndrome in TCM): Fritillaria tea: Fritillaria cirrhosa, green tea 3g, Zingiber officinalis, and sugar, infused with boiling water (Wu, 1995).

Common cold (so-called Wind Cold Syndrome in TCM): Allium-Soja tea: Allium fistulosum, Glycine soja fermentatum, Schizonepeta tennuifolia, Mentha haptocalyx, Gardenia jasminoides, Gypsum fibrosum, and green tea, decoct and take daily (Yang, 1991).

Common cold (Wind Heat Syndrome in TCM): Lophatherum tea: Lophatherum gracile, and green tea, boil the former herb for 5 min, then infuse it with green tea (Huang, 1995).

3.2.10. Urinary diseases

Urolithiasis (so-called Spleen and Stomach Deficiency Syndrome in TCM): Lysimachia tea: Lysimachia christina, Stigma Zea mays, and green tea, take the decoction (Huang, 1995).

Acute nephritis (so-called Edema in TCM): Imperata tea: Imperata cylindrica, Stigma Zea mays, and green tea, decoct and take as drink (Wu, 1995).

3.2.11. Eye diseases

Senile cataract (so-called Kidney Yin Deficiency Syndrome in TCM): Lycium-Chrysanthemum tea: Lycium barbarum, Chrysanthemum morifolium, and green tea, infuse with boiling water for 10 min (Huang, 1995).

Glaucoma (Early stage): Cassia tea: Semen Cassia obtussifolia, and Green tea (Cheng et al. 1991).

3.2.12. Tooth caries and ozostomia

Tea (black tea, green tea or Oolong tea), infuses with boiling water, used it for gargling (Lin et al. 1996).

Asarum tea: Asarum sieboldi, Glycyrrhiza uralensis, green tea, decoct former two herbs with water, then add tea, take after meal (Lin & Chen, 1996).

For Ozostomia. Osmanthus tea: Flos Osmanthus fragrans, boil with water, then add black tea, for gargling (Lin & Chen, 1996).

3.2.13.   Other diseases

Carbuncle (Early stage):   Hedyotis-Glycyrrhiza tea: Fresh Hedyotis diffusa, Glycyrrhiza uralensis, and ɢʀᴇᴇɴ ᴛᴇᴀ, boil the former two herbs, and then add tea, take 4 times daily (Huang, 1995).

Eczema (so-called Dampness Heat Syndrome in TCM):   Alumen-Sophora tea: Alumen, Sophora flavescens, and green tea, add water and boil for 10 min, wash the skin lesion (Huang, 1995).

Dysmenorrhea (so-called Qi Stagnation and Blood Stasis Syndrome in TCM): Ligusticum tea: Ligusticum wallichii and green tea, boil with water, take before meal (Huang, 1995).

3.2.14.   Stop smoking tea

Picrorhiza tea: Picrorhiza scrofulariflora, and green tea, boil former with water, and then add tea, take 1 dose daily (Wu, 1995).

4.   SUMMARY

Tea is the most popular drink. It not only relieves thirst, but also serves as herbal medicine, so-called herbal tea. Tea is a kind of non-specific wide-spectrum physiological active agent that may be used to treat many disorders, and is indicated for cancer, coronary heart disease, hyperlipidemia, hypertension, obesity, diabetes mellitus, dysentery, caries, headache, fatigue, anti-aging, and even stop smoking. Tea is mild in action, good in taste; it can be repeatedly even daily used, and is convenient to take. In addition, tea is cheap in price and rich in resource. It is most indicated for chronic diseases, particularly for aged and weak people. New forms of tea such as instant tea and teabags have been available. Some active principles have been isolated, e.g., tea polyphenols, tea pigments and others. In traditional Chinese medicine, tea therapy is apparently one of the important ways to maintaining good health.

REFERENCES

Chen, K.J., Ma, X.C., Shi, D.Z., Li, C.S., Zhang, W.G., Zhou, W.Q. *et al.* (1994) *Essence of Qing Palace Tea Drink*, People's Health Publishing House, Beijing. **11**, p. 6.
Chen, W.Y., Lin, B.H., and Chen, L. (1997) Clinical hypotensive and lipid regulatory effect of Fujian Oolong tea and green tea. *China Tea*, **1**, 32–33.

Cheng, L., Chen, S.C., Dai, B.Y., Wei, J.W., Chen, F., and Wu, Q.R. (1991) *Treat Hundred Diseases with Tea and Wine*, Shanghai Science and Technology Archive Press. **4**, 9–91.

Cheng, Q. (1990) Anti-cancer effect of tea. *China Tea*, **5**, 19.

Du, Q.Z. Jiang, H.Y. (1997) Pharmacology of tea pigment and its application. *China Tea*. **5**, 76.

Editorial Committee. (1979) *Concise Dictionary of Traditional Chinese Medicine*, People's Health Publishing House, Beijing, **3**, 585.

Huang, C.C. (1995) *Diet, Wine, Tea Therapy Treats Hundred Diseases*, Modern China Press. **12**, 93–299.

Li, R.L. (1997) Progress on molecular level to study pharmacologic effect of tea polyphenol. *Fujian Tea*, **2**, 24.

Lin, Q.L., and Chen, X.Y. (1996) Tea therapy on caries. *J. Tea*, **22**(2), 49–50.

Lu, Tong. "*To Xie Meng*". 775–835 A.D.

Lu, Yu. *Tea Classic*, 758 A.D.

Luo, Q.F. (1992) *A Complete Works of Chinese Herbal Tea*, Guizhou People's Press, **8**, 1–203.

Mao, J.L. (1996) *Health Consultant of Tea Drink*. Sichuan Dictionary Press, **9**, 10–281.

Oguni, I., Nasu, K., and Nomura, T. (1987) Epidemiological and physiological studies on the anti-tumor activity of the fresh green tea. *Abstracts of International Symposium on Tea-Quality-Human Health,* pp. 222–226.

Poem Classic. 1100-476 B.C. Quoted from H.J. Zhang (ed.), (1989) *Chinese Herbal Tea*, People's Health Publishing House, **2**, 2.

Ruan, Y.C. (1987) Selenium in tea and its relationship with human health. *China Tea*, **5**, 10.

Tang, D.H., Xia, B., Wu, H.Q. (1997) Effect of tea drink on urban middle and old aged. *China Tea*, (5), 28.

Teng, J. (1992) *An Introduction to Japanese Tea Ceremony Culture*, Oriental Press, Beijing, p. 11.

Wang, D.F., Xie, X.F., Cai, C.L., Yang, M., and Zhang, Y.C. (1995) Analysis of pharmacological ingredients of old tea in treating diabetes mellitus. *Chin. Herbal Med.*, **26**(5), 255–257.

Wu, S.N. (1995) *Tea Therapy Treats Hundred Diseases*, Anhui Science and Technology Press, Hefei, China, **7**, 10–262.

Yang, S.S. (1991) *Herbal Tea Therapy of Chinese Medicine to Treat Hundred Diseases*, Xue Yuan Press, Beijing, **1**, 81–194.

Yang, X.Q. (1996) New contribution of tea to mankind. *J. Tea*, **22**(3), 42.

Yang, X.Q., Wang, Y.F., Gao, Y.G., and Shen, S.R. (1997) Effect of tea in coming century. *Fujian Tea*. **3**, 12.

Zhou, X. (1993) *Herbal Tea:A Treasure to Benefit Health and Life Expectance*, China Construction Material Industry Press, p. 62.

13. AGRONOMY AND COMMERCIAL PRODUCTION OF TEA

ZONG-MAO CHEN AND NING XU

Tea Research Institute, Chinese Academy of Agricultural Sciences, 1 Yunqi Road, Hangzhou 310008, China

1. DEVELOPMENTAL HISTORY OF TEA INDUSTRY

Tea as a drink originated from the China Shennong period, around 4000–5000 years ago. Originally, tea was used as a medicine for various ills. It was recorded that tea drinking in ancient China was popular in the 5th century AD and tea production has been developed rapidly since the Tang dynasty (618–907 AD) and accepted as a beverage. In the year of 780 AD, there was the first authentic account of tea. This famous book, Cha Ching (Tea Classic) was written by Lu Yu in which described the morphology of tea plant, tea drinking and tea manufacture.

Tea was firstly appeared for medical purpose in many ancient publications, such as the "Shi-Lun" (Dissertation on foods) written by famous doctor Hua Tuo in the period of East Han Dynasty and the "Ming-Yi-Bi-Lou" (Alternative record of famous doctor) written by Dao Hongjing in Nang dynasty. During that time, tea was infused with the fresh leaves plucked from the wild tea plant. From the viewpoint of modern technology, this kind of tasting was not good enough because there was no manufacturing process existed. The active ingredients within tea fresh leaves were only partly extracted by the water. The manufacture of tea has developed rapidly since the Tang dynasty. Although green tea was created firstly in ancient China; Oolong tea and black tea have appeared successively since the Ming dynasty (around 15th century) (Chen, 1992).

According to the legend, the spreading of tea from China to other countries started during the period of East Han dynasty (25 AD). However, the more dependable record was in Tang dynasty (815 AD). Japan was the first country where tea was introduced from China. Although it was in fashion in the palace and in the noble class; however, it declined after a short time period. The real source of Japanese tea was usually attributed to a famous Buddhist monk Ese in the period of Song dynasty. He had a deep investigation on the pharmacological function of tea and published the famous monograph of "Description on tea drinking for keeping in good health" in 1211 AD, two years before his death. The introduction of China tea to Korea was recorded in 823 AD, several years later than to Japan. The spreading of tea from China to western countries was around 800 years later than that introduced to Japan. The first tea

243

reached Europe around 1607 AD on Dutch ships from Macao to Holland. It was green tea! Black tea did not replace it till the middle of the eighteenth century. As early as 1641 AD, a Dutch doctor Nikolas Dirk stated that tea was the medicine that could cure many human disorders. Within a few years, tea had become very popular in Dutch high society. However, it was extremely expensive and sold in medicine shops. The introduction of tea to England was from Holland, around ten years later than to the continent. The first solid evidence of the sale of tea in England was a newspaper advertisement for the coffeehouse Thomas Garway in London in 1658 AD. It listed tea as effective in curing 14 human disorders. Afterwards, the habit of drinking tea gradually became popular in high society in England and moved into the family within around 30 years. Therefore, it can be seen from the above statements that the introduction of tea into England also started from medical usage. The introduction of tea into France was recorded in 1625 AD from Holland. It was said that the Palace doctor of King Louis XIV used tea to treat headaches of the King. Then the therapeutic effect was noticed by the high society of France (Chen, 1996).

The first tea reached Russia via the Chinese ambassador in 1618 as a gift to Czar Alexis. By 1700 AD Russia received over 600 camel loads of tea annually and in 1796, Russia was consuming over 6000 camel loads of tea per year. In Russia, tea was also used for dysentery treatment and detoxification and was reported to be extremely effective. Tea plantations were established in Russia in 1913.

India, the largest tea-producing country in the world nowadays, received tea from English merchants after tea was popularized in Europe. According to records, it happened around 1780, 170 years later than to England. The first tea garden in India appeared in 1862. Tea came to Sri Lanka from India in 1939. However, the first tea garden in Sri Lanka appeared in 1872, ten years later than that in India (Chen, 1996).

The tea cultivation history in Africa is relatively short. The first record of cultivation in Africa was in 1850; however, the tea industry did not develop until the middle of the 20th century. The development of tea plantations in South America was around 1900. Now, tea plants are distributed worldwide ranging from 42°N to 34°S. Tea is now grown commercially in tropical and subtropical regions of Asia, Africa and South America, and also in limited areas in North America and Australia.

2.   COMMERCIAL TEA PRODUCTION IN THE 20TH CENTURY

Since the beginning of this century, the world tea acreage, tea production and tea exportation increased steadily and continuously during most years of the century, and stagnated only in the period of the Second World War. The world acreage increased 1.44 times from 1910 to 1950, and increased 87% from 1950 to 1990. After 1990, tea acreage in the world became more stabilized (ITC, 1996; 1998). The increase in tea production was even more obvious than that of tea acreage. Tea production increased 1.13 times from 1900 to 1950, and doubled every 20 years from 1950 to 1990. However, the productive level has moved up and down in the range of 2.50 to 2.70

Table 13.1 World Tea Acreage, Tea Production and Tea Exportation in 20ᵗʰ Century.

Year	Tea acreage (× 10,000 ha)	Tea production (× 10,000 t)	Tea exportation (× 10,000 t)
1900	–	30.00	28.00
1910	52.59*	40.25	34.23
1920	65.78*	53.53	29.95
1930	86.03*	70.79	40.65
1940	99.79*	61.73	41.68
1950	128.68	64.15	39.66
1960	138.08	93.33	52.97
1970	163.77	125.17	64.82
1980	236.00	184.80	87.39
1990	240.94	252.29	113.45
1995	249.17	261.64	107.97
1996	240.09	262.90	110.66
1997	248.50	266.70	115.60

Tea acreage in China is not included due to the lack of information.

million tons since 1990. Tea production in 1997 created a historical record – 2.66 million tons, a 5.05% increase on 1990 (ITC, 1996; 1998). Tea acreage in major tea producing countries in this century is listed in Table 13.1.

Of the total world tea production in 1997, the Asian tea producing countries produced 83.6%. The African and South American tea producing countries produced 13.5% and 2.6%, respectively (Figure 13.1). This portion has changed to a certain extent in this century. Asian and African countries produced 99.9% and 0.1% in 1900 and 93.8% and 3.3% in 1950, respectively, illustrating that the contribution of African countries to the world tea production was increasing.

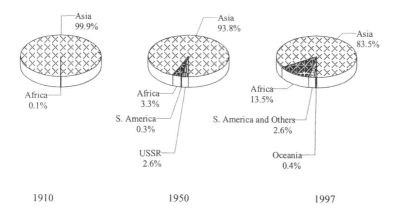

Figure 13.1 Proportion of World Tea production from various continents

Table 13.2 World Tea Production in Major Tea-producing Countries in 1997.

Country	Production in 1997 (× 10,000 t)	Percentage of the world total production (%)
India	81.06	30.39
China*	61.33	23.00
Sri Lanka	27.74	10.40
Kenya	22.07	8.27
Indonesia	13.10	4.91
Turkey	11.00	4.12
Japan	9.12	3.42
Iran	6.00	2.25
Argentina	5.50	2.06
Bangladesh	5.35	2.00
Malawi	4.39	1.64
Vietnam	4.20	1.57

*China Taiwan is not included.

In the major tea producing countries in the world, according to information in 1997, India took first place (810,600 tons), producing 30.4% of the total world tea production. China occupied the second position, producing 613,300 tons (excluding Taiwan), around 23.0% of the total production. Then followed Sri Lanka (277,400 tons), Kenya (220,700 tons) and Indonesia (131,000 tons), producing 10.4%, 8.2% and 4.9% of the total world production respectively. The above mentioned 5 countries produced 76.9% of the world tea (Table 13.2) (ITC, 1998).

The plucking acreage in the world has increased continuously since the beginning of this century; however, it stabilized in the last 10 years. The world plucking acreage in 1997 was around 2,485,000 hectares. The average yields per hectare have increased steadily since the 1950s, and the increase was most rapid in the 1980s. The average yields per hectare in the world in 1990 were 33.7% higher than those in 1980 and 1.1 times higher than those in 1950. The average yields per hectare in 1997 were 1073 kg, only 1.9% higher than those in 1990 (Table 13.3). There are large differences in the average yields per hectare in various tea-producing countries (Table 13.4). According to information in 1996, Kenya had the highest, the average yields per hectare reached 2285 kg. The average yields in those countries including Malawi, India, Japan, Iran and Turkey ranged between 1450–2000 kg per hectare. Those in Sri Lanka, Bangladesh, Indonesia and Argentina ranged between 1000–1400 kg. The average yields per hectare in Tanzania, Uganda, Vietnam and China ranged from 500–1000 kg (Chen, 1998b).

The world tea exportation amounted to 280,000 tons at the beginning of this century and only increased to 390,000 tons in 1950. The amounts exported increased only 39% during the 50 years period. This 0.78% low annual increasing rate could be attributed to the outbreak of World War I and World War II. However, it has shown an obvious increasing tendency since 1950. The export amounts in 1997 reached 1,150,000 tons, and 1.91 times higher than those in 1950 (Table 13.1). The average annual rate of increase was 4.06%. Among the total world tea exportation, more than 65% is exported from Asia and 28% from Africa. The tea production of those Asian

Table 13.3 World Average Unit Yield.

Year	Average yield per hectare (kg made tea/ha)
1940	537.76
1950	498.50
1960	675.90
1970	766.10
1980	783.10
1990	1047.10
1995	1950.00
1996	1120.80
1997	1073.60

tea-producing countries forms 83.2% of the total world tea production in 1997, but the export volumes form only 67% of the total world exporting volumes. However, the tea production of those African tea-producing countries is only 13.5% of the total world tea production, but the export amounts form 27.0% of the world tea exporting volumes. This also reflects the fact that domestic consumption in those Asian tea-producing countries is increasing. The percentage of the tea export amounts from Asia in the total world exportation has decreased around 5% during the last 15 years, while that from Africa has increased around 10% (Figure 13.2). In the 19th century, China almost monopolized the world tea export market. The export amounts from China reached 134,000 tons in 1886. The export amounts from India surpassed those from China in 1900 and occupied first place in the world tea market. This position has been maintained for a 90 year period. The tea exportation from Sri Lanka surpassed that from India from 1990–1994 to take first place in the world. Kenya caught up with Sri Lanka in 1995 and 1996. However, the export amounts from Sri Lanka reached

Table 13.4 Average Yield Per Hectare in Various Tea-producing Countries in 1996.

Country	Average yield per hectare (kg made tea/ha)
Kenya	2285
Malawi	1961
India	1795
Japan	1683
Turkey	1493
Iran	1470
Sri Lanka	1460
Bangladesh	1150
Indonesia	1120
Argentina	1102
Tanzania	961
Uganda	849
China	537
Vietnam	503

*Calculated by the quotient of total production and tea acreage.

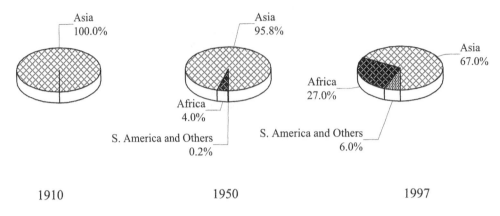

Figure 13.2   World Tea Exports

268,000 tons, a historic record, in 1997, and took first place in the world once again (Chen, 1998b). Figure 13.2 shows the tea exportation in major tea-producing countries.

The world tea import amounts were around 330,000 tons in 1910 and 392,700 tons in 1950, increasing nearly 20% in 40 years. It increased to 636,700 tons in 1970, an increase of 62% in 20 years. The world tea import amounts in 1997 were 1,152,400 tons, 1.9 times higher than those in 1950 and 2.46 times higher than those at the beginning of this century. The total tea import amounts have been up and down

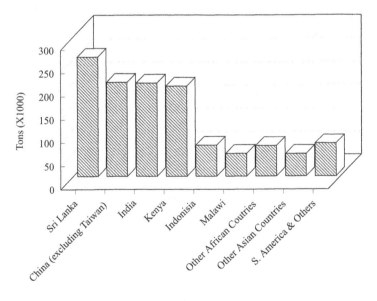

Figure 13.3   Tea Exports in Major Tea-Producing Countries

Table 13.5 Proportion of the Tea Importation of Various Continents in the World Total Amounts.

Year	Europe	America	Latin America	Africa	Asia	Oceania
1910	64.4	18.5	1.4	2.1	3.1	5.8
1920	63.2	18.3	1.7	3.3	3.7	6.7
1930	64.6	14.9	1.3	6.0	4.2	6.2
1940	56.8	16.7	1.5	7.1	7.6	6.9
1950	49.8	19.7	0.7	13.4	7.4	8.4
1960	53.9	13.6	1.1	13.1	11.0	6.9
1970	48.6	12.8	1.6	17.2	12.0	5.2
1980	39.2	12.2	1.6	16.2	25.2	3.5
1990	43.4	8.4	1.2	15.9	25.5	2.1
1996	37.2	9.3	1.4	16.8	27.9	2.1
1997	39.8	8.6	1.6	18.2	24.9	1.7

between 1.01 million tons to 1.15 million tons since 1990. The first 5 countries to import tea in 1997 were CIS/Russia (197,300 tons), UK (150,500 tons), Pakistan (86,800 tons), USA (81,200 tons) and Egypt (77,900 tons). The ratio of tea importation between the five continents varied significantly over the past century. At the beginning of this century, the import amounts in European countries occupied two-thirds of the world total import, and decreased to one half of the total. There was a further decrease to 40% in the 1980s. The percentage of Asian and African countries increased from 3.1% and 2.1% at the beginning of this century to 7.44% and 13.4% in the middle of this century, respectively; and further increased to more than 25% and more than 16% respectively in the 1980s (ITC, 1996; 1998). The import percentage of American and Oceanic countries has gradually decreased since the beginning of this century. The percentage of European, Asian, African, American (including South American) and Oceanic countries of world total import amounts was 39.8%, 24.9%, 18.2%, 15.3% and 1.8% in 1997 respectively, reflecting the amount of tea consumption in Asian and African countries which was gradually increasing (Table 13.5).

Corresponding with the continuous increase of world tea production, has been a gradual decreasing tendency of world tea price in the last 50 year period, except in the years of 1956–1958, 1976–1978, 1983–1985 and 1996–1997. The world tea price in 1994–1995 was only 40% of the price in 1960 (ITC, 1996) (Figure 13.4). The world tea price in 1996–1997 was raised in most of the world tea auction market due to the unfavorable weather in those major tea-producing countries including India, Sri Lanka, Kenya and Indonesia. It is a good opportunity which has seldom been seen in recent years. Unfortunately, the world tea price has dropped sharply since the third quarter of 1998.

To summarize the developments in the world tea industry: world tea acreage continuously increased in the first 80 years of this century, and stabilized in the most recent 15 years. Tea production increased 2.26% annually in the first half of the century, and it doubled every 20 years since 1950 to 1990, and showed a slowly increasing tendency after 1990. The annual increasing percentage in the average tea yields per hectare was 2.75% between 1950–1990, and the average was 0.35%

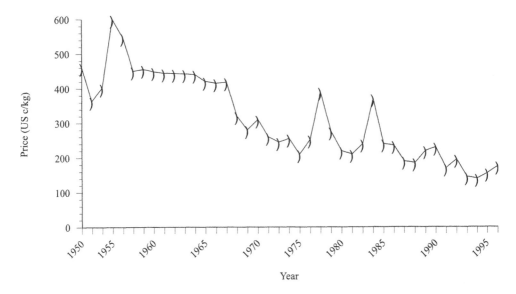

Figure 13.4   World Tea Price 1950–1996

between 1990–1997. The annual increasing rate in tea export was less than 1% in the first 50 years of this century and 4.07% in the latter 47 years period. The increasing rate in tea import was only 0.38% in the first 50 years of this century and 4% in the latter 40 year period from 1950–1990; notably, it increased slowly in the 1990s.

3.   AGROBOTANICAL CHARACTERISTICS

Although the tea plant is an ancient plant with a long history, the confusion and modification in nomenclature have continued for almost two centuries. As early as 1753, Linnaeus described the tea plant as *Thea sinensis*, and it was modified to *Camellia sinensis* in August of the same year. Since then, the genus name of *Thea* and *Camellia* has had a checkered history. Masters (1844) described the small-leaved China plant as *Thea sinensis* and the large-leaved Assam plant as *T. assamica*. For a long time, *Thea* and *Camellia* were considered to be separate genera. Fujita *et al.* (1973) used the presence of eugenol glycoside as a major criterion to separate the two genera. It was present in the essential oil of *Camellia*, but not in *Thea*. However, these two genera are so alike that it is not enough to differentiate them according to their morphological, anatomical and biochemical characteristics. Hence, Wight (1962) put forward that *Thea* should be synonymous with *Camellia*. Since then, the name *Camellia* prevailed. However, the species name is still in confusion. Watt in India named *Camellia thea* in 1907; Cohen-Stuart in Indonesia used a new name of *Camellia theifera* (Banerjee, 1992; Chen, 1992). In 1950, Qian Chongshu, a famous Chinese botanist, used the nomenclature *Camellia sinensis*. Sealy in the UK (1958) also gave the same name and included two varieties: var. *sinensis* (small-leaf variety) and var. *assamica* (large-leaf variety). Since then, despite some papers contributing to the botanical name of tea plant, uniformity has been achieved. Tea is the most important of all *Camellia* spp. both commercially and taxonomically.

Botanically, tea plant belongs to the order Theales, family Theaceae, genus *Camellia*. All varieties and cultivars of tea plant belong to a single species, *Camellia sinensis*. Although some scientists in China reported 17 new species and 1 new variety in the genus *Camellia* besides the *C. sinensis* in 1984 (Tan *et al.*, 1984). However, it was not generally accepted. Tea plant is a perennial evergreen; the aerial portion of tea plant is grown as a tree, semi-tree or shrub depending on the influence of the external environment. In general, the China variety (var. *sinensis*) tea plant usually grows into a shrub about 1–2 m high, characterized by more or less virgate stems. Leaves are small, hard, dark green in color with a dull surface. The *Assam* variety (var. *assamica)* is described as an erect tree with many branches, 3–10 m high. Leaves are 15–20 cm long, light green in color, with a glossy surface.

The fresh shoots are the economic harvest of tea plant. The phyllotaxy of leaves on the shoot is alternate. The leaf pose on the stem includes erect, semi-erect, horizontal, and drooping according to the variety. Leaves are leathery in texture, with silvery or light-colored hairs on the undersurface of tender leaves. There are 7–15 pairs of veins on the leaf. The lateral veins curve upward and connect with the upper veins, forming a close transporting network, which is characteristic of the leaves of tea plant. Leaves are serrated at the margin. The first few new leaves in the flushing period of the tea shoot usually have a characteristic small size, being thick and brittle with a blunt apex. The petiole is wider and flat, and called fish-leaf or in Indian terminology the *Janam*. Its position on the shoot is of very great importance when considering the standard of plucking. Sometimes the leaf primodium differentiates from the vegetative bud of tea plant ceasing growth prematurely instead of developing into the normal leaf. It is termed the dormant bud or in Indian terminology the *Banjhi*. Normal tea shoots show a distinct periodicity of growth, i.e., after the development of several normal leaves, the *Banjhi* bud forms, thus completing a full periodic shoot growth rhythm.

Tea flowers are bisexual with a light fragrance and are white in color. Their diameter is 20–55 mm. The morphology of the flower is one of the important indices in the classification of the tea plant. The fruit on the tea plant is green in color, three-celled, thick-walled, and shiny at first, but then duller and slightly rough later. Tea seed is brown in color, thin-shelled, about 1 cm in diameter, and semi-globose in shape.

4. ECOLOGY AND AGRONOMY

Tea plants can be grown over a considerable range of conditions from temperate climates to hot, humid subtropics and tropics. At present, commercial cultivation of tea plant extends from 42°N to 34°S latitude. Spread over such a vast landmass of diverse topography, the tea areas differ considerably in their soil-climatic environments. However, the optimum average daily ambient temperature for tea growth ranges between 20° and 30°C. When the temperature is higher than 30°C or lower than 13°C, the growth of the tea plant is retarded. The tolerance of tea plant to the minimum temperature varies with the varieties, it generally ranges between −3° and −15°C. Tea plant is a rain-fed crop. The quantity and distribution are equally important in the cultivation of tea plant. Rainfall in the tea area of the world varies from less than 1000 mm to 6000 mm in a year. Irrespective of the total quantity of rainfall, it is important that the rain is distributed evenly throughout the year, especially during the growing season. Persistent rainless conditions cause drought and prolonged drought could be lethal to the plant. The minimum quantity of rain required to sustain healthy growth depends on the soil and other environmental conditions and cultural factors which are connected with the loss of water from the soil and tea plant. Evaporative loss of water in a whole year from many tea areas exceeds 1270 mm, which is believed to be the minimum annual requirement of water by the tea plant. In a large number of tea growing countries, the rainfall exceeds evaporative losses in the course of a year, but its uneven distribution leads to deficit of water in the soil during periods of scant rain, such as in a large number of African tea-growing countries.

Tea plants are not overcritical to soil. The range of soil types on which tea is grown in the major tea producing countries in the world is remarkably wide including the latosols, red-yellow podzolic and reddish-brown lateritic, alluvial, andosols, volcanic soils, *etc* (Othieno, 1992). Tea plants are very sensitive to the soil acidity. They cannot survive in alkaline soil. The optimum pH of soil for growing tea ranges between 4.5 and 6.5. In recent years the soil acidity in the tea producing countries has lowered due to the heavy application of inorganic nitrogenous fertilizers. In acidic soils where tea is grown, large amounts of aluminum occur in the exchangeable form. The availability of aluminum decreases as the pH of the soil rises and at pH 7 and above none or very little aluminum is found in the exchangeable form.

The tea output per unit area is proportional to the garden coverage. To obtain maximum productivity within a short period, a density of more than 12,000–20,000 bushes per hectare in large-leaf varieties and 45,000–60,000 bushes (20,000 bushes × 3 rows) in small- to medium-leaf varieties is recommended (Chen, 1992). The economic lifespan of tea plant is generally around 40 years. It is recommended to pull out and replace plants with new clones when they reach this age. However, techniques such as collar-pruning and heavy-pruning of old bushes are adopted in China, Sri Lanka, and other countries in order to maximize the benefits during the early stages.

The principle of fertilization is to compensate the nutrients taken away by the crop and eluted by rainfall in a timely manner. Ordinarily, the schedule of fertilization is determined according to the nutritional status in the soil, the yield level in the previous pruning cycle, and the yield predicted by agrometeorological conditions. Nitrogen is the first nutrient selected for tea growing, because the shoots are the objective of tea cultivation. The tea crop is therefore sensitive to application of nitrogen fertilizer. In favorable circumstances a yield response commences 3–4 weeks after application of fertilizer. Usually, the increase of tea yield per unit area is directly proportional to the amount of nitrogen applied up to levels of 150 kg nitrogen per hectare. However, this linear response is reduced at higher application but remains positive. Very high levels of nitrogen application may therefore be uneconomic and may induce the acidification of the soil, with an adverse effect on tea quality and influence on the absorption of other nutrients. On this basis, the level of nitrogen application is controlled around 200–300 kg (Chen, 1992). Next to nitrogen, phosphorus and potassium are the most important nutrients in tea production. The quantity of phosphorus removed by the crop is small in comparison with nitrogen or potassium. The availability of phosphorus in the soil is highest when the soil has a pH between 5.0–7.0. The availability declines rapidly as the soil pH falls below 5.5 or rises above 7.0. In very acid soils phosphorus is combined with hydroxides of iron and aluminum to form compounds which are insoluble in water and unavailable to tea plant. The recommended dosage of phosphorus ranges from 75 kg to 100 kg P_2O_5 per hectare (Chen, 1992). Potassium is indispensable for young tea plants due to the stimulation of growth and the formation of a strong frame. The mature tea plant also responds to potassium, although a visible response does not appear immediately. However, in plants deficient in potassium a response to potassium fertilizer often appears very quickly. The recommended dosage of potassium application ranges from 100 kg to 150 kg per hectare (Chen, 1992). A balanced fertilization is emphasized from the viewpoint of increasing efficiency. Thus, it is recommended to apply a compound fertilizer consisting of the optimum dosage of nitrogen, phosphorus, potassium and other micronutrients to obtain the maximum effect. Organic manure is often neglected nowadays. So, it is worth stressing the importance of increasing the application of organic manure. The organic materials could halt or, at least, limit the deterioration of the soil. In addition, it could minimize the outbreak of tea pests and diseases as well as improve the quality of manufactured tea.

The requirements for microelements are few. However, a deficiency of some microelements occurs in some tea areas of the world. The increasing incidence of magnesium and zinc deficiency in tea plants was reported in some tea areas in India, Sri Lanka and east African countries (Chen, 1998a). Copper deficiency was also reported in some tea areas in Malawi and Japan; and zinc deficiency in some areas in Sri Lanka and Kenya. So, the application of microelement fertilizer produced significant effects in some instances.

The effect of shading is to modify light, airflow, temperature, and humidity as well as to decrease the physical damage of solar radiation and to minimize the excessive evaporation of water from leaves. The fallen leaves of the shading tree increase the source of organic matter; however, shading increases the incidence of tea blister blight (*Exobasidium vexans*). In some tropical and subtropical tea growing countries including India, Sri Lanka, Indonesia and African countries, shading of the plant is a popular technique used in tea production. However, in countries situated in the northern latitudes, such as China, Japan, Iran, Georgia, Argentina, *etc*, where the temperature drops very low during winter, shade trees are rarely grown. Shade is also considered unnecessary in the tropical and subtropical areas above an elevation of approximately 1300 m as the temperature remains low at such altitudes. So, the benefit and risk analysis of shading is a disputed issue, possibly because of the wide geographical distribution and various climate conditions. It is regarded that shading is necessary in tea areas with a maximum temperature higher than 35°C and relative humidity lower than 40% (Chen, 1994).

Pruning is a "necessary evil" for tea plants. The objective of pruning is to maintain the plant permanently in the younger phase, to stimulate the growth of shoots, and to build a rational frame height. In mature tea gardens, light-pruning and heavy-pruning should be done alternately. The best time for pruning is during a dormant period, because this is the time that the carbohydrate reserves within the tea plant are at a high level. Pruning during the drought season is not suitable. The rationale for the plucking system is based on the fact that certain amounts of re-growth leaves remain on the plucking table, thus guaranteeing enough supply of carbohydrates to the tea shoots. Generally, the terminal bud is removed together with one to three leaves for manufacture. The interval of the plucking cycle of tea plants mainly depends on the growth rate of the plant, generally 5–14 days. In most areas of Japan and Georgia, plucking has been fully mechanized; however, in most areas of the world tea is plucked by hand.

Tea plant is a C3 plant with high photorespiration; the utilization ratio of the tea plant of solar radiation energy is far lower than that of other crops. According to a study in India, only 7% of the photosynthetic products are used in the growth of the tea shoot, 9% in the formation of frame branches, and 84% is exhausted during respiration and other metabolic processes. How to improve the harvesting index via the breeding route or through cultivation techniques is a problem to be solved in the future.

5. CONSUMPTION AND CUSTOM

There are 125 countries and regions in the world that import tea. The average consumption *per capita* was only 0.19 kg at the beginning of this century. It increased rather slowly due to intense hostilities. The average tea consumption *per capita* was 0.21 kg in 1950, only a 20 g increase in a 50 year period. It increased more rapidly in the 40 years after 1950. The average consumption *per capita* in 1997 was 0.445 kg, 1.34 times higher than that in 1900 and 1.11 times higher than that in 1950. However, it slightly decreased compared with that in 1990 (0.51 kg) (ITC, 1996; 1998). Table 13.6 shows the average tea consumption *per capita* and the world tea population. Among the various countries in the world, Ireland had the highest average annual tea consumption *per capita* (3.23 kg) in the world according to statistics from 1995–1997, followed by the United Kingdom (2.46 kg), Kuwait (2.41 kg), Qatar (2.08 kg) and Turkey (1.77 kg). The following characteristics can be summarized about world tea consumption in recent years: (1) consumption in the United Kingdom, the largest tea-importing country historically, shows a decreasing tendency, (2) internal consumption in the major tea-producing countries (India and China) increased rapidly, (3) the proportion of imports from the Asian and African countries increased from 35.5% in 1980 to 43.1% in 1997, (4) the proportion of CTC black tea in the total world black tea trade increased significantly from 39% in 1980 to 59.4% in 1997. The proportion of tea bags and instant tea in the total tea trade has increased significantly in the last 10 years.

Table 13.6 World Average Annual Tea Consumption *Per Capita*.

Year	1900	1950	1960	1970	1980	1990	1997
World population (× 10,000)	160,800	251,300	302,700	367,800	441,500	490,000	600,000
Tea consumption (kg/person/year)	0.19	0.21	0.31	0.34	0.42	0.51	0.45

The consumptive preferences of tea in the world vary widely, depending on the drinking customs in a long historical period, and also varied with the age, sex, nationality, *etc.* Black tea is the most popular consuming tea kind in the world. Almost all the people in European, American, Oceanic countries and most of the African countries as well as the Southeast Asian countries (including India, Sri Lanka, Pakistan, Bangladesh, *etc.*) consume black tea. Green tea is mainly consumed in northeast Asian countries (including China, Japan, Korea, *etc.*) and the northwest African countries (including Morocco, Algeria, Tunisia, Libya, Mauritania, *etc.*). Perfumed tea is popular in South American countries and mint green tea is popular in northwest African countries. Even for the same kind of tea, people from different countries have a different quality preference. For instance, Sri Lanka black tea is preferred in Russia and CIS countries. CTC-style black tea is favored largely in the United Kingdom, Pakistan and Egypt. In North America, the low-priced and light-liquoring black teas are demanded. In China, people in Fujian and Taiwan consume mostly the Oolong tea (a semi-fermented tea). However, people in North China prefer Jasmine tea (a scented tea), and those from Southeast China like to drink green tea. In China, the people of minorities (including Tibet, Xinjiang, and Mongolia) prefer the compressed tea. The drinking style in China and Japan mainly adopts the brewing form. In India, Sri Lanka and European countries the traditional style of tea drinking is mainly the cooking method. In modern society, some new drinking styles have been created, such as the instant tea, tea bag, iced tea, ready-to-drink tea (liquid tea), which are simple and convenient, so suitable for the fast rhythm of the modern life style.

Many of the things considered pleasures of life, such as sweets, rich food, alcohol and smoking have turned out to have deleterious effects on human health. Tea, however, is one of those rare treasures, enjoyed throughout the world and for a long time in history. Tea has been shown to benefit health and even to counteract some of the bad effects of our other favorite pleasures. It could be predicted that the consumption of tea will increase with the development of research on the beneficial effect on human health. We hope everyone will drink more tea for enjoyment and for health.

REFERENCES

Banerjee, B. (1992) Botanical Classification of Tea. In: *Tea: Cultivation to Consumption,* Willson, K.C. & Clifford, M.N. (Ed.), Chapman & Hall, London, pp. 25–51.

Chen, Z.M. (Ed.) (1992) *China Tea Classics*, Shanghai Civilization Press, Shanghai, pp. 786 (Chinese).

Chen, Z.M. (1994) Tea. In: *Encyclopedia of Agricultural Science*, Vol. 4. Academic Press, New York, pp. 281–288.

Chen, Z.M. (1996) The spreading of tea and the development of tea therapy. *Seminar papers of the 4th International Tea Culture Festival,* May 25–28, 1996, Seoul, Korea, pp. 93–96.

Chen, Z.M.(1998a) A review and prospect on tea science at the crossing of century (Chinese). *J. Tea Sci.,* **18**(2), 81–88.

Chen, Z.M. (1998b) World Tea Industry in the Crossing of Century (Chinese). In: *Scientific Progress and Discipline Development,* Zhuo, G.Z. (Ed.), China Sci. & Technol. Publishers, Beijing, pp. 768–772.

Eden, T. (1976) *Tea.* Longman, London, pp. 236.

International Tea Committee (1996) *World Tea Statistics, 1910–1990.* London.

International Tea Committee (1998) *Annual Bulletin of Statistics,* London.

Othieno, C.O. (1992) Soils. In: *Tea: Cultivation to Consumption.* Willson, K.C. & Clifford, M. N. (Ed.), Chapman & Hall, London. pp. 137–172.

Tan, Y.J., Chen, B.H., Yu, F.L. *et al.* (1984) New species and variety of tea plants found in Yunnan, China (Chinese). *J. Tea Sci.,* **4**(1), 19–30.

INDEX

Printed and bound by CPI Group (UK) Ltd, Croydon, CR0 4YY

23/10/2024

01778254-0005